Texts and Monographs in Physics

Series Editors: R. Balian W. Beiglböck H. Grosse E. H. Lieb
N. Reshetikhin H. Spohn W. Thirring

Springer
Berlin
Heidelberg
New York
Barcelona
Hong Kong
London
Milan
Paris
Singapore
Tokyo

Texts and Monographs in Physics

Series Editors: R. Balian W. Beiglböck H. Grosse E. H. Lieb
N. Reshetikhin H. Spohn W. Thirring

Manfred Salmhofer

Renormalization

An Introduction

With 24 Figures

Springer

Prof. Dr. Manfred Salmhofer

Mathematik
ETH-Zentrum
CH-8092 Zürich, Switzerland

Editors

Roger Balian

CEA
Service de Physique Théorique de Saclay
F-91191 Gif-sur-Yvette, France

Nicolai Reshetikhin

Department of Mathematics
University of California
Berkeley, CA 94720-3840, USA

Wolf Beiglböck

Institut für Angewandte Mathematik
Universität Heidelberg
Im Neuenheimer Feld 294
D-69120 Heidelberg, Germany

Herbert Spohn

Theoretische Physik
Ludwig-Maximilians-Universität München
Theresienstraße 37
D-80333 München, Germany

Harald Grosse

Institut für Theoretische Physik
Universität Wien
Boltzmanngasse 5
A-1090 Wien, Austria

Walter Thirring

Institut für Theoretische Physik
Universität Wien
Boltzmanngasse 5
A-1090 Wien, Austria

Elliott H. Lieb

Jadwin Hall
Princeton University, P. O. Box 708
Princeton, NJ 08544-0708, USA

Library of Congress Cataloging-in-Publication Data

Salmhofer, Manfred, 1964– Renormalization: an introduction / Manfred Salmhofer. p. cm. –
(Texts and monographs in physics, ISSN 0172-5998) Includes bibliographical references and index.
1. Renormalization (Physics) I. Title. II. Series.
QC174.17.R46S35 1998 530.1'43–dc21 98-40042 CIP

ISBN 978-3-642-08430-0

© Springer-Verlag Berlin Heidelberg 2010
Printed in Germany

Cover design: *design & production* GmbH, Heidelberg

Preface

Why another book on the renormalization of field theory? This book aims to contribute to the bridging of the gap between the treatments of renormalization in physics courses and the mathematically rigorous approach. It provides a simple but rigorous introduction to perturbative renormalization, and, in doing so, also equips the reader with some basic techniques which are a prerequisite for studying renormalization nonperturbatively. Beside these technical issues, it also contains a proof of renormalizability of ϕ^4 theory in $d \leq 4$ dimensions and a discussion of renormalization for systems with a Fermi surface, which are realistic models for electrons in metals.

Like the two courses on which it is based, the book is intended to be easily accessible to mathematics and physics students from the third year on, and after going through it, one should be able to start reading the current literature on the subject, in particular on nonperturbative renormalization.

Chapter 1 provides a brief motivation for studying quantum theory by functional integrals, as well as the setup. In Chap. 2, the techniques of Gaussian integration and Feynman graph expansions are introduced. I then give simple proofs of basic results, such as the theorem that the logarithm of the generating functional is a sum of values of connected Feynman graphs.

In Chap. 3, the Wilson renormalization flow is defined, and perturbative renormalizability of ϕ^4 theory in $d \leq 4$ dimensions is proven using a renormalization group differential equation. The Feynman graph expansion of Chap. 2 is the explicit solution to this equation, but the analysis of the differential equation leads to a very simple renormalizability proof. It also brings out clearly that the renormalization subtractions really amount to a change of boundary conditions. In Chap. 4, a similar renormalization flow is applied to an infrared problem, that of many-fermion systems. Using the method of overlapping loops, I determine the leading contributions to the renormalization group flow to all orders of perturbation theory. This leading behaviour is then calculated in the simplest cases and its physical implications are discussed heuristically. Readers with a little experience in field theory can read Chap. 4 independently of Chaps. 1–3.

I would like to thank Horst Knörrer for the suggestion to write this text and for many valuable comments. I also thank László Erdős and Pirmin Lemberger for reading carefully through Chaps. 1–3 and for suggesting var-

ious improvements. Finally, I would like to thank Joel Feldman and Eugene Trubowitz for the collaboration which brought about many of the results about many-fermion systems described in Chap. 4 and for many discussions, Christian Lang and Erhard Seiler for teaching me a good part of what I know about quantum field theory, and Volker Bach, Georg Keller, Christoph Kopper, Walter Metzner, Edwin Langmann, and Christian Wieczerkowski for discussions.

A saying attributed to Einstein goes 'Everything should be made as simple as possible, but not simpler.' I have made the proofs as clear and concise as I could without oversimplifying matters or discarding important details. Certainly, not everything here is as simple as it could possibly be, but I hope that it is at least readable.

Zurich, September 1998 Manfred Salmhofer

Table of Contents

1. Field Theory

1.1 A Motivation for Path Integrals

To motivate the study of field theory via the functional integral, we review the path integral of quantum mechanics, as discovered by Dirac [1] and Feynman [2]. We shall be brief and refer to the literature for many details because a number of excellent accounts (e.g. [3]) deal with this topic in depth (and because these details will not be required for understanding the rest of the text). The presentation given here is taken from [3]; it is included for the convenience of the reader and because, even though beautiful mathematical work has given a mathematically rigorous foundation for the functional integrals of quantum mechanics [3] *and* quantum field theory [6, 7, 5], many people still seem to believe that this is not the case. In fact, however, the Hamiltonian formalism of quantum field theory is so singular in more than two dimensions that the path integral seems a much better starting point.

The classical Hamiltonian

$$H_{cl} = \frac{p^2}{2m} + V(x) \tag{1.1}$$

becomes a quantum-mechanical Hamiltonian

$$H = -\frac{\hbar^2}{2m}\Delta + V(x) \tag{1.2}$$

under the replacement $p \to \hbar/i\ \partial/\partial x$. States of spinless particles on \mathbb{R}^d are described by vectors $\psi \in \mathcal{H} = L^2(\mathbb{R}^d, \mathbb{C})$, with the interpretation that for any measurable $R \subset \mathbb{R}^d$, $\int_R |\psi(t, x)|^2 d^d x$ is the probability for finding the particle at time t in the region R. Given an initial state ψ_0 in the domain $D(H)$ of the Hamiltonian H at $t = 0$, the state at later times $t > 0$ is determined by the Schrödinger equation

$$i\hbar\frac{\partial\psi}{\partial t} = H\psi \tag{1.3}$$

with initial condition $\psi(0, x) = \psi_0(x)$. If V is such that H is self-adjoint, then the time evolution $t \mapsto \exp(-itH)$, $t \in \mathbb{R}$, is a strongly continuous semigroup, and for an $\psi_0 \in \mathcal{H}$, the state at time $t > 0$ is given by

$$\psi(t, x) = \left(e^{-itH/\hbar}\, \psi_0\right)(x)$$

$$= \int_{\mathbb{R}^d} dy\, K(t; x, y)\, \psi_0(y). \tag{1.4}$$

In Dirac's notation,

$$K(t; x, y) = \langle x \mid e^{-itH/\hbar} \mid y \rangle. \tag{1.5}$$

Feynman's path integral formula is a rewriting of K in terms of an integral over paths, obtained from the Trotter product formula.

Lemma 1.1. (Trotter product formula) *If A and B are bounded operators, then*

$$e^{A+B} = \lim_{N \to \infty} \left(e^{A/N}\, e^{B/N}\right)^N, \tag{1.6}$$

with the limit in the operator norm.

Proof. This proof is taken from [3]. Let $C = \exp((A + B)/N)$ and $D = e^{A/N}e^{B/N}$. We have to show that $\left\|C^N - D^N\right\| \to 0$ as $N \to \infty$. By telescoping,

$$C^N - D^N = \sum_{k=0}^{N-1} C^k (C - D) D^{N-1-k}. \tag{1.7}$$

Using the product inequality of the operator norm and

$$\max\{\|C\|, \|D\|\} \le e^{(\|A\|+\|B\|)/N}, \tag{1.8}$$

we get

$$\left\|C^N - D^N\right\| \le \sum_{k=0}^{N-1} \|C\|^k \|C - D\| \|D\|^{N-1-k} \tag{1.9}$$

$$\le \|C - D\|\, N\, e^{(\|A\|+\|B\|)\,(N-1)/N}. \tag{1.10}$$

Expanding the exponentials, we see that $C - D$ starts with the term of order N^{-2}. More precisely

$$C - D = \sum_{m \ge 2} \frac{N^{-m}}{m!} \left((A + B)^m - \sum_{m'+m''=m} \frac{m!}{m'!m''!} A^{m'} B^{m''}\right), \tag{1.11}$$

so, taking norms and using the product and triangle inequality, we get

$$\|C - D\| \le N^{-2}\, e^{(\|A\|+\|B\|)/N}\, (\|A\| + \|B\|)^2, \tag{1.12}$$

hence

$$\left\|C^N - D^N\right\| \le N^{-1}\, e^{\|A\|+\|B\|}\, (\|A\| + \|B\|)^2. \tag{1.13}$$

∎

The Hamiltonian H given by (1.2) is unbounded, so the Lemma does not apply to it. It is proven in [3] that in the physically interesting cases, the limit exists in the strong sense, that is, if both sides are applied to a vector $\psi \in \mathcal{H}$, the resulting vector converges in \mathcal{H} as $N \to \infty$. More precisely [3]:

- if A and B are self-adjoint and $A + B$ is essentially self-adjoint on $D(A) \cap D(B)$, then

$$e^{i(A+B)} = \underset{N\to\infty}{\text{s–lim}} \left(e^{iA/N} \, e^{iB/N} \right)^N \qquad (1.14)$$

- if A and B are bounded below, then

$$e^{-(A+B)} = \underset{N\to\infty}{\text{s–lim}} \left(e^{-A/N} \, e^{-B/N} \right)^N . \qquad (1.15)$$

(see also [4], Sect. X.11). These are the physically relevant cases because both the time evolution e^{-itH} and the heat kernel $e^{-\beta H}$ are covered.

The free time evolution has the kernel

$$K_0(t; x, y) = \langle x \mid e^{-itH_0/\hbar} \mid y \rangle = \left(\frac{m}{2\pi i\hbar t} \right)^{d/2} e^{im|x-y|^2/2\hbar t}, \qquad (1.16)$$

and ([4], Sect. IX.7)

$$\left(e^{-itH_0/\hbar}\phi \right)(x) = \int_{\mathbb{R}^d} dy \, K_0(t; x, y)\phi(y), \qquad (1.17)$$

where $\int_{\mathbb{R}^d} dx$ is understood as a limit $\lim_{M\to\infty} \int_{|x|\leq M} dx$ in L^2 norm ([4], Sect. IX.2).

Because

$$\langle x \mid e^{-i\Delta t/\hbar \, H_0} e^{-i\Delta t/\hbar \, V} \mid y \rangle = K_0(\Delta t; x, y) \, e^{-i\Delta t/\hbar \, V(y)}, \qquad (1.18)$$

we have

$$\left(e^{-i\Delta t/\hbar \, (H_0+V)}\psi \right)(x_0) = \lim_{N\to\infty} \left(\frac{m}{2\pi i\hbar\Delta t} \right)^{Nd/2} \int_{\mathbb{R}^{dN}} dx_1 \dots dx_N$$

$$e^{i/\hbar \, S_N(x_0,\dots,x_N;t)} \, \psi(x_N) \qquad (1.19)$$

with $\Delta t = t/N$ and

$$S_N(x_0, \dots, x_N; t) = \sum_{k=1}^{N} \Delta t \left(\frac{m}{2} \left(\frac{x_k - x_{k-1}}{\Delta t} \right)^2 - V(x_k) \right). \qquad (1.20)$$

A classical particle in a potential V moving along a path $X(t)$ has the action

$$S(X) = \int_0^t ds \left(\frac{m}{2}\dot{X}(s)^2 - V(X(s)) \right). \qquad (1.21)$$

For a polygonal path $X^{(P)}$ given by x_0, \ldots, x_N, one gets

$$
\begin{aligned}
S(X^{(P)}) &= \sum_{k=1}^{N} \Delta t \left(\frac{m}{2} \left(\frac{x_k - x_{k-1}}{\Delta t} \right)^2 - \int_0^t ds\, V(X(s)) \right) \\
&\approx S_N(x_0, \ldots, x_N; t).
\end{aligned}
\tag{1.22}
$$

Thus one may interpret the right hand side of (1.19) as an integral over all polygonal paths, with S_N approximating the classical action associated to each path.

One would expect that in the limit $N \to \infty$, the set of all polygonal paths becomes the set of all paths, that $S_N(X^{(P)}) \to S(X)$, and that [1, 2]

$$
\left(e^{-it(H_0 + V)/\hbar} \psi \right)(x_0) = \int_{\Omega_{x_0}} e^{iS(X)/\hbar} \psi(X(t)) dX
\tag{1.23}
$$

where Ω_{x_0} is the set of all paths $X : [0, t] \to \mathbb{R}^d$ with $X(0) = x_0$ and dX is a measure on Ω_{x_0}. This is *Feynman's path integral formula*. One of its most beautiful consequences is that the variational condition for the classical solution X_{cl},

$$
\left. \frac{\delta}{\delta X} S(X) \right|_{X = X_{\text{cl}}} = 0
\tag{1.24}
$$

appears as the stationary phase approximation to the integral, which should be valid in the limit $\hbar \to 0$.

The problem is, unfortunately, that the measure dX does not exist, and that therefore, the path integral formula makes mathematical sense only at finite N. Thus all manipulations have to be done before taking the limit, which is messy and removes the elegance of the approach. A concrete indication that the limit dX does not exist is that the prefactor in (1.19) diverges as $N \to \infty$.

However, if one takes e^{-tH} instead of e^{itH}, the exponential of the kinetic term decays exponentially and can be included in the measure: replacing it by τ, that is, going to imaginary time, one gets instead of K_0 the kernel

$$
p(\tau, x, y) = \left(\frac{m}{2\pi\hbar\tau} \right)^{d/2} e^{-m|x-y|^2/2\hbar\tau}.
\tag{1.25}
$$

Since

$$
\langle x \mid e^{-\Delta\tau/\hbar\, H_0} e^{-\Delta\tau/\hbar\, V} \mid y \rangle = p(\Delta t; x, y)\, e^{-\Delta\tau/\hbar\, V(y)},
\tag{1.26}
$$

we have

$$
\begin{aligned}
\left(e^{-\Delta\tau/\hbar\,(H_0 + V)} \psi \right)(x_0) = \lim_{N \to \infty} \left(\frac{m}{2\pi\hbar\Delta\tau} \right)^{Nd/2} \int_{\mathbb{R}^{dN}} dx_1 \ldots dx_N \\
e^{-S_E(x_0, \ldots, x_N)/\hbar}\, \psi(x_N)
\end{aligned}
\tag{1.27}
$$

with $\Delta\tau = \tau/N$ and

$$S_E(x_0,\ldots,x_N;\tau) = \sum_{k=1}^{N} \Delta\tau \left(\frac{m}{2} \left(\frac{x_k - x_{k-1}}{\Delta\tau} \right)^2 + V(x_k) \right). \qquad (1.28)$$

Since

$$\int_{\mathbb{R}^d} \mathrm{d}y\, p(\tau;x,y) = 1, \qquad (1.29)$$

the *Gaussian measure* defined by

$$\mathrm{d}\mu_{x_0}^{(N)}(X) = \prod_{k=1}^{N} \exp\left(-\frac{m}{2\hbar} \frac{(x_k - x_{k-1})^2}{\Delta\tau} \right) \mathrm{d}x_1\ldots\mathrm{d}x_N \qquad (1.30)$$

is normalized, i.e. $\int \mathrm{d}\mu_{x_0}^{(N)}(X) = 1$, independently of N. And indeed, one can show that the limit $N \to \infty$ of $\mathrm{d}\mu_{x_0}^{(N)}$ exists as a measure on the set of all paths starting at x_0. It is the *conditional Wiener measure* $\mathrm{d}\mu_{x_0}(X)$ describing Brownian motion, that is, the analogue of free quantum mechanical motion, but with the Schrödinger equation replaced by the heat equation. For suitable V, e.g. $V \in L^p \cap L^\infty$, where $p > \frac{d}{2}$ [4], (1.27) converges as $N \to \infty$ to the (convergent) integral

$$\left(e^{-\beta H}\psi \right)(x_0) = \int_{\Omega_{x_0}} \mathrm{d}\mu_{x_0}(X) e^{-\int_0^\beta V(X(\tau))\mathrm{d}\tau} \psi(X(\beta)). \qquad (1.31)$$

Note that the improper integral $\int \mathrm{d}x\, \exp(ix^2)$ is convergent, but not absolutely convergent, which prevents a similar procedure in the real-time case. In fact, complex-valued Gaussian measures on infinite-dimensional spaces exist only under very restricted conditions, e.g. if the imaginary part of the covariance is relatively trace class to the real part.

For determining spectral properties of H, $e^{-\beta H}$ is as good as e^{itH}. Moreover, the path integral representation for the partition function for the canconical ensemble (in a finite volume) $\mathrm{tr}\, e^{-\beta H}$ is obvious from (1.31). The *Feynman-Kac formula* (1.31) has many applications in quantum mechanics [3].

In quantum field theory, one is not dealing with a single particle, but with infinitely many particles, because one has to account for the creation and annihilation of particles. One can formally write down a Hamiltonian, but it becomes very difficult to give a mathematical definition of it. We shall simply define the theory by the functional integral. The lesson from the quantum mechanical case is that free particles are described by a Gaussian measure. We shall therefore put the quadratic part of the action into the measure before taking the continuum limit that defines the field theory.

In the Minkowski spacetime, there are further problems: because of the indefinite metric, the kinetic term is not semibounded. For this reason, one studies the Euclidean version of the theory, i.e. the theory with imaginary time, and then tries to do the analytical continuation to real time.

A historical introduction and physical motivations for doing quantum field theory, as well as a thorough development of the theory is, e.g., in the recent book by Weinberg [8].

1.2 Gaussian Integrals and Random Variables

The fields in quantum field theory can be considered as random variables associated to a Gaussian measure that is perturbed by an interaction term. The Gaussian measure determines the properties of free particles. In this section, we give the elementary definitions and properties of Gaussian random variables indexed by a finite set. This is of relevance for quantum field theory because we shall define a regularized field theory on the lattice with finitely many points and then study the continuum limit.

1.2.1 Preliminaries

All identities for Gaussian integration are based on the following two elementary formulas.

Remark 1.2. Let $a > 0$. Then

$$\int_{-\infty}^{\infty} e^{-ax^2} dx = \sqrt{\frac{\pi}{a}} \tag{1.32}$$

and for all $b \in \mathbb{C}$

$$I(b) = \int_{\mathbb{R}} e^{-\frac{a}{2}x^2 + bx} dx = e^{b^2/2a} \sqrt{\frac{2\pi}{a}}. \tag{1.33}$$

Proof. Since $-\frac{a}{2}x^2 + bx = -\frac{a}{2}(x - \frac{b}{a})^2 + \frac{b^2}{2a}$,

$$I(b) = e^{b^2/2a} \int_{\mathbb{R}} dx\, e^{-a(x-b/a)^2/2}. \tag{1.34}$$

For $b \in \mathbb{R}$, a shift in the integration variable proves (1.33). For $\beta = \text{Im}\, b \neq 0$, the shift gives

$$I(b) = e^{b^2/2a} \int_{\mathbb{R}} dx\, e^{-a(x-i\beta/a)^2/2}. \tag{1.35}$$

Since $\left| e^{-a(x-i\beta/a)^2/2} \right| = e^{-ax^2/2}\, e^{\beta^2/2a}$,

$$\left| \int_{|x|>L} dx\, e^{-a(x-i\beta/a)^2/2} \right| \le e^{\beta^2/2a} \int_{|x|>L} dx\, e^{-ax^2/2} \xrightarrow[L\to\infty]{} 0, \quad (1.36)$$

so it suffices to show that

$$\int_{-L}^{L} dx\, e^{-a(x-i\beta/a)^2/2} \xrightarrow[L\to\infty]{} \int_{\mathbb{R}} dx\, e^{-ax^2/2}. \quad (1.37)$$

Since the integrand is analytic in x, we may deform the contour to consist of a piece $[-L, L] + i\beta/a$ and two vertical pieces, going from $\pm L$ to $\pm L + i\beta/a$. The contribution from $[-L, L] + i\beta/a$ equals $\int_{-L}^{L} e^{-ax^2/2} dx$, which converges to the desired result as $L \to \infty$. The contribution from the two vertical pieces vanishes as $L \to \infty$ because the integrand is exponentially small in L and the integration path is of fixed length $|\beta/a|$. ∎

1.2.2 Gaussian Integrals in Finitely Many Variables

From now on, let $N \in \mathbb{N}$, $A \in M_N(\mathbb{R})$, $A = A^t > 0$ (i.e., all eigenvalues of A are positive) and denote $C = A^{-1}$. Denote the scalar product on \mathbb{R}^N by

$$(\phi, \psi) = \sum_{i=1}^{N} \phi_i \psi_i \quad (1.38)$$

and $d^N \phi = d\phi_1 \ldots d\phi_N$.

Remark 1.3.

$$\int e^{-\frac{1}{2}(\phi, A\phi)} d^N \phi = (2\pi)^{\frac{N}{2}} \det A^{-\frac{1}{2}} = \det(2\pi C)^{\frac{1}{2}} \quad (1.39)$$

and for all $J \in \mathbb{C}^N$

$$\int e^{-\frac{1}{2}(\phi, A\phi)+(J,\phi)} d^N \phi = \det(2\pi C)^{\frac{1}{2}} e^{\frac{1}{2}(J,CJ)}. \quad (1.40)$$

Proof. Since A is symmetric, there is a $V \in SO(N)$ such that

$$V^{-1} A V = D = \text{diag}\{a_1, \ldots, a_N\} \quad (1.41)$$

where a_i are the eigenvalues of A. Since A is strictly positive, $a_i > 0$ for all $i \in \{1, \ldots, N\}$. Let $\psi = V\phi$, then

$$(\phi, A\phi) = (\psi, D\psi) = \sum_{i=1}^{N} a_i \psi_i^2 \quad (1.42)$$

and $d^N\phi = |\det V| d^N\psi = d^N\psi$. Thus by (1.32)

$$
\begin{aligned}
\int e^{-\frac{1}{2}(\phi, A\phi)} d^N\phi &= \int d^N\psi\, e^{-\frac{1}{2}\sum\limits_{i=1}^{N} a_i\psi^2} \\
&= \prod_{i=1}^{N} \int d\psi\, e^{-\frac{1}{2}a_i\psi^2} = \prod_{i=1}^{N} \sqrt{\frac{2\pi}{a_i}} \qquad (1.43) \\
&= \frac{(2\pi)^{\frac{N}{2}}}{(\det A)^{\frac{1}{2}}} = \det(2\pi C)^{\frac{1}{2}}.
\end{aligned}
$$

Similarly, $(J, \phi) = (J, V^{-1}\psi) = (VJ, \psi)$, so, with $b = VJ \in \mathbb{C}^N$, and by (1.33)

$$
\begin{aligned}
\int d^N\phi\, e^{-\frac{1}{2}(\phi, a\phi) + (J, \phi)} &= \prod_{i=1}^{N} \int d\psi\, e^{-a_i/2\,\psi^2 + b_i\psi} \\
= \prod_{i=1}^{N} \sqrt{\frac{2\pi}{a_i}}\, e^{b_i^2/2a_i} &= (\det 2\pi C)^{\frac{1}{2}}\, e^{\frac{1}{2}(J, CJ)} \qquad (1.44)
\end{aligned}
$$

because

$$
\sum_{i=1}^{N} \frac{b_i^2}{2a_i} = \frac{1}{2}(b, D^{-1}b) = \frac{1}{2}(J, V^{-1}D^{-1}VJ) = \frac{1}{2}(J, CJ). \qquad (1.45)
$$

∎

Definition 1.4. *Let* $C \in GL_N(\mathbb{R})$, $C > 0$. *The measure*

$$
d\mu_C(\phi) = (\det 2\pi C)^{-\frac{1}{2}}\, e^{-\frac{1}{2}(\phi, C^{-1}\phi)} d^N\phi \qquad (1.46)
$$

is called the Gaussian measure on \mathbb{R}^N *with mean zero and covariance matrix* C.

By definition, μ_C has total mass equal to one, i.e. $\int d\mu_C(\phi) = 1$. By Remark 1.3,

$$
\int d\mu_C(\phi)\, e^{i(J, \phi)} = e^{-\frac{1}{2}(J, CJ)} \qquad (1.47)
$$

holds for all $J \in \mathbb{C}^N$. This Fourier transform is often called the characteristic function of the measure μ_C. All moments of the measure μ_C can be calculated using (1.47) as polynomials in the covariance C, e.g.

$$
\begin{aligned}
\int d\mu_C(\phi)\phi_1\phi_2 &= \left[\frac{\partial^2}{\partial J_1 \partial J_2} \int d\mu_C(\phi)\, e^{(J, \phi)} \right]_{J=0} \\
&= \left[\frac{\partial^2}{\partial J_1 \partial J_2} e^{\frac{1}{2}(J, CJ)} \right]_{J=0} \qquad (1.48) \\
&= \frac{1}{2}(C_{21} + C_{12}) = C_{12}.
\end{aligned}
$$

We shall see much more of that when calculating more complicated moments of μ_C as sums of values of Feynman graphs.

Equation (1.47) already suggests that the limit $N \to \infty$ may be taken under suitable conditions on C and J, e.g. that J is in the form domain of C. The existence of moments of μ_C in that limit then depends on further properties of C; this will be central in the discussion of renormalization.

A natural interpretation from the probabilistic point of view (as well as that of field theory) is to regard not only the measure $d\mu_C$ as interesting, but also the variables ϕ_1, \dots, ϕ_N. The random variables ϕ_1, \dots, ϕ_N distributed according to (1.46) are called Gaussian random variables indexed by $\{1, \dots, N\}$, that is, Gaussian fields.

The right hand side of (1.47) is defined for $C \geq 0$ as well (in fact, for any $C \in M_N(\mathbb{C})$). As we shall show now, the $\int d\mu_C$ on the left hand side of (1.47) is a measure for any $C \geq 0$ (i.e., where some of the eigenvalues vanish), but not for $C < 0$. To see this, let us go back to the one-dimensional case where the matrix C is a number $c > 0$, and where $d\mu_C(\phi) = d\phi_1\, g_C(\phi_1)$ with density

$$g_C(x) = \frac{1}{\sqrt{2\pi c}}\, e^{-x^2/2c}. \tag{1.49}$$

Since $\int_{\mathbb{R}} g_C(x)dx = 1$ and $g_C(x) \underset{c \to 0}{\longrightarrow} 0$ for all $x \neq 0$, g_C is an approximate identity, i.e. $g_C \to \delta$ in $\mathcal{S}'(\mathbb{R})$. Thus

$$d\mu_C(\phi) \underset{c \to 0+}{\longrightarrow} \delta(\phi)d\phi. \tag{1.50}$$

Similarly, the vanishing of an eigenvalue of C in the N-dimensional case simply implies that the measure contains a δ function that restricts to the linear subspace orthogonal to the corresponding eigenvector.

In this way, the measure $d\mu_C$ has a natural meaning also for $C \geq 0$, which we adopt from now on. It is natural because $d\mu_C$ for $C \geq 0$ can always be obtained by taking $\varepsilon \to 0^+$ of a measure $d\mu_{C_\varepsilon}$ with $C_\varepsilon > 0$ for $\varepsilon > 0$. With this definition, (1.47) holds for all $C \geq 0$, with a genuine measure μ_C on the left hand side. We shall need nonnegative covariances all the time in what follows because we shall impose strict cutoffs in momentum space.

Finally, note that for C with some negative eigenvalues, the left hand side of (1.47) diverges: although for any $C \in GL_N(\mathbb{R})$ the linear functional

$$L_C(F) = (\det 2\pi C)^{-\frac{1}{2}} \int d^N\phi\, e^{-\frac{1}{2}(\phi, C\phi)} F(\phi) \tag{1.51}$$

can be defined on a suitably restricted space of functions, such as $C_0^\infty(\mathbb{R}^N)$, the inequality

$$L_C(F) \leq \text{const } \|F\|_\infty \tag{1.52}$$

holds only if $C \geq 0$. So there is no measure associated to the linear functional L_C if C has negative eigenvalues.

1.2.3 The Covariance Splitting Formula

Theorem 1.5. *Let $C_1, C_2 \in M_N(\mathbb{R})$, $C_i = C_i^t$ for $i \in \{1, 2\}$ and $C_1 \geq 0$, $C_2 \geq 0$. Define $C = C_1 + C_2$. Then for all $F \in \mathcal{L}^1(\mu_C)$,*

$$
\int d\mu_C(\phi) F(\phi) = \int d\mu_{C_1}(\phi_1) \int d\mu_{C_2}(\phi_2) F(\phi_1 + \phi_2)
$$
$$
= \int d\mu_{C_1}(\phi_1) \int d\mu_{C_2}(\phi - \phi_1) F(\phi). \qquad (1.53)
$$

In other words, if $C = C_1 + C_2$, the Gaussian random variable ϕ is a sum of two independent Gaussian random variables, $\phi = \phi_1 + \phi_2$, and the Gaussian measure factors, $d\mu_C(\phi) = d\mu_{C_1}(\phi_1) d\mu_{C_2}(\phi_2)$.

Proof. Let $F(\phi) = e^{i(J, \phi)}$. Then by (1.47)

$$
\int d\mu_C(\phi) e^{i(J, \phi)} = e^{-\frac{1}{2}(J, CJ)} = e^{-\frac{1}{2}(J, C_1 J) - \frac{1}{2}(J, C_2 J)}
$$
$$
= \int d\mu_{C_1}(\phi_1) e^{i(J, \phi_1)} \int d\mu_{C_2}(\phi_2) e^{i(J, \phi_2)} \quad (1.54)
$$
$$
= \int d\mu_{C_1}(\phi_1) \int d\mu_{C_2}(\phi_2) e^{i(J, \phi_1 + \phi_2)},
$$

so (1.53) holds for this case. Since a measure is uniquely defined by its Fourier transform, the measures $d\mu_C$ and $d\mu_{C_1} * d\mu_{C_2}$ are identical, therefore (1.53) holds. ∎

Remark 1.6. One can derive (1.53) from (1.54) also more explicitly. By the explicit formula (1.54), both sides are analytic in J. Thus we may take derivatives with respect to J. If F is a polynomial, $F(\phi) = P(\phi)$, with

$$
P(\phi) = \sum_{\nu=0}^{n} \sum_{j_1, \ldots, j_\nu} f_{j_1 \ldots j_\nu} \phi_{j_1} \cdots \phi_{j_\nu}, \qquad (1.55)
$$

where the inner sum runs over all sequences (j_1, \ldots, j_ν) with $j_k \in \{1, \ldots, N\}$, we have

$$
\int d\mu_C(\phi) P(\phi) e^{i(J, \phi)} = \int d\mu_C(\phi) P(\frac{1}{i} \frac{\partial}{\partial J}) e^{i(J, \phi)}
$$
$$
= P(\frac{1}{i} \frac{\partial}{\partial J}) \int d\mu_C(\phi) e^{i(J, \phi)} \qquad (1.56)
$$

where the notation means that every ϕ_{j_l} gets replaced by $\frac{1}{i} \frac{\partial}{\partial J_{j_l}}$. Using (1.54), this becomes

$$= P(\frac{1}{i}\frac{\partial}{\partial J}) \int d\mu_{C_1}(\phi_1) \int d\mu_{C_2}(\phi_2)\, e^{i(J,\phi_1+\phi_2)}$$

$$= \int d\mu_{C_1}(\phi_1) \int d\mu_{C_2}(\phi_2)\, P(\phi_1+\phi_2)\, e^{i(J,\phi_1+\phi_2)}. \tag{1.57}$$

The exchange of integration and differentiation with respect to J is justified because all integrals are still absolutely convergent. For the same reason, the limit $J \to 0$ exists, and it implies that (1.53) holds for all polynomials F. More general F can be approximated by polynomials. This technique of differentiating with respect to the *source terms* J will be a Leitmotiv in most of the technical discussion about Feynman graph representations.

Definition 1.7. *Let Γ be a finite set, $|\Gamma| = N$, and let $F = \{\phi : \Gamma \to \mathbb{R}\}$ (then $F \cong \mathbb{R}^N$). For $\phi, \psi \in F$ let $(\phi, \psi)_\Gamma = \sum_{x \in \Gamma} \phi(x)\psi(x)$, let C be a linear map on F such that $(\phi, C\phi) > 0$ for all $\phi \neq 0$, and let A be its inverse. The Gaussian measure on F is defined by*

$$d\mu_C(\phi) = (\det 2\pi C)^{\frac{1}{2}}\, e^{-\frac{1}{2}(\phi, A\phi)} \prod_{x \in \Gamma} d\phi(x). \tag{1.58}$$

Remark 1.8. If $\nu : \{1, \ldots, n\} \to \Gamma$ is bijective, $d\mu_C(\phi)$ can be written as

$$d\mu_C(\phi) = (\det 2\pi(C_{\nu(i),\nu(j)}))^{\frac{1}{2}}\, e^{-\frac{1}{2}\sum_{i,j=1}^{N} \phi(\nu(i)) A_{\nu(i)\nu(j)} \phi(\nu(j))} \prod_{i=1}^{N} d\phi_{\nu(i)}. \tag{1.59}$$

$d\mu_C(\phi)$ is well-defined because all factors occuring in its definition are basis independent: the right hand side of (1.59) does not depend on ν.

1.3 Field Theory on a Lattice

We want to define a regularized field theory corresponding to the Euclidean action

$$S = \int_{\mathbb{R}^d} d^d x \left(\frac{1}{2}\nabla\phi(x)^2 + \frac{m_\phi^2}{2}\phi(x)^2 + \lambda\phi(x)^4\right)$$

$$= \int_{\mathbb{R}^d} d^d x \left[\phi(x)\frac{1}{2}(-\Delta + m_\phi^2)\phi(x) + \lambda\phi(x)^4\right] \tag{1.60}$$

where, in going from the first to the second line, we have assumed that ϕ vanishes at infinity so fast that there are no boundary terms – this is justified because we have to assume enough decay of ϕ and its derivatives to make the integral for S converge.

1.3.1 Discretization

We use the lattice regularization only as a method to define the field theory. To analyze perturbation theory, we shall later take the continuum limit at a fixed ultraviolet cutoff in momentum space, to restore the Euclidean symmetries of the theory. The advantage of using the lattice is for us that we can use ordinary derivatives instead of functional derivatives (the latter can easily be defined in the continuum, but we shall not need that here and it loads the presentation with unnecessary discussions).

However, the importance of the lattice theory goes well beyond the purpose for which it is used here. Up to now, the lattice regularization provides the only nonperturbative regularization of functional integrals that works independently of the coupling strength. Also, to this day it remains the only nonperturbative regularization that does not break gauge-invariance.

We discretize space and time and introduce a finite volume. For convenience in using the Fourier transformation, we introduce the finite volume as a torus of sidelength L. We take a lattice of spacing $\varepsilon > 0$, and define

$$\Gamma = \Gamma_{\varepsilon,L} = \varepsilon \mathbb{Z}^d / L\mathbb{Z}^d \tag{1.61}$$

where $\varepsilon > 0$ is to be small, $L \gg 1$ and $\frac{L}{2\varepsilon} \in \mathbb{N}$. When no confusion can arise, we simply denote $\Gamma_{\varepsilon,L}$ by Γ. A particular choice of coordinates is to write Γ as

$$\{x \in \varepsilon\mathbb{Z}^d : -\frac{L}{2} \le x_k < \frac{L}{2} \quad \forall k \in \{1,\ldots,d\}\} \tag{1.62}$$

with periodic boundary conditions. The number of points of this lattice is

$$|\Gamma_{\varepsilon,L}| = (\frac{L}{\varepsilon})^d. \tag{1.63}$$

In the end, we are interested in the field theory on \mathbb{R}^d which means that we want to take the thermodynamic limit $L \to \infty$, and the continuum limit $\varepsilon \to 0$. The theory where only $\varepsilon \to 0$ is taken is a continuum field theory in a finite volume, the theory where only $L \to \infty$ is taken is a system of statistical mechanics. Since the number of lattice points goes to infinity in both of these limits by (1.63), both limits are systems with an infinite number of degrees of freedom, and hence interesting by themselves. The theory of critical phenomena, i.e. of the long-distance behaviour of the correlations of an infinite-volume system at fixed ε is closely related to studying the continuum limit of field theory [14].

The continuum derivative of a field $\phi : \mathbb{R}^d \longrightarrow \mathbb{R}$ is discretized by

$$d_\varepsilon\phi(x) = \left(\frac{1}{\varepsilon}(\phi(x + \varepsilon e_k) - \phi(x))\right)_{k \in \{1,\ldots,d\}} \tag{1.64}$$

where e_k is the unit vector in k-direction, and the discretized action is

$$S_\Gamma(\phi) = \varepsilon^d \sum_{x \in \Gamma} \left(\frac{1}{2} (d_\varepsilon \phi(x))^2 + \frac{m_\phi{}^2}{2} \phi(x)^2 + \lambda \phi(x)^4 \right) \tag{1.65}$$

where $\lambda \geq 0$ and $m \geq 0$. m is the mass and λ the self-coupling. In all detail,

$$S_\Gamma(\phi) = \varepsilon^d \sum_{x \in \Gamma} \left(\frac{1}{2\varepsilon^2} \sum_{k=1}^{d} (\phi(x + \varepsilon e_k) - \phi(x))^2 + \frac{m_\phi{}^2}{2} \phi(x)^2 + \lambda \phi(x)^4 \right) . \tag{1.66}$$

If $\phi_\Gamma : \Gamma \longrightarrow \mathbb{R}$ is obtained by restricting $\phi \in C^1(\mathbb{R}^d/L\mathbb{Z}^d, \mathbb{R})$ to the lattice Γ, $\phi_\Gamma = \phi|_\Gamma$, then

$$S_{\Gamma_{\varepsilon,L}}(\phi_{\Gamma_{\varepsilon,L}}) \xrightarrow[\varepsilon \to 0]{} S(\phi). \tag{1.67}$$

Many other choices for S have the same property, so the lattice action is not uniquely determined by the continuum action one wants to regularize. Whether different actions give the same continuum limit for the correlation functions of the statistical mechanical system given by S (defined in the next section) is a much more difficult question.

We now streamline the notation to avoid too many ε's and sums. We introduce the notation

$$\int_\Gamma dx\, F(x) = \varepsilon^d \sum_{x \in \Gamma} F(x), \tag{1.68}$$

which is suggested by the convergence

$$\int_{\Gamma_{\varepsilon,L}} dx\, F(x) \xrightarrow[\varepsilon \to 0]{} \int_{\mathbb{R}^d/L\mathbb{Z}^d} dx\, F(x) \tag{1.69}$$

for any $F \in C^1(\mathbb{R}^d/L\mathbb{Z}^d, \mathbb{C})$, and define a scalar product on \mathbb{C}^Γ by

$$(\phi, \psi)_\Gamma = \int_\Gamma dx\, \overline{\phi(x)} \psi(x). \tag{1.70}$$

The backward derivative d_ε^*, defined by

$$(d_\varepsilon^* \phi)_k(x) = \frac{1}{\varepsilon} (\phi(x - \varepsilon e_k) - \phi(x)), \qquad k \in \{1, \ldots, d\} \tag{1.71}$$

is the adjoint of the forward derivative d_ε:

$$(\phi, d_\varepsilon \psi)_\Gamma = (d_\varepsilon^* \phi, \psi)_\Gamma. \tag{1.72}$$

Equation (1.72) is proven by summation by parts. There are no boundary terms because we chose the lattice as a torus. Introducing the lattice Laplacian

$$\Delta = -d_\varepsilon^* d_\varepsilon \tag{1.73}$$

we can sum by parts in the sum for S_Γ, and get

$$S_\Gamma(\phi) = \epsilon^d \sum_{x \in \Gamma} \left[\phi(x) \frac{1}{2} ((-\Delta + m_\phi{}^2)\phi)(x) + \lambda \phi(x)^4 \right]$$

$$= \int_\Gamma dx \left(\frac{1}{2} \phi(x) ((-\Delta + m_\phi{}^2)\phi)(x) + \lambda \phi(x)^4 \right) \quad (1.74)$$

$$= \frac{1}{2} (\phi, (-\Delta + m_\phi{}^2)\phi)_\Gamma + \lambda \int_\Gamma dx\, \phi(x)^4.$$

The action of Laplacian is, more explicitly,

$$(-\Delta \phi)(x) = \frac{1}{\epsilon^2} \sum_{k=1}^d (2\phi(x) - \phi(x + \epsilon e_k) - \phi(x - \epsilon e_k)). \quad (1.75)$$

Since for all ϕ,

$$(\phi, -\Delta \phi)_\Gamma = (d_\epsilon \phi, d_\epsilon \phi)_\Gamma \geq 0, \quad (1.76)$$

$-\Delta$ is a nonnegative operator, just like its continuum counterpart.

Every linear operator on \mathbb{C}^Γ is, when represented in a basis, simply a matrix in $M_{|\Gamma|}(\mathbb{C})$. Choosing the orthonormal basis $(E_x)_{x \in \Gamma}$ by $E_x(y) = \delta_\Gamma(x, y)$, where

$$\delta_\Gamma(x, y) = \epsilon^{-d} \delta_{xy}, \quad (1.77)$$

we have $\phi(x) = (E_x, \phi)_\Gamma$ and

$$\int_\Gamma dy\, E_y(x) E_y(x') = \delta_\Gamma(x, x'), \quad (1.78)$$

so

$$(A\phi)(x) = \int_\Gamma dy\, A(x, y)\phi(x) \quad (1.79)$$

with the matrix representation (or "integral kernel")

$$A(x, y) = (E_x, AE_y) \quad (1.80)$$

of A. In particular, the unit matrix has kernel $1(x, y) = \delta_\Gamma(x, y)$. The corresponding distributional kernel in the continuum is, of course, $\delta(x - y)$. If A is invertible, the kernel of the inverse satisfies

$$\int_\Gamma dz\, A(x, z)(A^{-1})(z, y) = \delta_\Gamma(x, y). \quad (1.81)$$

We also define a derivative

$$\frac{\delta}{\delta J(x)} = \epsilon^{-d} \frac{\partial}{\partial J(x)} \quad (1.82)$$

so that

$$\frac{\delta}{\delta J(x)} J(y) = \delta_\Gamma(x, y). \quad (1.83)$$

This also agrees with the formal continuum convention

$$\frac{\delta}{\delta J(x)} J(y) = \delta(x - y).\tag{1.84}$$

With this, we have

$$\frac{\delta}{\delta J(x)} e^{(J,\phi)_\Gamma} = J(x) e^{(J,\phi)_\Gamma}.\tag{1.85}$$

The Laplacian has the kernel

$$-\Delta(x,y) = \varepsilon^{-d-2} \sum_{k=1}^{d} (2\,\delta_{xy} - \delta_{x+\varepsilon e_k, y} - \delta_{x-\varepsilon e_k, y})\tag{1.86}$$

$$= \varepsilon^{-2} \sum_{k=1}^{d} (2\delta_\Gamma(x,y) - \delta_\Gamma(x + \varepsilon e_k, y) - \delta_\Gamma(x - \varepsilon e_k, y))$$

(in the continuum, the distributional kernel would be $(-\Delta\delta)(x - y)$).

1.3.2 The Partition Function and Correlations

With $D_\Gamma\phi = \prod_{x\in\Gamma} d\phi(x)$, the functional integral with sources $J : \Gamma \longrightarrow \mathbb{R}$ is defined as

$$Z_{\Gamma,\lambda}(J) = \int D_\Gamma\phi\, e^{-S_\Gamma(\phi)+(J,\phi)_\Gamma}.\tag{1.87}$$

This integral converges if $\lambda \geq 0$ and $m_\phi > 0$, or if $\lambda > 0$ and $m_\phi{}^2 = \mu \in \mathbb{R}$ is arbitrary.

Expectation values of observables \mathcal{O}, that is, functions of the field ϕ, are defined by

$$\langle\mathcal{O}\rangle = \frac{1}{Z_{\Gamma,\lambda}(0)} \int D_\Gamma\phi\, e^{-S_\Gamma(\phi)} \mathcal{O}(\phi).\tag{1.88}$$

Only those \mathcal{O} are admitted for which this integral converges; this includes, e.g., all polynomials in the fields, i.e., all moments of the measure with density $e^{-S_\Gamma(\phi)}$.

Definition 1.9. *The correlation functions, also called Green functions, are*

$$\langle\phi(x_1)\ldots\phi(x_n)\rangle = \frac{1}{Z_{\Gamma,\lambda}(0)} \int D_\Gamma\phi\, e^{-S_\Gamma(\phi)} \prod_{k=1}^{n} \phi(x_k).\tag{1.89}$$

These moments can be calculated as derivatives of $Z_{\Gamma,\lambda}(J)$ with respect to J, via

$$\langle\phi(x_1)\ldots\phi(x_n)\rangle = \left[\frac{1}{Z_{\Gamma,\lambda}(J)} \frac{\delta^n}{\delta J(x_1)\ldots\delta J(x_n)} Z_{\Gamma,\lambda}(J)\right]_{J=0}.\tag{1.90}$$

Thus the partition function $Z_{\Gamma,\lambda}(J)$ with source terms J contains all the information about the correlation functions. The derivatives

$$
\left[\frac{\delta^n}{\delta J(x_1) \dots \delta J(x_n)} \log Z_{\Gamma,\lambda}(J) \right]_{J=0}
\tag{1.91}
$$

are the *connected correlation functions*. In Chap. 2, we shall prove that the connected correlation functions are sums of values of connected Feynman graphs.

The significance of the Green functions is that the S-matrix can be constructed from them by the LSZ-formulas [9].

Note that in (1.88), we are integrating over all field configurations ϕ : $\Gamma \longrightarrow \mathbb{R}$ in the same way as we were integrating over all polygonal paths in the quantum-mechanical case. In the limit $\varepsilon \longrightarrow 0$, there is no reason to assume that these configurations arise from C^1 fields ϕ on $\mathbb{R}^d/L\mathbb{Z}^d$. On the contrary, the typical field configurations are not C^1, just as the typical path contributing to the Wiener integral is not C^1.

A simple geometrical interpretation of the ϕ field in $d = 2$ is that of the height of a surface, so $Z_\Gamma(J)$ describes a statistical mechanical ensemble of random surfaces. The probability density is given by the Boltzmann weight e^{-S}/Z, so the action S effectively appears as an energy function. The form of S shows that large values of $(d_\varepsilon \phi)^2$, of $m_\phi{}^2 \phi^2$, and of $\lambda \phi^4$ are supressed. Note that, if $\varepsilon \to 0$, the scale of smoothness, which is ε, goes to zero, so there is no reason to expect the random surfaces to be C^1 in the continuum limit.

The particle physics interpretation of Z is that of a massive field theory with a repulsive self-interaction $\lambda \geq 0$. For $\lambda = 0$, the particles are free. Note that while the measure $D_\Gamma \phi$ has no limit as $|\Gamma| \to \infty$, the Gaussian measure obtained by combining the quadratic part of the action with $D_\Gamma \phi$, and normalizing suitably, has a limit. This is a natural procedure from the physical point of view because the propagation of free particles is then given by the covariance of the Gaussian measure.

The lattice Laplacian generates a random walk ensemble, which is yet another way of viewing the statistical mechanical system. Important nonperturbative results, like triviality of ϕ^4 in $d \geq 5$, are based on the random walk representation [10, 11]. In that language, $m_\phi{}^2$ is the killing rate and $\lambda > 0$ means that the random walks interact: they repel each other.

For $m_\phi{}^2 > 0$, $-\Delta + m_\phi{}^2 \geq m_\phi{}^2 > 0$, so the Gaussian integral converges. Since $\lambda \geq 0$, the ϕ^4 term only improves the convergence. The case $m_\phi = 0$ (massless field) can also be defined (one has to take care of the zero modes of Δ, the "torons" (=constant configurations). This is no problem on $\Gamma_{\varepsilon,L}$, but the limit $L \to \infty$ becomes more difficult in that case. The slow decay of $(-\Delta)^{-1}(x,y)$ in $d = 2$ causes the Mermin-Wagner theorem [15]. The limit $\varepsilon \to 0$ and $L \to \infty$ does not exist for $d = 2$: there is no massless field in $d = 2$ [16]. We shall keep $m_\phi > 0$ when analyzing perturbative field theory.

1.3.3 The Ising Model

We show that the Ising model is a limit of the statistical mechanical system defined by our lattice field theory. The local term in the action is drawn in Fig. 1.1.

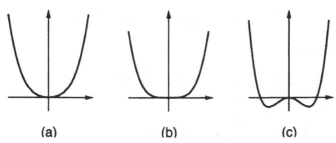

Fig. 1.1. The ϕ^4-potential for different values of $m_\phi{}^2$. (a) $m_\phi{}^2 > 0$, (b) $m_\phi{}^2 = 0$, (c) $m_\phi{}^2 < 0$.

Suppose that $m_\phi{}^2 = \mu < 0, \lambda > 0$, and let $\varepsilon = 1$ for simplicity. Since

$$\frac{\mu}{2}\phi^2 + \lambda\phi^4 = \lambda(\phi^2 + \frac{\mu}{4\lambda})^2 - \frac{\mu^2}{16\lambda}, \tag{1.92}$$

the local term in the action is of the form drawn in Fig. 1.1(c).

We use

$$-\frac{1}{2}\sum_{x,k}(\phi(x + e_k) - \phi(x))^2 = \sum_{x,k}\phi(x)\phi(x + e_k) - d\sum_x \phi(x)^2 \tag{1.93}$$

to rewrite the action in the exponent of the integrand for $Z_\Gamma(J)$ as

$$-S_\Gamma(\phi) = \sum_{x,k}\phi(x)\phi(x + e_k) - \sum_x((\frac{\mu}{2} + d)\phi(x)^2 - \lambda\phi(x)^4)$$

$$= \sum_{x,k}\phi(x)\phi(x + e_k) - \lambda\sum_x\left(\phi(x)^2 + \frac{\mu + 2d}{4\lambda}\right)^2 + \frac{(\mu + 2d)^2}{16\lambda}|\Gamma|. \tag{1.94}$$

If $\mu + 2d < 0$,

$$Z_\Gamma(J) = K^{|\Gamma|}\int \prod_{x \in \Gamma}\left(d\phi(x)\, e^{J(x)\phi(x) - \lambda(\phi(x)^2 - \alpha^2)^2}\right)\prod_{x,k} e^{\phi(x)\phi(x + e_k)} \tag{1.95}$$

with

$$K = \exp(\frac{(\mu + 2d)^2}{16\lambda}), \quad \alpha = \sqrt{\frac{|\mu + 2d|}{4\lambda}}. \tag{1.96}$$

The last factor in the integrand suppresses values of $|\phi(x)|$ far away from α. In the limit $\lambda \to \infty$, α kept fixed, the last factor becomes proportional to a sum of δ functions because it is peaked at $\pm\alpha$, and because

$$\int e^{-\lambda(\phi^2-\alpha^2)}d\phi \;\approx\; \int e^{-2\alpha^2\lambda(\phi-\alpha)^2}d\phi + \int e^{-2\alpha^2\lambda(\phi+\alpha)^2}d\phi$$

$$\approx\; 2\sqrt{\frac{\pi}{2\alpha^2\lambda}}. \tag{1.97}$$

Therefore, the rescaled partition function

$$\tilde{Z}_\Gamma(J) = \lambda^{|\Gamma|/2}\, Z_\Gamma(J) \tag{1.98}$$

is a constant times

$$Z_\Gamma^I(J) = \sum_{\phi:\Gamma\to\{-\alpha,\alpha\}} e^{(J,\phi)_\Gamma}\, e^{\sum_{x,k}\phi(x)\phi(x+e_k)}. \tag{1.99}$$

Rescaling the fields, $s(x) = \frac{\phi(x)}{\alpha}$, and the sources, $H(x) = \frac{J(x)}{\alpha}$, we recognize Z_Γ^I as the partition function of the Ising model

$$Z_\Gamma^I(H) = \sum_{s:\Gamma\to\{-1,+1\}} e^{-\beta E(s)} \tag{1.100}$$

with the interaction

$$E(s) = -\sum_{x,k} s(x)s(x+e_k) - \sum_x H(x)s(x)) \tag{1.101}$$

at inverse temperature

$$\beta = \alpha^2 = \lim_{\lambda\to\infty} \frac{|\mu(\lambda) + 2d|}{4\lambda}. \tag{1.102}$$

The source term now appears as a spatially varying magnetic field H. The interaction is between nearest neighbours, and it is ferromagnetic, i.e. equal spins have bigger Boltzmann weight $e^{\beta s(x)s(x+e_k)}$ than opposite neighbouring spins. For this reason, the Laplacian is also called ferromagnetic.

The quantities that can be put in relation to physical measurements are the correlation functions defined in Definition 1.9, i.e. the functions

$$\langle F(s)\rangle = \frac{1}{Z} \sum_{s:\Gamma\to\{1,-1\}} F(s)\, e^{-\beta E(s)}. \tag{1.103}$$

For instance, $\langle s_x\rangle$ is the local magnetization and $\sum_y(\langle s_x s_y\rangle - \langle s_x\rangle\langle s_y\rangle)$ is the magnetic susceptibility of the spin system.

This is not the place to go into details about the Ising model and statistical mechanics, since there are many excellent references in the literature [17, 18, 19, 20, 21, 22]. We only mention very briefly some of the main results. The behaviour of the Ising model at $H = 0$ depends on β: for small β (high temperature), the spins are disordered, and $\langle s_x s_y\rangle \to 0$ exponentially fast as $|x - y| \to \infty$. For large β,

$$\langle s_x s_y \rangle \xrightarrow[|x-y|\to\infty]{} \sigma(\beta)^2 \neq 0. \tag{1.104}$$

$\sigma(\beta)$ vanishes at a critical point β_c. Note that for $\beta > \beta_c$, the correlation functions depend on the boundary conditions. With our choice, we pick the symmetric phase where $\langle s_x \rangle = 0$. Had we chosen boundary conditions that favour spin alignment, such as the condition that all spins on the boundary are $+1$, we could have achieved $\langle s_x \rangle = \sigma(\beta)$ so that

$$\langle s_x s_y \rangle - \langle s_x \rangle \langle s_y \rangle \xrightarrow[|x-y|\to\infty]{} 0, \tag{1.105}$$

again, exponentially fast for $\beta > \beta_c$. Only at $\beta = \beta_c$, one has a power law decay. It is at this point that one has to look for a continuum limit if one wants to construct a continuum theory without a momentum space cutoff directly from a lattice theory. As mentioned, we shall take $\varepsilon \to 0$ while keeping a finite cutoff Λ_0 in momentum space, and then study the limit $\Lambda_0 \to \infty$ directly in the continuum (which is a different limit).

1.4 Free Fields

For $\lambda = 0$, the integral is Gaussian, and this corresponds to the case of free (noninteracting) particles. The free action is diagonal in Fourier space. We give only the main definitions of the Fourier transform here; details about the Fourier transform are in Appendix A.2.

The dual lattice to Γ is $\Gamma^* = \frac{2\pi}{L}\mathbb{Z}^d / \frac{2\pi}{\varepsilon}\mathbb{Z}^d$. We define $p \cdot x = p_1 x_1 + \ldots + p_d x_d$, $\int_{\Gamma^*} dp \hat{F}(p) = L^{-d} \sum_{p\in\Gamma^*} \hat{F}(p)$, and $\delta_{\Gamma^*}(p,q) = L^d \delta_{pq}$. The Fourier transform of $f : \Gamma \to \mathbb{C}$ is $\hat{f} : \Gamma^* \to \mathbb{C}$, given by

$$\hat{f}(p) = \int_\Gamma dx \, e^{-ip\cdot x} f(x). \tag{1.106}$$

The inverse relation is

$$f(x) = \int_{\Gamma^*} dp \, e^{ip\cdot x} \hat{f}(p). \tag{1.107}$$

We denote the scalar product for functions $\hat{\phi}, \hat{\psi} \in \mathbb{C}^{\Gamma^*}$ by

$$(\hat{\phi}, \hat{\psi})_{\Gamma^*} = \int_{\Gamma^*} dp \, \overline{\hat{\phi}(p)} \hat{\psi}(p). \tag{1.108}$$

Fourier transformation is unitary: $(\hat{\phi}, \hat{\psi})_{\Gamma^*} = (\phi, \psi)_\Gamma$. The integral kernel of a linear operator transforms as

$$\hat{A}(p,q) = \int_\Gamma dx \int_\Gamma dy \, e^{-ip\cdot x} A(x,y) \, e^{iq\cdot y}. \tag{1.109}$$

By(1.86) and (1.109), the Fourier transform of the kernel of the lattice Laplacian is

$$- \hat{\Delta}(p,q) = \delta_{\Gamma^\bullet}(p,q) D_\varepsilon(p) \qquad (1.110)$$

with

$$D_\varepsilon(p) = \frac{2}{\varepsilon^2} \sum_{k=1}^{d} (1 - \cos(\varepsilon p_k)). \qquad (1.111)$$

That is, the Laplacian is diagonal in the Fourier basis. Thus

$$(-\hat{\Delta} + m_\phi^2)^{-1}(p,q) = \delta_{\Gamma^\bullet}(p,q) \frac{1}{m_\phi^2 + D_\varepsilon(p)} = \delta_{\Gamma^\bullet}(p,q) \, \hat{C}_\varepsilon(p). \qquad (1.112)$$

In the limit $\varepsilon \to 0$, $D_\varepsilon(p) \to p^2$, so

$$(-\Delta + m_\phi^2)^{-1}(p,q) \xrightarrow[\varepsilon \to 0]{} L^d \delta_{pq} \frac{1}{p^2 + m_\phi^2} = \delta_{\Gamma^\bullet}(p,q) \, \hat{C}(p). \qquad (1.113)$$

The quadratic form defined by the action is

$$\begin{aligned}
(\phi, (-\Delta + m_\phi^2)\phi)_\Gamma &= \int_{\Gamma^\bullet} dp \, \overline{\hat{\phi}(p)} \, (m_\phi^2 + D_\varepsilon(p)) \hat{\phi}(p) \\
&= \int_{\Gamma^\bullet} dp \, \hat{\phi}(-p)(m_\phi^2 + D_\varepsilon(p)) \hat{\phi}(p) \quad (1.114)
\end{aligned}$$

because $\phi(x) \in \mathbb{R} \; \forall x$ implies $\overline{\hat{\phi}(p)} = \hat{\phi}(-p)$. Thus, the partition function is

$$Z_{\Gamma,\lambda}(J) = Z_{\Gamma,0}(0) \int d\mu_{C_\Gamma}(\phi) \, e^{-\lambda \int_\Gamma \phi(x)^4 dx + (J,\phi)_\Gamma} \qquad (1.115)$$

where the normalization constant of the Gaussian measure

$$Z_{\Gamma,0}(0) = \det(2\pi C_\Gamma)^{1/2} = \prod_{p \in \Gamma^\bullet} \left(\frac{2\pi}{\varepsilon^d} L^d \frac{1}{m_\phi^2 + D_\varepsilon(p)} \right)^{1/2} \qquad (1.116)$$

is unimportant because it drops out in the quotient defining the correlation functions. We can therefore redefine $Z_{\Gamma,\lambda}(J)$ such that it does not contain that constant, and get

$$Z_{\Gamma,\lambda}(J) = \int d\mu_{C_\Gamma}(\phi) \, e^{-\lambda V(\phi) + (J,\phi)_\Gamma} \qquad (1.117)$$

with $V(\phi) = \int_\Gamma \phi(x)^4 dx$.

The quadratic part of the action, which defines the covariance C_Γ of the Gaussian measure, describes free particles that propagate from x to y (graphically later denoted by a line connecting x and y). The interaction term describes the influence of particles on one another; for the ϕ^4 theory one may imagine it as an interaction between two particles, denoted by a vertex with four legs. The interaction is local. This is, in absence of other fields, the only way to maintain Lorentz invariance in the real-time theory.

1.5 Properties of the Free Covariance

The properties of $C_\Gamma(x, y)$ in the limits $L \to \infty$ and $\varepsilon \to 0$ are central to the question why renormalization is necessary. For $\varepsilon > 0$ and $L < \infty$,

$$C_\Gamma(x, y) = \int_{\Gamma^*} dp \, \frac{e^{ip(x-y)}}{m_\phi{}^2 + D_\varepsilon(p)}. \tag{1.118}$$

For $m_\phi > 0$, the denominator is always nonzero, so the summand is continuous in p on $\mathbb{R}^d / \frac{2\pi}{\varepsilon} \mathbb{Z}^d$ and one can take the limit $L \to \infty$, in which the Riemann sum $\int_{\Gamma^*} dp$ becomes the integral $\frac{1}{(2\pi)^d} \int_B dp$ where $B = \mathbb{R}^d / \frac{2\pi}{\varepsilon} \mathbb{Z}^d$. In other words, the integral is taken over $[-\frac{\pi}{\varepsilon}, \frac{\pi}{\varepsilon})^d$. Thus

$$\lim_{L \to \infty} C(x, y) = \frac{1}{(2\pi)^d} \int_B d^d p \, \frac{e^{ip(x-y)}}{m_\phi{}^2 + D_\varepsilon(p)}. \tag{1.119}$$

The limit $\varepsilon \to 0$ exists if $x - y \neq 0$, and is given by the integral

$$C(x, y) = \frac{1}{(2\pi)^d} \int_{\mathbb{R}^d} d^d p \, \frac{e^{ip(x-y)}}{m_\phi{}^2 + p^2}. \tag{1.120}$$

One can calculate $C(x, y)$ in terms of a modified Bessel function

$$C(x, y) = (2\pi)^{-d/2} \left(\frac{m_\phi}{|x - y|} \right)^{\frac{d}{2}-1} K_{\frac{d}{2}-1}(m|x - y|). \tag{1.121}$$

In particular, for $d = 3$,

$$C(x, y) = \frac{1}{4\pi|x - y|} e^{-m_\phi|x-y|}. \tag{1.122}$$

Most properties of C can, however, be found without any explicit formulas. For instance, the exponential decay as $|x - y| \to \infty$ follows directly from the analyticity of $1/(p^2 + m_\phi{}^2)$ for $|p| < m$. By translation and rotation invariance, $C(x, y)$ is a function of $|x - y|$. Scaling out the factor m_ϕ gives

$$C(x, y) = \int \frac{d^d p}{(2\pi)^d} \frac{e^{ip(x-y)}}{m_\phi{}^2 + p^2} = m_\phi{}^{d-2} \tilde{C}_d(m_\phi|x - y|) \tag{1.123}$$

with

$$\tilde{C}_d(t) = \int \frac{d^d p}{(2\pi)^d} \frac{e^{ip_1 t}}{1 + p^2}. \tag{1.124}$$

Lemma 1.10. *(i) For all $t > 0$: $\tilde{C}_d(t) > 0$*
(ii) For $t \geq 1$,

$$\tilde{C}_d(t) \leq \text{const } t^{-\frac{d-1}{2}} e^{-t} \tag{1.125}$$

(iii) For $t \leq 1$,

$$\tilde{C}_d(t) = \text{const} \begin{cases} t^{-(d-2)} & \text{if } d \geq 3 \\ \log t & \text{if } d = 2 \end{cases} \tag{1.126}$$

The bounds become sharp as $t \to 0$. The constant depends only on d.

Proof. This is a standard proof [6]. For convenience of the reader, it is reproduced in Appendix A.3. ∎

Note that $C(x, x)$ does not exist: there is a singularity in the propagator at short distances. This is the usual behaviour of a Green function (cf. the quantum mechanical propagator), but also the reason why renormalization is necessary. For large $|x - y|$, $|C(x, y)| \leq \text{const } e^{-m_\phi |x-y|}$. If m_ϕ were zero, this would be replaced by a power law decay, and one would have an infrared problem as well. For $\varepsilon > 0$, there is no singularity, but the propagator grows as $\varepsilon \to 0$.

Remark 1.11. The lattice propagator obeys

$$C_{\Gamma_{\varepsilon,L}}(x, x) \sim \text{const} \begin{cases} \log \varepsilon & \text{for } d = 2 \\ \varepsilon^{-(d-2)} & \text{for } d \geq 3 \end{cases} \tag{1.127}$$

as $\varepsilon \to 0$. For $d = 2$, the constant depends on $m_\phi > 0$.

Proof. Because

$$p^2 \geq \frac{2}{\varepsilon^2}(1 - \cos(p\varepsilon)) \geq \frac{2}{\pi^2}p^2, \tag{1.128}$$

we have

$$\int\limits_{[-\frac{\pi}{\varepsilon}, \frac{\pi}{\varepsilon}]^d} \frac{d^d p}{p^2 + m_\phi^2} \leq \int\limits_{[-\frac{\pi}{\varepsilon}, \frac{\pi}{\varepsilon}]^d} \frac{d^d p}{m_\phi^2 + D_\varepsilon(p)} \leq \int\limits_{[-\frac{\pi}{\varepsilon}, \frac{\pi}{\varepsilon}]^d} \frac{d^d p}{m_\phi^2 + \frac{2}{\pi^2}p^2}. \tag{1.129}$$

Because

$$\left\{p : |p| \leq \frac{\pi}{\varepsilon}\right\} \subset \left[-\frac{\pi}{\varepsilon}, \frac{\pi}{\varepsilon}\right]^d \subset \left\{p : |p| \leq \frac{\sqrt{2}\pi}{\varepsilon}\right\} \tag{1.130}$$

we have

$$\int\limits_{\{p : |p| \leq \frac{\pi}{\varepsilon}\}} \frac{d^d p}{p^2 + m_\phi^2} \leq C_{\Gamma_{\varepsilon,L}}(x, x) \leq \int\limits_{\{p : |p| \leq \frac{\sqrt{2}\pi}{\varepsilon}\}} \frac{d^d p}{m_\phi^2 + \frac{2}{\pi^2}p^2}. \tag{1.131}$$

Both the lower and the upper bound behave in the same way, namely as given in (1.127) (see also (1.137)). ∎

1.6 Problems With and Without Cutoffs

To see why the short-distance singularity of the continuum covariance is a problem, consider $\langle \phi(x)^2 \rangle_{\Gamma, \lambda=0}$. This function diverges for $\varepsilon \to 0$,

$$\int d\mu_{C_\Gamma}(\phi)\, \phi(x)^2 = C_\Gamma(x, x) \xrightarrow[\varepsilon \to 0]{} \infty. \tag{1.132}$$

So, the second moment $\langle \phi(x)\phi(y) \rangle = C(x, y)$ has a singularity at $x = y$ (coinciding points). This by itself would not be a problem. The problem is that these singularities cause divergences in the continuum limit of the partition function, and of the correlations, defined above. For instance, the integral

$$\int d\mu_{C_\Gamma}(\phi)\, \phi(x)^4 = 3\, C_\Gamma(x, x)^2 \tag{1.133}$$

occurs in the expansion

$$\int d\mu_{C_\Gamma}(\phi)\, e^{\lambda \int_\Gamma \phi(x)^4 dx} = 1 + \lambda \int_\Gamma dx \int d\mu_{C_\Gamma}(\phi)\, \phi(x)^4 + \ldots \tag{1.134}$$

already in first order, and it diverges as $\varepsilon \to 0$.

There are two ways out of this problem – renormalize, or never remove the cutoff. Chaps. 2 and 3 are concerned with renormalization. In the remainder of this section, we briefly discuss why keeping a cutoff is a problem for the particle physics interpretation of the model, and then motivate Wick ordering.

We don't know if the world is a lattice or not, but on measurable scales, it is Lorentz invariant. A lattice model is not even rotation invariant, so one has to take $\varepsilon \to 0$ to get back Euclidean invariance. But one does not have to take a lattice to regularize the propagator. One could, instead, also take the following continuum limit with a rotation-invariant cutoff Λ_0. Let $K \in C^\infty(\mathbb{R}_0^+, [0, 1])$ obey

$$K(x) = \begin{cases} 1 & \text{if } 0 \le x \le 1 \\ 0 & \text{if } x \ge 4, \end{cases} \tag{1.135}$$

$K'(x) < 0$ for $x \in (1, 4)$, and let $\Lambda_0 \gg 1$. Define

$$C_{0, \Lambda_0}(x, y) = \frac{1}{(2\pi)^d} \int d^d p\, \frac{e^{ip(x-y)}}{p^2 + m_\phi^2}\, K\!\left(\frac{p^2}{\Lambda_0^2}\right). \tag{1.136}$$

Then $C_{0, \Lambda_0}(x, y)$ has no short-distance singularity because, as we shall show in (3.7), for $\Lambda_0 \ge 2$,

$$|C_{0, \Lambda_0}(0, 0)| \le \text{const} \begin{cases} \log(\Lambda_0) & \text{if } d = 2 \\ \Lambda_0^{d-2} & \text{if } d \ge 3. \end{cases} \tag{1.137}$$

The constant depends on m_ϕ, and grows like $|\log m_\phi|$ as $m_\phi \to 0$ in $d = 2$, and it is independent of m_ϕ for $d \ge 3$. Hence all moments of the measure with this

covariance are finite (but Λ_0-dependent). Moreover, $C_{0,\Lambda_0}(x,y) = \tilde{C}_{\Lambda_0}(|x-y|)$ is rotation invariant.

One could also have taken a Pauli-Villars cutoff

$$\frac{1}{p^2 + m_\phi{}^2} - \frac{1}{p^2 + \Lambda_0^2} \tag{1.138}$$

instead of $\frac{1}{p^2+m_\phi{}^2}$: this removes the singularity, at least for $d \le 3$, and it is a less restrictive cutoff. But all these cutoffs violate physical positivity: a Euclidean invariant field theory with a genuine Hilbert space (i.e., where the scalar product is positive definite, as it should be), the two-point function has the Källen-Lehmann representation

$$\langle \hat{\phi}(p)\hat{\phi}(-p) \rangle = \int_0^\infty d\rho(\mu) \frac{1}{\mu^2 + p^2} \tag{1.139}$$

where $d\rho(\mu)$ is a positive measure. But then the two-point function cannot decay faster than $1/p^2$ for $|p| \to \infty$. Thus any cutoff that makes \hat{C} integrable leads to problems with physical positivity. We therefore have to remove the cutoff in the end.

Another reason for not keeping Λ_0 is that such a momentum space is not gauge invariant.

Since there seems to be no way of smoothing the propagator without losing some physical requirement, one can only try to change the interaction. The first step, which suffices to overcome the above divergence, but not others, is to do Wick ordering. It means replacing the monomials in the action by their Wick ordered counterparts (which will be defined in the next chapter), for instance replacing $\phi(x)^2$ by

$$: \phi(x)^2 := \phi(x)^2 - C_\Gamma(x,x). \tag{1.140}$$

Then, by definition,

$$\int : \phi(x)^2 : d\mu_{C_\Gamma}(\phi) = 0 \tag{1.141}$$

even though $\int \phi(x)^2 d\mu_{C_\Gamma}(\phi)$ diverges as $\varepsilon \to 0$. Similarly, one replaces $\phi(x)^4$ by

$$: \phi(x)^4 := \phi(x)^4 - 6\,C_\Gamma(x,x)\phi(x)^2 + 3\,C_\Gamma(x,x)^2 \tag{1.142}$$

in the interaction term, which then satisfies

$$\int : \phi(x)^4 : d\mu_{C_\Gamma}(\phi) = 0. \tag{1.143}$$

For the moment we just note that the problems with the limit $\varepsilon \to 0$ in (1.134) disappear to first order, i.e. the first order expansion in λ of the redefined

$$Z_{\Gamma,\lambda}(J) = \int d\mu_{C_\Gamma}(\phi)\,e^{-\lambda \int_\Gamma :\phi(x)^4:dx + (J,\phi)_\Gamma} \tag{1.144}$$

is bounded uniformly in ε. There is no claim that this cures higher order divergences in general, and it doesn't. For $\varepsilon > 0$, let us write this out as

$$Z_{\Gamma,\lambda}(J) = \frac{1}{Z_{\Gamma,0}(0)} \int D_\Gamma \phi \, e^{-(\phi,(-\Delta+m_\phi{}^2)\phi)_\Gamma + (J,\phi)_\Gamma}$$
$$\cdot e^{-\lambda \int_\Gamma dx (\phi(x)^4 - 6C_\Gamma(x,x)\phi(x)^2 + 3C_\Gamma(x,x)^2)}. \qquad (1.145)$$

By the lattice translation invariance, $C_\Gamma(x,x) = C_\Gamma(0,0)$ is a constant, so we have added the ε-dependent term

$$6\,\lambda C_\Gamma(0,0) \int_\Gamma dx\, \phi(x)^2 - 3\,\lambda\, C_\Gamma(0,0)^2 \int_\Gamma dx \qquad (1.146)$$

to the action.

The second of these terms is just a constant, which can be absorbed in a redefinition of Z. The first term is a correction of the mass term $-6\lambda C_\Gamma(0,0)$, which diverges as $\varepsilon \to 0$ by (1.127). The interpretation of this term is the main point of field-theoretic renormalization: the extra term does not change the functional form of the action $S_\Gamma(\phi)$, but only the value of the *bare mass*, that is, the mass parameter in the action, from $m_\phi{}^2$ to $m_\phi{}^2 - 6\lambda C_\Gamma(0,0)$. The bare mass is not observable because the particles are not free. The observed mass contains the interaction effects. There is, as we shall see, a one-to-one relationship between the bare and the observed mass, so one may fix the latter by experiment. This means that it is fixed as the value of a correlation function at a particular momentum. The simplest choice is

$$\int dx \, \langle \phi(0)\phi(x) \rangle = \frac{1}{m_\phi{}^2}. \qquad (1.147)$$

Adjusting everything this way, one has removed the divergences in first order perturbation theory.

The additional terms in S_Γ are *local*, so there is no problem with Lorentz invariance, once the correlation functions have been shown to converge.

If one can absorb all divergent terms into a redefinition of finitely many parameters of the action (preferably, only few of them), such as $m_\phi{}^2$ and λ, to any order in the expansion, then the theory is called perturbatively renormalizable. In such a theory, a finite number of parameters have to be fixed by experiment. Once this is done, one can predict the results of other experiments, by *renormalized perturbation theory*, in which no divergences appear, and compare them to measurements.

Of course, the above discussion has not really proved anything of that kind. The divergences may reappear in higher orders, in particular because the terms added to make the first order finite may actually cause infinities in higher orders. It requires more insight into the structure of field theory to see that this is not so and to understand how renormalization really works (also, to say it again, Wick ordering is not sufficient for $d > 2$).

It will turn out that Wilson's form of the renormalization group, (see Chap. 3), provides a different point of view of this subtraction procedure, as a change of initial conditions for the sequence of effective actions. In particular, it can be formulated without any explicit reference to the bare parameters of the action, that is, to the very high energy behaviour, once the form of the action at that energy is determined. It relates only quantities at energy scales that are independent of, and much smaller than, the cutoff.

In the next two chapters, we shall develop the formalism and prove some basic results about perturbative renormalization, that is, statements about the Green functions that hold in a *formal* power series expansion in λ. The question about convergence or divergence of the expansion in λ for the ϕ^4 theory is much harder and is discussed here only in a simplified case, namely the flow generated by "bubble" graphs, which leads to the so-called Landau pole in four dimensions (see the last section of Chap. 3).

2. Techniques

To get a deeper understanding, as well as the ability to do field theoretic calculations, we need to learn the standard techniques of Gaussian integration. Not all of this will be needed in Chap. 3, but these standard techniques are a prerequisite for reading most of the current literature in field theory.

Most of this chapter is concerned with Feynman graph techniques. Emphasis is put on defining Feynman graphs in a simple but precise way and on giving an unambiguous meaning to the "sum over values of Feynman graphs". Although every statement about Feynman graphs can in principle be restated without a reference to a picture, their value for thinking about field-theoretic problems goes far beyond mnemotechnics. We shall see, for instance, that some combinatorial factors are very easily – and safely – calculated with the help of Feynman graphs. Needless to say, they are also the central tool of the detailed calculations that are the basis for the great success of perturbative quantum field theory in particle physics.

2.1 Integration by Parts

Let A be a positive operator on \mathbb{R}^Γ and C its inverse. Recall that $\frac{\delta}{\delta\phi(x)} = \varepsilon^{-d}\frac{\partial}{\partial\phi(x)}$. Since $\int_\Gamma dz\, C(x,z)A(z,y) = \varepsilon^{-d}\delta_{xy}$, and $\frac{\delta}{\delta\phi(x)}\phi(y) = \varepsilon^{-d}\delta_{xy}$, we have

$$\phi(x) = \int_\Gamma dy\, C(x,y)\,\frac{\delta}{\delta\phi(y)}\,\frac{1}{2}(\phi, A\phi)_\Gamma, \tag{2.1}$$

so

$$\phi(x)\, e^{-\frac{1}{2}(\phi, A\phi)_\Gamma} = -\int_\Gamma dy\, C(x,y)\frac{\delta}{\delta\phi(y)}e^{-\frac{1}{2}(\phi, A\phi)_\Gamma}. \tag{2.2}$$

With this we get by integration by parts

$$\begin{aligned}
& \int D_\Gamma\,\phi\, F(\phi)\,\phi(x)\, e^{-\frac{1}{2}(\phi, A\phi)} \\
& = -\int_\Gamma dy\, C(x,y)\int D_\Gamma\,\phi\, F(\phi)\frac{\delta}{\delta\phi(y)}e^{-\frac{1}{2}(\phi, A\phi)} \\
& = \int_\Gamma dy\, C(x,y)\int D_\Gamma\,\phi\, e^{-\frac{1}{2}(\phi, A\phi)}\frac{\delta}{\delta\phi(y)}F(\phi).
\end{aligned} \tag{2.3}$$

Inserting the normalization, we get the integration by parts formula

$$\int d\mu_{C_\Gamma}(\phi)\, \phi(x) F(\phi) = \int_\Gamma dy\, C(x,y) \int d\mu_C(\phi)\, \frac{\delta F}{\delta \phi(y)}. \tag{2.4}$$

2.2 Wick Ordering

We have already seen that Wick ordering can cancel some divergences in the continuum limit. Since Wick ordering alone does not remove all divergences, this may not seem a big deal (and one can do renormalization without Wick ordering). However, Wick ordering is natural, because it simply corresponds to introducing orthogonal polynomials associated to the measure $d\mu_C$ given by the weight function $e^{-\frac{1}{2}(\phi, A\phi)_\Gamma}$, and it leads to substantial simplifications both in the Feynman graph representation and the renormalization group equation.

2.2.1 Definition and Main Properties

Let $C = C_\Gamma$ be a nonnegative symmetric operator on \mathbb{C}^Γ. For $J : \Gamma \to \mathbb{C}$, let

$$\mathcal{W}_\Gamma(J, \phi) = e^{i(J,\phi)_\Gamma + \frac{1}{2}(J, C\, J)_\Gamma}. \tag{2.5}$$

Definition 2.1. *Let \mathcal{P}_Γ be the algebra of polynomials in $(\phi(x))_{x\in\Gamma}$. Wick ordering is the \mathbb{C}-linear map $\Omega_C : \mathcal{P}_\Gamma \to \mathcal{P}_\Gamma$ that takes the following values on the monomials:*

$$\Omega_C(1) = 1, \tag{2.6}$$

and for $n \geq 1$ and $x_1, \ldots, x_n \in \Gamma$ (not necessarily distinct)

$$\Omega_C(\phi(x_1)\ldots\phi(x_n)) = \left[\prod_{k=1}^n \frac{1}{i}\frac{\delta}{\delta J(x_k)}\, \mathcal{W}_\Gamma(J,\phi)\right]_{J=0}. \tag{2.7}$$

Remark 2.2. Ω_C *is well-defined because for all permutations $\pi \in \mathcal{S}_n$*

$$\Omega_C(\phi(x_1)\ldots\phi(x_n)) = \Omega_C(\phi(x_{\pi(1)})\ldots\phi(x_{\pi(n)})). \tag{2.8}$$

The term of highest degree in the polynomial $\Omega_C(\phi(x_1)\ldots\phi(x_n))$ is $\phi(x_1) \ldots \phi(x_n)$.

Proof. Let $\pi \in \mathcal{S}_n$. Since $\frac{\delta}{\delta J(x)} = \varepsilon^{-d}\frac{\partial}{\partial J(x)}$, and since the partial derivatives commute,

$$\begin{aligned}
\Omega_C(\phi(x_{\pi(1)})\ldots\phi(x_{\pi(n)})) &= \left[\prod_{k=1}^n \frac{1}{i}\frac{\delta}{\delta J(x_{\pi(k)})}\, \mathcal{W}_\Gamma(J,\phi)\right]_{J=0} \\
&= \left[\prod_{k=1}^n \frac{1}{i}\frac{\delta}{\delta J(x_k)}\, \mathcal{W}_\Gamma(J,\phi)\right]_{J=0} \\
&= \Omega_C(\phi(x_1)\ldots\phi(x_n)). \tag{2.9}
\end{aligned}$$

Thus $\Omega_C(\phi(x_1)\ldots\phi(x_n))$ is well-defined on the set of all monomials, a subset of which is a basis of the linear space \mathcal{P}_Γ. Ω_C therefore defines a unique \mathbb{C}-linear map from \mathcal{P}_Γ to \mathcal{P}_Γ. ∎

Usually, the Wick ordered monomial $\Omega_C(\phi(x_1)\ldots\phi(x_p))$ is denoted by $:\phi(x_1)\ldots\phi(x_p):$. In the following, it will be important to keep track of what covariance C is used for Wick ordering, therefore we use the notation Ω_C. Of course, a Wick ordered monomial is a polynomial, and not a monomial, in the fields.

Theorem 2.3. *Define, as a formal power series in J,*

$$\Omega_C\left(e^{i(J,\phi)_\Gamma}\right) = \sum_{n=0}^{\infty} \frac{i^n}{n!}\, \Omega_C\left((J,\phi)_\Gamma^n\right). \tag{2.10}$$

Then

$$\mathcal{W}_\Gamma(J,\phi) = \Omega_C\left(e^{i(J,\phi)_\Gamma}\right). \tag{2.11}$$

Let $\alpha : \Gamma \to \mathbb{C}$ be defined by

$$\alpha(x) = \sum_{k=1}^{n}\alpha_k\delta_\Gamma(x,x_k) = \begin{cases} \varepsilon^{-d}\alpha_k & \text{if } x = x_k \\ 0 & \text{otherwise.} \end{cases} \tag{2.12}$$

Then

$$\Omega_C(\phi(x_1)\ldots\phi(x_n)) = \left[\left(\prod_{k=1}^{n}\frac{1}{i}\frac{\partial}{\partial\alpha_k}\right)e^{i(\alpha,\phi)_\Gamma+\frac{1}{2}(\alpha,C\alpha)_\Gamma}\right]_{\alpha=0}. \tag{2.13}$$

Proof. By Taylor expansion in J, and by definition of Ω_C,

$$\mathcal{W}_\Gamma(J,\phi) - 1$$

$$= \sum_{n\geq 1}\frac{1}{n!}\prod_{k=1}^{n}J(x_k)\sum_{x_1,\ldots,x_n\in\Gamma}\left[\prod_{k=1}^{n}\frac{\partial}{\partial J(x_k)}\mathcal{W}_\Gamma(J,\phi)\right]_{J=0}$$

$$= \sum_{n\geq 1}\frac{1}{n!}\prod_{k=1}^{n}J(x_k)\int dx_1\ldots dx_n\left[\prod_{k=1}^{n}\frac{\delta}{\delta J(x_k)}\mathcal{W}_\Gamma(J,\phi)\right]_{J=0}$$

$$= \sum_{n\geq 1}\frac{1}{n!}\prod_{k=1}^{n}J(x_k)\int dx_1\ldots dx_n\, i^n\Omega_C\left(\prod_{k=1}^{n}\phi(x_k)\right)$$

$$= \sum_{n\geq 1}\frac{i^n}{n!}\,\Omega_C\left((J,\phi)_\Gamma^n\right)$$

$$= \Omega_C\left(e^{i(J,\phi)_\Gamma}\right) - 1. \tag{2.14}$$

The last equality holds because Ω_C is continuous in the formal power series topology (see Appendix A.1). (2.13) is obvious from the definition if one sets $J(x) = \varepsilon^{-d}\alpha(x)$. ∎

We shall use the suggestive notation $\Omega_C\left(e^{i(J,\phi)_\Gamma}\right)$ for $\mathcal{W}_\Gamma(J,\phi)$ since the power series is convergent for all ϕ and J.

The next theorem contains an alternative formula for the Wick ordered monomials that will be useful in Chap. 3.

Theorem 2.4. *Let*

$$\Delta_C = \frac{1}{2}\left(\frac{\delta}{\delta\phi}, C\,\frac{\delta}{\delta\phi}\right) = \frac{1}{2}\int_\Gamma dx \int_\Gamma dy\, \frac{\delta}{\delta\phi(x)}\, C(x,y)\frac{\delta}{\delta\phi(y)}\,. \qquad (2.15)$$

Then

$$\Omega_C\big(\phi(x_1)\ldots\phi(x_n)\big) = e^{-\Delta_C}\,\phi(x_1)\ldots\phi(x_n). \qquad (2.16)$$

In particular, if C depends differentiably on a parameter t, then

$$\frac{\partial}{\partial t}\Omega_C\big(\phi(x_1)\ldots\phi(x_n)\big) = -\frac{\partial\Delta_C}{\partial t}\Omega_C\big(\phi(x_1)\ldots\phi(x_n)\big). \qquad (2.17)$$

Proof. Consider C as a formal parameter. Then for any power series $f(z) = \sum_k f_k z^k$

$$f(\Delta_C)\,e^{i(J,\phi)_\Gamma} = f\left(\frac{1}{2}(iJ, C\,iJ)_\Gamma\right)\,e^{i(J,\phi)_\Gamma}, \qquad (2.18)$$

so

$$e^{-\Delta_C}\,e^{i(J,\phi)_\Gamma} = e^{\frac{1}{2}(J,CJ)_\Gamma + i(J,\phi)_\Gamma}, \qquad (2.19)$$

from which (2.16) follows by definition of Ω_C, because derivatives with respect to J commute with Δ_C. If C depends differentiably on a parameter t, (2.17) follows by taking a derivative of (2.16) because Δ_C commutes with $\frac{\partial\Delta_C}{\partial t}$. ∎

2.2.2 Further Results

Section 2.2.1 contains all the statements about Wick ordering that we shall need for doing calculations. For completeness, and to verify that Wick ordering indeed gives the results stated in Sect. 1.6, we derive here the recursion relation and prove the orthogonality relations.

Theorem 2.5. *The Wick ordered monomials satisfy the recursion relation*

$$\Omega_C\left(\prod_{k=1}^{n+1}\phi(x_k)\right) = \Omega_C\left(\prod_{k=1}^{n}\phi(x_k)\right)\phi(x_{n+1})$$

$$- \sum_{k=1}^{n} C(x_k, x_{n+1})\,\Omega_C\left(\prod_{\substack{l=1\\l\neq k}}^{n}\phi(x_l)\right). \qquad (2.20)$$

Moreover,

$$\int d\mu_C(\phi)\,\Omega_C\left(e^{i(J,\phi)_\Gamma}\right) = 1, \qquad (2.21)$$

for all $n \geq 1$,

$$\int \mathrm{d}\mu_C(\phi)\, \Omega_C\big(\phi(x_1)\ldots\phi(x_n)\big) = 0, \qquad (2.22)$$

and if $n_1 \neq n_2$,

$$\int \mathrm{d}\mu_C(\phi)\, \Omega_C\left(\prod_{k=1}^{n_1}\phi(x_k)\right) \Omega_C\left(\prod_{l=1}^{n_2}\phi(y_l)\right) = 0. \qquad (2.23)$$

Proof. Since

$$\frac{1}{\mathrm{i}}\frac{\delta}{\delta J(x_{n+1})}\mathcal{W}_\Gamma(J,\phi) = \left(\phi(x_{n+1}) - \mathrm{i}\int_\Gamma \mathrm{d}x\, C(x_{n+1},x)J(x)\right)\mathcal{W}_\Gamma(J,\phi),$$
$$(2.24)$$

(2.20) follows if we can show that

$$\left[\prod_{k=1}^{n}\frac{1}{\mathrm{i}}\frac{\delta}{\delta J(x_k)}\, J(x)\, \mathcal{W}_\Gamma(J,\phi)\right]_{J=0} = \sum_{k=1}^{n}\delta_\Gamma(x,x_k)\, \Omega_C\left(\prod_{\substack{l=1 \\ l\neq k}}^{n}\phi(x_l)\right).$$
$$(2.25)$$

By the product rule, every derivative can act on $J(x)$ or on $\mathcal{W}_\Gamma(J,\phi)$. If $\frac{\delta}{\delta J(x_k)}$ acts on $J(x)$, we get $\delta_\Gamma(x,x_k)$, and all other derivatives must act on \mathcal{W}_Γ. This gives the sum on the right side of (2.25). If no derivative acts on $J(x)$, evaluating at $J=0$ gives zero, so (2.25) holds.

(2.21) is obvious from doing the Gaussian integral: by (1.47),

$$\int \mathrm{d}\mu_C(\phi)\, \Omega_C\left(\mathrm{e}^{\mathrm{i}(J,\phi)_\Gamma}\right) = \int \mathrm{d}\mu_C(\phi)\, \mathrm{e}^{\mathrm{i}(J,\phi)_\Gamma}\, \mathrm{e}^{\frac{1}{2}(J,CJ)_\Gamma}$$
$$= 1. \qquad (2.26)$$

Equation (2.22) follows from (2.21) by expansion in J. By (2.13),

$$\int \mathrm{d}\mu_C(\phi)\Omega_C\left(\prod_{k=1}^{n_1}\phi(x_k)\right) \Omega_C\left(\prod_{l=1}^{n_2}\phi(y_l)\right)$$
$$= \left[\prod_{k=1}^{n_1}\frac{1}{\mathrm{i}}\frac{\partial}{\partial\alpha_k} \prod_{l=1}^{n_2}\frac{1}{\mathrm{i}}\frac{\partial}{\partial\beta_l} F(\alpha,\beta)\right]_{\alpha=\beta=0} \qquad (2.27)$$

where

$$F(\alpha,\beta) = \int \mathrm{d}\mu_C(\phi)\, \mathrm{e}^{\mathrm{i}(\alpha+\beta,\phi)_\Gamma}\, \mathrm{e}^{\frac{1}{2}(\alpha,C\alpha)_\Gamma+\frac{1}{2}(\beta,C\beta)_\Gamma}$$
$$= \mathrm{e}^{-\frac{1}{2}\left(\alpha+\beta,C(\alpha+\beta)\right)_\Gamma+\frac{1}{2}(\alpha,C\alpha)_\Gamma+\frac{1}{2}(\beta,C\beta)_\Gamma} \qquad (2.28)$$
$$= \mathrm{e}^{-(\alpha,C\beta)_\Gamma}.$$

Expanding the exponential, we see that every term has the same degree in α of as in β. Therefore taking derivatives and evaluating at $\alpha = \beta = 0$ gives zero if $n_1 \neq n_2$. ∎

The recursion relation gives

$$
\begin{aligned}
\Omega_C(\phi(x)^2) &= \phi(x)\Omega_C(\phi(x)) - C(x,x)\Omega_C(1) \\
&= \phi(x)^2 - C(x,x)
\end{aligned}
\tag{2.29}
$$

and

$$
\begin{aligned}
\Omega_C(\phi(x)^4) &= \phi(x)\Omega_C(\phi(x)^3) - 3C(x,x)\Omega_C(\phi(x)^2) \\
&= \phi(x)\left(\phi(x)\Omega_C(\phi(x)^2) - 2C(x,x)\Omega_C(\phi(x))\right) \\
&\quad - 3C(x,x)\Omega_C(\phi(x)^2) \\
&= \phi(x)^4 - 6C(x,x)\phi(x)^2 + 3C(x,x)^2,
\end{aligned}
\tag{2.30}
$$

which are the identities used in the Sect. 1.6.

The integration by parts formula gives

$$
\int d\mu_C(\phi) \prod_{k=1}^{n+1} \phi(x_k) = \sum_{k=1}^{n} C(x_k, x_{n+1}) \int d\mu_C(\phi) \prod_{\substack{l=1 \\ l \neq k}}^{n} \phi(x_l)
\tag{2.31}
$$

which is similar to the definition of the Wick ordered monomial, only that the terms that get added when one does integration by parts get subtracted in the Wick monomial.

In one variable, the Wick ordered monomials are scaled Hermite polynomials: Calling $\phi(x_1) = \phi$, $C(x_1, x_1) = C$, one gets $\Omega_C(1) = 1$, $\Omega_C(\phi) = \phi$, and for $n \geq 1$, $\Omega_C(\phi^{n+1}) = \phi\,\Omega_C(\phi^n) - nC\,\Omega_C(\phi^{n-1})$. Calling $P_n(\phi) = \Omega_C(\phi^n)$, this is the recursion relation

$$
P_{n+1}(\phi) = \phi P_n(\phi) - nC\, P_{n-1}(\phi).
\tag{2.32}
$$

Thus

$$
P_n(\phi) = \left(\frac{C}{2}\right)^{n/2} H_n\!\left(\frac{\phi}{\sqrt{2C}}\right)
\tag{2.33}
$$

where $H_0(x) = 1$, $H_1(x) = 2x$,

$$
H_{n+1}(x) = 2xH_n(x) - 2nH_{n-1}(x)
\tag{2.34}
$$

are the usual Hermite polynomials. The generating function of the H_n is

$$
e^{2xs - s^2} = \sum_{n=0}^{\infty} \frac{H_n(x)}{n!} s^n.
\tag{2.35}
$$

2.3 Evaluation of Gaussian Integrals

2.3.1 Labelled Feynman Graphs

In this section we evaluate Gaussian integrals of the form

$$\tilde{I}(V_1, \ldots, V_p; \psi) = \int d\mu_C(\phi) \prod_{k=1}^{p} V_k(\phi + \psi) \qquad (2.36)$$

where

$$V_k(\phi) = \int_\Gamma dx_1 \ldots \int_\Gamma dx_{m_k} v_k(x_1, \ldots, x_{m_k}) \phi(x_1) \ldots \phi(x_{m_k}) \qquad (2.37)$$

and

$$I(V_1, \ldots, V_p; \psi) = \int d\mu_C(\phi) \prod_{q=1}^{p} \Omega_C(V_q(\phi + \psi)). \qquad (2.38)$$

We shall do the function I in detail; the changes in the representation for \tilde{I} will then be obvious. The result is going to be a sum of values of "Feynman graphs", which will be defined in detail below. However, we briefly indicate now how these graphs are going to arise, to make it easier to follow the derivation below. We shall visualize every v_k as a vertex shown in Fig. 2.1, with m_k external, numbered, legs. These legs will be joined to an external field ψ or to another vertex by a line, which has an associated factor $C(x_i, x_j)$.

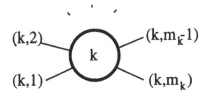

Fig. 2.1. A fully labelled vertex

A precise definition will follow below at the end of the derivation, but we shall discuss the graphical significance of various objects already when we introduce them. We introduce the notation

$$X_k = (x_{k,1}, \ldots, x_{k,m_k}), \qquad \int dX_k = \int_\Gamma dx_{k,1} \ldots \int_\Gamma dx_{k,m_k} \qquad (2.39)$$

and

$$\phi^{X_k} = \phi(x_{k,1}) \ldots \phi(x_{k,m_k}) \qquad (2.40)$$

so that

$$V_k(\phi) = \int dX_k v(X_k) \phi^{X_k}. \qquad (2.41)$$

The v_k are assumed to be symmetric under permutations

$$v_k(x_1,\ldots,x_{m_k}) = v_k(x_{\pi(1)},\ldots,x_{\pi(m_k)}) \qquad \forall \pi \in S_{m_k}, \tag{2.42}$$

because any nonsymmetric part of v_k would cancel in (2.41). Also, we define the *set of legs of vertices determined by* V_1,\ldots,V_p as

$$B_0(p; m_1,\ldots,m_p) = \{(k,i_k) : k \in \{1,\ldots,p\}, i_k \in \{1,\ldots,m_k\}\} \tag{2.43}$$

and a partial ordering \prec on B_0 by

$$(k,i) \prec (l,i) \Leftrightarrow k < l. \tag{2.44}$$

Note that B_0 specifies the set of vertices completely: every vertex has a number $k \in \{1,\ldots,p\}$, and every one of its legs is numbered as well.

We now drop the arguments of B_0, keeping in mind that it depends on p and m_1,\ldots,m_p. For $\emptyset \neq B \subset B_0$ let

$$\begin{aligned} L_0(B) &= \{(b,b') : b, b' \in B, b \prec b'\} \\ &= \{((k,i),(l,j)) : (k,i) \in B, (l,j) \in B \text{ and } k < l\}. \end{aligned} \tag{2.45}$$

We shall see that $L_0(B')$ is the set of all internal lines that can be formed by pairing legs of vertices. Obviously, $I(V_1,\ldots,V_p;\psi)$ is linear in every V_k. Since $(X_k)_k = (x_b)_{b \in B_0}$,

$$I(V_1,\ldots,V_p;\psi) = \int_\Gamma \prod_{b \in B_0} dx_b \prod_{k=1}^p v(X_k) \, J(X_1,\ldots,X_p;\psi) \tag{2.46}$$

with

$$J(X_1,\ldots,X_p;\psi) = \int d\mu_C(\phi) \prod_{k=1}^p \Omega_C((\phi + \psi)^{X_k}). \tag{2.47}$$

Since the X_i will be fixed in the following, we write $J(\psi) = J(X_1,\ldots,X_p;\psi)$. We now use (2.13) in the form

$$\Omega_C\left((\phi + \psi)^{X_k}\right) = \left[\prod_{j=1}^{m_k} \frac{1}{i} \frac{\partial}{\partial \alpha_{k,j}} \, \Omega_C\left(e^{i(\alpha^{(k)}, \phi + \psi)_\Gamma}\right)\right]_{\alpha^{(k)}=0} \tag{2.48}$$

with

$$\alpha^{(k)} = \alpha_{k,1}\, \delta_\Gamma(x, x_{k,1}) + \ldots + \alpha_{k,m_k}\, \delta_\Gamma(x, x_{k,m_k}). \tag{2.49}$$

Note that even if the $x_{k,r}$ are not distinct, the α's are; this is a typical example of the combinatorial method of "giving identities" to identical objects that appear several times. Doing this for all $k \in \{1,\ldots,p\}$, we arrive at

$$J(\psi) = \left[\prod_{b \in B_0} \frac{1}{i} \frac{\partial}{\partial \alpha_b} \int d\mu_C(\phi) \prod_{k=1}^p \Omega_C(e^{i(\alpha^{(k)}, \phi + \psi)})\right]_{\alpha_b=0 \ \forall b \in B_0}. \tag{2.50}$$

Since $\Omega_C(e^{i(\alpha^{(k)},\phi)}) = e^{i(\alpha^{(k)},\phi)+\frac{1}{2}(\alpha^{(k)},C\alpha^{(k)})}$, we get

$$J(\psi) = \left[(\prod_{b\in B_0} \frac{1}{i}\frac{\partial}{\partial\alpha_b}) e^{\frac{1}{2}\sum_{k=1}^{p}(\alpha^{(k)},C\alpha^{(k)})} \int d\mu_C(\phi)e^{i\sum_{k=1}^{p}(\alpha^{(k)},\phi+\psi)} \right]_{\alpha=0}. \quad (2.51)$$

By (1.40), the Gaussian integral is

$$\int d\mu_C(\phi)e^{i\sum_{k=1}^{p}(\alpha^{(k)},\phi+\psi)} = e^{i\sum_{k=1}^{p}(\alpha^{(k)},\psi)} e^{-\frac{1}{2}(\sum_{k=1}^{p}\alpha^{(k)},C\sum_{l=1}^{p}\alpha^{(l)})}. \quad (2.52)$$

By the symmetry of C, $(\alpha^{(k)},C\alpha^{(l)}) = (\alpha^{(l)},C\alpha^{(k)})$, so

$$J(\psi) = \left[(\prod_{b\in B_0} \frac{1}{i}\frac{\partial}{\partial\alpha_b}) e^{-\sum_{k<l}(\alpha^{(k)},C\alpha^{(l)})} e^{i\sum_{k=1}^{p}(\alpha^{(k)},\psi)} \right]_{\alpha=0}. \quad (2.53)$$

Note that the Wick ordering factor has cancelled the $k = l$ terms in the sum. This is the reason why after Wick ordering, no two legs of a single vertex will be joined to form a line that connects this vertex to itself (a "self–contraction").

We rewrite $\sum_{k=1}^{p} i(\alpha^{(k)},\psi) = \sum_{b\in B_0} i\alpha_b\,\psi(x_b)$ and

$$-\sum_{k<l}(\alpha^{(k)},C\alpha^{(l)}) = \sum_{(b,b')\in L_0(B_0)} i\alpha_b C(x_b,x_{b'})i\alpha_{b'}. \quad (2.54)$$

Note that the sum in (2.54) is the reason why $L_0(B_0)$ was defined with the partial ordering \prec. It arises in a natural way from the restriction $k < l$, which in turn is caused by Wick ordering and by the symmetry of C.

Since we evaluate at $\alpha = 0$ to get J, only a polynomial linear in every α_b contributes to J. Thus we may replace the exponential by a polynomial, and $i\alpha_b$ by α_b everywhere, so that

$$J(\psi) = \left[(\prod_{b\in B_0} \frac{\partial}{\partial\alpha_b}) e^{\sum_{b\in B_0}\alpha_b\psi(x_b) + \sum_{(b,b')\in L_0(B_0)}\alpha_b C(x_b,x_{b'})\alpha_{b'}} \right]_{\alpha=0}$$

$$= \left[(\prod_{b\in B_0} \frac{\partial}{\partial\alpha_b}) \prod_{b\in B_0}(1+\alpha_b\psi(x_b)) \prod_{\substack{(b,b')\in \\ L_0(B_0)}}(1+\alpha_b C(x_b,x_{b'})\alpha_{b'}) \right]_{\alpha=0}. \quad (2.55)$$

It is convenient to use the following lemma.

Lemma 2.6. *Let $n \geq 1$ and f and g be polynomials in $\xi = (\xi_1,\ldots,\xi_n)$. Then*

$$\left[f\left(\frac{\partial}{\partial\xi}\right) g(\xi) \right]_{\xi=0} = \left[g\left(\frac{\partial}{\partial\xi}\right) f(\xi) \right]_{\xi=0}. \quad (2.56)$$

Proof. The identity is linear in f and g, so it suffices to verify it for monomials. For monomials, it is obvious. ∎

By Lemma 2.6,

$$
J(\psi) = \left[\exp\left(\sum_{b \in B_0} \psi(x_b)\frac{\partial}{\partial\alpha_b} + \sum \frac{\partial}{\partial\alpha_b}C(x_b, x_{b'})\frac{\partial}{\partial\alpha_{b'}} \right) \prod_{b \in B_0} \alpha_b \right]_{\alpha=0} \tag{2.57}
$$

$$
= \left[\prod_{\substack{(b,b')\in \\ L_0(B_0)}} \left(1 + \frac{\partial}{\partial\alpha_b}C(x_b, x_{b'})\frac{\partial}{\partial\alpha_{b'}}\right) \prod_{b \in B_0} \left(1 + \psi(x_b)\frac{\partial}{\partial\alpha_b}\right) \prod_{b \in B_0} \alpha_b \right]_{\alpha=0}.
$$

We now apply the identity

$$
\prod_{r \in M} (1 + a_r) = \sum_{N \subset M} \prod_{r \in N} a_r, \tag{2.58}
$$

which holds for any finite family $(a_r)_{r \in M}$ of commuting objects, to (2.57), take the derivatives and evaluate at zero. Since every α_b appears only once in $\prod_{b \in B_0} \alpha_b$, every derivative can act at most once. If we apply

$$
\prod_{b \in B_0} \left(1 + \psi(x_b)\frac{\partial}{\partial\alpha_b}\right) = \sum_{E \subset B_0} \prod_{b \in E} \psi(x_b)\frac{\partial}{\partial\alpha_b} \tag{2.59}
$$

first, this leaves $\prod_{b \in B_0 \backslash E} \alpha_b$ for the other derivatives to act on. Because we evaluate at $\alpha = 0$, all of these factors must be used up by derivatives (that is, all non-external legs of the vertices must be used to make internal lines of the graph), and we obtain

$$
J(\psi) = \sum_{E \subset B_0} \prod_{b \in E} \psi(x_b) \, \mathcal{J} \tag{2.60}
$$

with

$$
\mathcal{J} = \sum_{L \subset L_0(B_0 \backslash E)} \left[\prod_{(b,b')\in L} \frac{\partial}{\partial\alpha_b}C(x_b, x_{b'})\frac{\partial}{\partial\alpha_{b'}} \prod_{b \in B_0 \backslash E} \alpha_b \right]_{\alpha=0}
$$

$$
= \sum_{E \subset B_0} \prod_{b \in E} \psi(x_b) \sum_{L \in \mathcal{L}(B_0 \backslash E)} \prod_{(b,b')\in L} C(x_b, x_{b'}) \tag{2.61}
$$

where $L \in \mathcal{L}(B)$ iff

$$
L = \{(b,b') \in L_0(B) : \pi_1|_L \text{ and } \pi_2|_L \text{ are bijective}\}. \tag{2.62}
$$

Here π_1 and π_2 are the projections on the first and second components of a pair (b,b'): $\pi_1(b,b') = b$ and $\pi_2(b,b') = b'$.

In other words, if $L \in \mathcal{L}(B)$, then every $e \in B_0$ appears exactly once as a first entry b or a second entry b' in a pair $(b,b') \in L$.

The procedure of differentiating with respect to the α_b and setting the α's equal to zero has the following natural graphical interpretation.

- The labelled vertex drawn in Fig. 2.1 represents the monomial $v_k(x_{k,1}, \ldots, x_{k,m_k})\, \alpha_{k,1} \ldots \alpha_{k,m_k}$, that is, leg number (k,l) corresponds to the factor $\alpha_{k,l}$.
- The action of $\psi(x_{ki})\frac{\partial}{\partial \alpha_{ki}}$ is to make leg (k,i) an external leg of the graph: the factor α_{ki} is replaced by $\psi(x_{ki})$.
- The differential operator $\frac{\partial}{\partial \alpha_{ki}}C(x_{ki}, x_{lj})\frac{\partial}{\partial \alpha_{lj}}$ removes the factors α_{ki} and α_{lj}. Accordingly, the legs (k,i) and (l,j) are joined to form a line of the graph.
- The restriction that $k < l$ following from the definition of $L_0(B)$ means that only legs of distinct vertices are joined to form lines.
- Evaluation at $\alpha = 0$ implies that all terms that still contain a factor α_b vanish. Thus all legs that are not assigned external fields must be paired to form internal lines of the graph.

Examples are given in the next subsection. The precise definition of a Feynman graph is as follows.

Definition 2.7. Let $B_0 = B_0(p; m_1, \ldots, m_p)$, as defined in (2.43), and for $\emptyset \neq B \subset B_0$, let $L_0(B)$ be defined as in (2.45) and $\mathcal{L}(B)$ be defined as in (2.62). Let $E \subset B_0$. A (labelled) Feynman graph for the vertex set determined by B_0, and with external legs in E, is an element $L \in \mathcal{L}(B_0 \setminus E)$.

Remark 2.8. By this definition, we have identifed a Feynman graph with its set L of internal lines. This is unambiguous since the vertices have already been fixed by specifying B_0. The Feynman graphs are called labelled because every vertex and every leg of each vertex is numbered, and hence distinguishable from every other one. We specify "internal" lines because we sometimes also say external lines for the external legs (people not used to graphs having 'external legs' can think of the endpoint of the external leg as a vertex with incidence number one; this is unambiguous because such vertices do not occur as internal ones in our models). Note, however, that the external legs do *not* count as lines of the Feynman graph under Definition 2.7. This is a natural convention, as we shall see later on.

Using the permutation symmetry (2.42) of the v_k we obtain the following representation for $I(\phi)$.

Theorem 2.9. Let $I(\phi) = I(V_1, \ldots, V_p, \phi)$ be given by (2.38). Then

$$I(\phi) = \sum_{m=0}^{\tilde{m}} \int_{\Gamma_m} d\xi_1 \ldots d\xi_m\, \phi(\xi_1) \ldots \phi(\xi_m)\, U_m(C; \xi_1, \ldots, \xi_m) \qquad (2.63)$$

with $\tilde{m} = \sum_{i=1}^{p} m_i$ and

$$U_m(C;\xi_1,\ldots,\xi_m) = \frac{1}{m!} \sum_{\eta_1,\ldots,\eta_m} \int_\Gamma \prod_{b\in B_0\backslash E} dx_b$$

$$\prod_{k=1}^p v_k(X_k) \sum_{L\in\mathcal{L}(B_0\backslash E)} \prod_{(b,b')\in L} C(x_b,x_{b'}) \qquad (2.64)$$

where the sum over $\eta_1,\ldots,\eta_m \in B_0$ is restricted such that the η's are pairwise distinct, $E = \{\eta_1,\ldots,\eta_m\}$, and where $x_{\eta_l} = \xi_l$ for all $l \in \{1,\ldots,m\}$. The U_m depend also on p and v_1,\ldots,v_p.

Remark 2.10. If we had not used Wick ordering, we would have had an extra term

$$\exp\left(\frac{1}{2}\sum_{k=1}^p (\frac{\partial}{\partial\alpha^{(k)}}, C\frac{\partial}{\partial\alpha^{(k)}})\right) \qquad (2.65)$$

which generates self-contractions of the vertices.

2.3.2 Symmetry Factors and Topological Feynman Graphs

As an example, consider the integral

$$J = \int d\mu_C(\phi)\, \Omega_C(\phi(x)^4)\Omega_C(\phi(y)^4). \qquad (2.66)$$

The graphical rules tell us that we have to join the legs of the two vertices shown in Fig. 2.2 in all possible ways to get the graphs. For instance, one can have the Feynman graphs in Fig. 2.3, with

$$L_1 = \{((1,1),(2,2)),((1,2),(2,1)),((1,3),(2,4)),((1,4),(2,3))\}$$
$$L_2 = \{((1,1),(2,2)),((1,2),(2,4)),((1,3),(2,1)),((1,4),(2,3))\}.(2.67)$$

Fig. 2.2. Two labelled vertices

The value of each such graph is $C(x,y)^4$, so to get the result we now have to count the number of Feynman graphs that one can make with these two vertices. There are four ways to join leg number $(1,1)$ to any leg of vertex

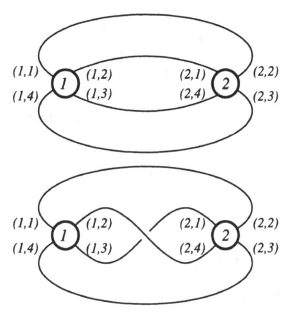

Fig. 2.3. Two labelled Feynman graphs, corresponding to L_1 and L_2

2. After that, there are three ways of joining $(1, 2)$ to any remaining leg of vertex 2, etc., so there are 4! different labelled graphs contributing to J. Thus

$$J(x, y) = 4! \, C(x, y)^4. \tag{2.68}$$

Generalizing this, we can also write the Feynman graph representation in terms of Feynman graphs (see Fig. 2.4) where the legs of the vertices are no longer labelled. Then one has to include the correct symmetry factors, like the 4! in (2.68). Unlabelled Feynman graphs are also called *topological Feynman graphs*. If a labelled Feynman graph G' has been obtained from another labelled Feynman graph G by a reordering of the legs of a vertex, then G and G' give rise to the same topological Feynman graph. For instance, the two graphs drawn in Fig. 2.3 give rise to the same topological graph, drawn in Fig. 2.4.

For small graphs, it is trivial to calculate the symmetry factors and thus to pass from one to the other representation. For large graphs, doing this can get rather involved. For the following graphical considerations, in particular the proof that taking the logarithm of the the partition function removes all disconnected Feynman graphs, working with labelled graphs is advantageous because every labelled Feynman graph appears exactly once in the sum and one does not have to think about any symmetry factors.

2.3.3 Motivations for Taking the Logarithm

The integral (2.66) occurs in the second order term of the expansion for

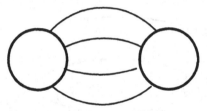

Fig. 2.4. The topological Feynman graph corresponding to the graphs given by L_1 and L_2.

$$\tilde{Z}_\Gamma = \int d\mu_C(\phi) \, e^{-\lambda \int_\Gamma dz \, \Omega_C(\phi(z)^4)}$$

$$= 1 + \frac{\lambda^2}{2} \int_\Gamma dx \int_\Gamma dy \int d\mu_C(\phi) \Omega_C(\phi(x)^4) \Omega_C(\phi(y)^4) + O(\lambda^4)$$

$$= 1 + \frac{\lambda^2}{2} 4! \int_\Gamma dx \int_\Gamma dy \, C_\Gamma(x,y)^4 + O(\lambda^4). \tag{2.69}$$

Since C_Γ depends only on $x - y$, and since $\int_\Gamma dx = L^d$,

$$\tilde{Z}_\Gamma = 1 + \frac{\lambda^2}{2} 4! \, L^d \int_\Gamma dy \, C_\Gamma(0,x)^4 + O(\lambda^4). \tag{2.70}$$

The factor L^d was to be expected: in a statistical mechanical system, $\tilde{Z}_\Gamma = \exp(L^d f_\Gamma)$ usually holds ("the free energy is a bulk quantity"), meaning that $\lim_{L\to\infty} f_{\Gamma_\epsilon,L} = f_\epsilon$: the free energy density stays finite in the thermodynamic limit. A proof that this is indeed the case requires decay assumptions on the interaction and is not trivial. We shall do this proof in the next sections.

One reason why this is nontrivial is that from fourth order on, disconnected graphs appear in the expansion for Z_Γ. This gives rise to terms of order $(L^d)^2$ and even higher powers of L^d in higher orders, which is of course consistent with what one gets when expanding $e^{L^d f_\Gamma}$, but it is clear that one has to prove finiteness of $\lim_{L\to\infty} f_\Gamma$. This proof is not by expansion in L^d, of course, and one also knows from cluster expansions that the corrections to the bulk behaviour are not $O(L^{-d})$ but more complicated functions of L [25]. By taking the logarithm, we shall be able to prove finiteness of the limit of the free energy density.

The other problem is the limit $\epsilon \to 0$, in which the question of the convergence of $\int_\Gamma dx \, C_\Gamma(0,x)^4$ for $\epsilon \to 0$ arises. If we take the continuum limit $\epsilon \to 0$ naively, replacing the Riemann sum by the integral $\int_{\mathbf{R}^d} dx \, C(0,x)^4$, and recall that

$$|C(0,x)| \leq \text{const } e^{-m|x|} \qquad \text{for} \quad |x| \geq 1, \tag{2.71}$$

we see that the integral over $\mathbf{R}^d \setminus \{x : |x| < 1\}$ converges because $m_\phi > 0$. But since

$$0 < C(0, x) \sim \text{const} \begin{cases} |x|^{-(d-2)} & \text{if } d \geq 3 \\ |\log(x)| & \text{if } d = 2 \end{cases} \tag{2.72}$$

as $x \to 0$, the integral

$$\int_{|x| \leq 1} d^d x\, C(0, x)^4 \sim \int_0^1 dr\, r^{d-1}\, r^{-4(d-2)}$$

$$= \int_0^1 dr\, r^{7-3d} \tag{2.73}$$

diverges for all $d \geq 3$. Only for $d = 2$,

$$\int_{|x| \leq 1} d^d x\, C(0, x)^4 \sim \int_0^1 r dr |\log r|^4 < \infty. \tag{2.74}$$

A more complicated, but still elementary, calculation shows that the limit $\varepsilon \to 0$ of the well-defined quantity $\int_\Gamma dx\, C_\Gamma(0, x)^4$ is indeed divergent for $d \geq 3$, and convergent for $d = 2$. Thus Wick ordering has not removed all divergences, at least if $d \geq 3$.

To remove these divergences, we shall have to analyze the singularities in more detail. For this purpose, it is inconvenient to deal with connected and disconnected graphs at the same time. The most efficient way to proceed is to use Wilson's effective action, defined by

$$G_{\text{eff}}^\Gamma(\psi) = \log \int d\mu_{C_\Gamma}(\phi) e^{-\lambda V(\phi + \psi)} \tag{2.75}$$

for $V(\phi)$ bounded below. We shall prove in the next sections that $G_{\text{eff}}^\Gamma(\psi)$ is the generating functional for the connected, amputated Green functions, and that $G_{\text{eff}}^\Gamma(\psi) = L^d g_{\text{eff}}^\Gamma(\psi)$ where $g_{\text{eff}}^\Gamma(\psi)$ has a finite limit as $L \to \infty$.

The renormalization group is then an analysis where C_Γ is split into many pieces, according to their distance from the singularity. The RG flow determines how singularities in Z and in the Green functions generated by g_{eff} build up in the limit $\varepsilon \to 0$.

2.4 Polymer Systems

In this section, we define the notion of an abstract polymer system and its partition function, and show that the logarithm of the partition function gets contributions from connected graphs only. This statement is of great use in statistical mechanics. It can be used to show that the free energy density indeed has a thermodynamic limit if the interaction decays sufficiently fast, and it is also useful to show exponential decay of correlations. The presentation in this section closely follows that in Erhard Seiler's book [20].

2.4.1 Preparation: Graphs and Partitions

Let $M \subset \mathbb{N}$ be a finite set and

$$\mathcal{G}(M) = \mathbb{P}(\{(i,j) \in M \times M : i < j\}) \tag{2.76}$$

where $\mathbb{P}(A)$ denotes the set of all subsets of A. In other words, every $G \in \mathcal{G}(M)$ is a set of ordered pairs $(i,j) \in M \times M$ with $i < j$. Every $G \in \mathcal{G}(M)$ defines a graph with vertex set M by the definition that the line between $k \in M$ and $l \in M$ is a line of the graph iff $(k,l) \in G$. We identify the graphical object with its set of lines $G \in \mathcal{G}(M)$ by the following

Definition 2.11. *A graph G is an element $G \in \mathcal{G}(M)$. A subset H of G is called a subgraph of G.*

For example, let $M = \{1, 2, 4, 5, 6, 7, 8\}$ and $G = \{(1,2), (1,4), (2,5), (6,8)\}$. The graph G is drawn in Fig. 2.5.

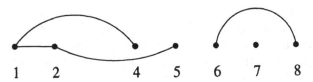

Fig. 2.5. An example of a graph

In graph theory, a graph on M is usually defined as a set of unordered pairs $\{i, j\}$ where $i < j$. This definition is obviously equivalent to the one given above. The notion of a graph is not the same as that of a Feynman graph because in a graph, two vertices can be joined by at most one line. The graph theoretical notation is, unfortunately, not completely standardized. In the nomenclature of [24], Feynman graphs would be called multigraphs and Feynman graphs without Wick ordering pseudo-multigraphs.

In the graphical representation, the concept that a graph can be disconnected (as in the example drawn in Fig. 2.5) or connected is obvious. If $G = \emptyset$, G is totally disconnected, i.e. it has $|M|$ connected components, each consisting of one vertex $i \in M$. Intuitively, a graph is connected if any two vertices $i \in M$ and $j \in M$ can be joined by a walk over lines of G. More formally, one defines for $M = \{m_1, \ldots, m_k\}$ with $m_1 < \ldots < m_k$ that $G \in \mathcal{G}(M)$ is connected iff there is a permutation $\pi \in \mathcal{S}(M)$ such that

$$\{(m_1, m_2), (m_2, m_3), \ldots, (m_{k-1}, m_k)\} \subset \{(\pi(i), \pi(j)) : (i,j) \in G\}. \tag{2.77}$$

An example for such a permutation is drawn in Fig. 2.6.

We denote the set of connected graphs on M by $\mathcal{G}_c(M)$.

If $G \in \mathcal{G}(M)$ is disconnected, G induces a partition $p(G)$ of M by grouping together the vertices in the connected components of G. More precisely:

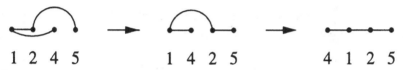

1 2 4 5 1 4 2 5 4 1 2 5

Fig. 2.6. Permuting a connected graph

Definition 2.12. *P is a partition of $M \neq \emptyset$ iff $P = (k, \{N_1, \ldots, N_k\})$ with $k \in \mathbb{N}$ and for all $l \in \{1, \ldots, k\}$: $\emptyset \neq N_l \subset M$, and $N_1 \cup \ldots \cup N_k = M$. We denote the set of partitions of M by $\mathcal{P}(M)$.*

If M is finite, $\mathcal{P}(M)$ is finite as well, because all $N_l \neq \emptyset$.

Lemma 2.13. *$p : \mathcal{G}(M) \to \mathcal{P}(M), G \to p(G)$ is a surjective mapping, therefore*

$$\sum_{G \in \mathcal{G}(M)} F(G) = \sum_{P \in \mathcal{P}(M)} \sum_{G : p(G) = P} F(G). \tag{2.78}$$

Proof. p is a well-defined map because the set $\{N_1, \ldots, N_k\}$, and not the sequence (N_1, \ldots, N_k), appears in the definition of the partition (otherwise the correspondence wouldn't be unique because for example both $\{N_1, N_2, \ldots, N_k\}$ and $\{N_2, N_1, \ldots, N_k\}$ are generated by the same G). Surjectivity is obvious. ∎

Remark 2.14. Let M be finite and $G \in \mathcal{G}(M)$. If $p(G) = (m, \{N_1, \ldots, N_m\})$, then

$$G = \bigcup_{k=1}^{m} G_k \tag{2.79}$$

with $G_k \in \mathcal{G}_c(N_k)$ for all k, and

$$G|_{N_k} = G_k = \{(i, j) \in G : i \in N_k \text{ and } j \in N_k\}. \tag{2.80}$$

Moreover,

$$\sum_{P \in \mathcal{P}(M)} F(P) = \sum_{m=1}^{|M|} \frac{1}{m!} \sum_{\substack{\emptyset \neq N_1, \ldots, N_m \subset M \\ N_1 \cup \ldots \cup N_m = M}} F(m, \{N_1, \ldots, N_m\}). \tag{2.81}$$

The $\frac{1}{m!}$ is there because we now sum over all sequences (N_1, \ldots, N_m) instead of all sets $\{N_1, \ldots, N_m\}$.

Lemma 2.15. *Let $A \subset \mathbb{N}$ be finite and nonempty. Then*

$$\prod_{\substack{i,j\in A\\ i<j}} (1+a_{ij}) = \sum_{G\in\mathcal{G}(A)} \prod_{(i,j)\in G} a_{ij} =$$

$$\sum_{m=1}^{|A|} \frac{1}{m!} \sum_{\substack{N_1,\ldots,N_m\subset A\\ N_k\neq\emptyset\forall k\\ N_1\dot\cup\ldots\dot\cup N_m=A}} \prod_{k=1}^{m} \left(\sum_{G\in\mathcal{G}_c(N_k)} \prod_{(i,j)\in G} a_{ij} \right). \qquad (2.82)$$

Proof. The first equation is a consequence of the multinomial theorem

$$\prod_{r\in R} (1+a_r) = \sum_{S\subset R} \prod_{r\in S} a_s$$

with $R = \{(i,j)\in A\times A : i<j\}$. By Lemma 2.13 and Remark 2.14

$$\sum_{G\in\mathcal{G}(A)} \prod_{(i,j)\in G} a_{ij} = \sum_{P\in\mathcal{P}(A)} \sum_{G:p(G)=p} \prod_{(i,j)\in G} a_{ij}$$

$$= \sum_{m=1}^{|A|} \frac{1}{m!} \sum_{N_1\dot\cup\ldots\dot\cup N_m=A} \sum_{\substack{G:p(G)=\\(m,\{N_1,\ldots,N_m\})}} \prod_{k=1}^{m} \prod_{(i,j)\in G|_{N_k}} a_{ij}$$

$$= \sum_{m=1}^{|A|} \frac{1}{m!} \sum_{N_1\dot\cup\ldots\dot\cup N_m=A} \prod_{k=1}^{m} \left(\sum_{G_k\in\mathcal{G}_c(N_k)} \prod_{(i,j)\in G_k} a_{ij} \right). \qquad (2.83)$$

∎

2.4.2 The Logarithm of the Polymer Partition Function

Definition 2.16. *Let Γ_0 be a finite set, and \sim be a symmetric relation on Γ_0 that is antireflexive, i.e. $\forall\gamma\in\Gamma_0$, $\gamma\not\sim\gamma$. Then we call the elements $\gamma\in\Gamma_0$ polymers, and γ and γ' compatible if $\gamma\sim\gamma'$. Let A be a commutative \mathbb{C}-algebra and $a : \Gamma_0 \to A$, $\gamma \to a(\gamma)$ a map. $a(\gamma)$ is called the activity of γ. A set $\Gamma\subset\Gamma_0$ is called compatible iff $\forall\gamma,\gamma'\in\Gamma : \gamma\neq\gamma' \Rightarrow \gamma\sim\gamma'$. The partition function of the polymer system is the finite sum*

$$Z_{\Gamma_0}(a) = \sum_{\substack{\Gamma\subset\Gamma_0\\ \Gamma\text{ compatible}}} \prod_{\gamma\in\Gamma} a(\gamma) = 1 + \sum_{\substack{\Gamma\subset\Gamma_0\\ \Gamma\neq\emptyset\\ \Gamma\text{ compatible}}} \prod_{\gamma\in\Gamma} a(\gamma). \qquad (2.84)$$

The name originates from chemistry. Polymers are long chain molecules with a hard-core property: no two polymers can occupy the same region of space. This disjointness is an example for a compatibility relation, and it is obvious that any polymer must be incompatible to itself.

Polymer systems arise in a natural way in expansions of the partition functions of statistical mechanics. For instance, the high-temperature expansion of

the Ising model can be brought in the form of a polymer expansion, with the polymers being connected clusters of bonds between nearest-neighbour sites on the lattice. Similarly, the low-temperature expansion of the Ising model can be rewritten that way, but with Bloch walls as the polymers [17, 18, 20].

One can make the compatibility constraint in (2.84) more explicit by introducing [20]

$$g(\gamma, \gamma') = \begin{cases} -1 & \gamma \not\sim \gamma' \\ 0 & \text{otherwise} \end{cases} \tag{2.85}$$

and

$$U(\gamma_1, \ldots, \gamma_n) = \prod_{1 \le i < j \le n} (1 + g(\gamma_i, \gamma_j)). \tag{2.86}$$

Since

$$U(\gamma_1, \ldots, \gamma_n) = \begin{cases} 1 & \text{for all } i, j: \ i \neq j \Rightarrow \gamma_i \sim \gamma_j \\ 0 & \text{otherwise,} \end{cases} \tag{2.87}$$

the constraint in (2.84) is enforced by including a factor $U(\gamma_1, \ldots, \gamma_n)$ in the summand. Thus $Z_{\Gamma_0}(z)$ is the polynomial

$$Z_{\Gamma_0}(a) = 1 + \sum_{n=1}^{\infty} \frac{1}{n!} \sum_{\gamma_1, \ldots, \gamma_n \in \Gamma_0} U(\gamma_1, \ldots, \gamma_n) \prod_{k=1}^{n} a(\gamma_k) \tag{2.88}$$

with

$$U(\gamma_1, \ldots, \gamma_n) = \sum_{G \in \mathcal{G}(\{1, \ldots, n\})} \prod_{(i,j) \in G} g(\gamma_i, \gamma_j). \tag{2.89}$$

Theorem 2.17. *Let A be a formal power series ring over \mathbb{C}, and the $a(\gamma)$ all have no constant part.*

$$\log Z_{\Gamma_0}(a) = \sum_{n=1}^{\infty} \frac{1}{n!} \sum_{\gamma_1, \ldots, \gamma_n \in \Gamma_0} U_c(\gamma_1, \ldots, \gamma_n) \prod_{k=1}^{n} a(\gamma_k) \tag{2.90}$$

with the Ursell function

$$U_c(\gamma_1, \ldots, \gamma_n) = \sum_{G \in \mathcal{G}_c(\{1, \ldots, n\})} \prod_{(i,j) \in G} g(\gamma_i, \gamma_j) \tag{2.91}$$

for $n > 1$ and $U_c(\gamma_1) = 1$ if $n = 1$.

Remark 2.18. 1. The only change in going from Z_{Γ_0} to $\log Z_{\Gamma_0}$ is that the 1 is left out and U is replaced by U_c. Also, the difference between U and U_c is that in (2.89), the sum runs over all graphs $G \in \mathcal{G}(\{1, \ldots, n\})$ and that in (2.91), the sum runs over all connected graphs $G \in \mathcal{G}_c(\{1, \ldots, n\})$. In this sense, taking the logarithm replaces graphs by connected graphs. The analogous statement for Feynman graphs will be proven in the next section.

2. $U_c(\gamma_1, \ldots, \gamma_n)$ is symmetric under permutations.

$$U_c(\gamma_1, \ldots, \gamma_n) = U_c(\gamma_{\pi(1)}, \ldots, \gamma_{\pi(n)}) \qquad \forall \pi \in S_n. \tag{2.92}$$

3. $U(\gamma_1, \ldots, \gamma_n) \neq 0$ only if $\{\gamma_1, \ldots, \gamma_n\}$ is compatible. By way of contrast, $U_c(\gamma_1, \ldots, \gamma_n) = 0$ if there are two nonempty sets $A, B \subset \{\gamma_1, \ldots, \gamma_n\}$ such that $A \sim B$ and $A \cup B = \{\gamma_1, \ldots, \gamma_n\}$.

4. A trivial polymer system is given by $\Gamma_0 = \{1\}$, $a(1) = a$. Then $Z = 1 + a$ and

$$\log Z = \sum_{n=1}^{\infty} \frac{(-1)^n}{n} a^n. \tag{2.93}$$

Proof. Denote $\underline{n} = \{1, \ldots, n\}$. By Lemma 2.15,

$$U(\gamma_1, \ldots, \gamma_n) = \sum_{m=1}^{n} \frac{1}{m!} \sum_{N_1 \dot{\cup} \ldots \dot{\cup} N_m = \underline{n}} \prod_{k=1}^{m} U_c((\gamma_r)_{r \in N_k}) \tag{2.94}$$

We now use

$$\prod_{r=1}^{n} a(\gamma_r) = \prod_{k=1}^{m} \left(\prod_{s \in N_k} a(\gamma_s) \right) \tag{2.95}$$

and get

$$Z_{\Gamma_0}(a) = 1 + \sum_{m=1}^{\infty} \frac{1}{m!} \sum_{n=m}^{\infty} \frac{1}{n!} \sum_{N_1 \dot{\cup} \ldots \dot{\cup} N_m = \underline{n}} \prod_{k=1}^{m} \psi(N_k) \tag{2.96}$$

where

$$\psi(M) = \sum_{(\gamma_r)_{r \in M}} U_c((\gamma_r)_{r \in M}) \prod_{r \in M} a(\gamma_r). \tag{2.97}$$

Since U_c and $\prod_{r \in M} a(\gamma_r)$ are permutation invariant, $\psi(M) = \tilde{\psi}(|M|)$. Thus we may rewrite the sum over N_1, \ldots, N_m by first summing over $n_1 = |N_1|, \ldots, n_m = |N_m|$, and then over N_1, \ldots, N_m with $|N_k| = n_k$ fixed. Then

$$Z_{\Gamma_0}(a) = 1 + \sum_{m=1}^{\infty} \frac{1}{m!} \sum_{n=m}^{\infty} \frac{1}{n!} \sum_{\substack{n_1, \ldots, n_m \geq 1 \\ n_1 + \ldots + n_m = n}} \prod_{k=1}^{m} \tilde{\psi}(n_k) \sum_{\substack{N_1 \dot{\cup} \ldots \dot{\cup} N_m = \underline{n} \\ |N_l| = n_l \;\; \forall l}} 1. \tag{2.98}$$

The last sum gives a multinomial factor,

$$\sum_{\substack{N_1 \dot{\cup} \ldots \dot{\cup} N_m = \underline{n} \\ |N_l| = n_l \;\; \forall l}} 1 = \frac{n!}{n_1! \ldots n_m!}. \tag{2.99}$$

The $n!$ cancels, so the summand is independent of n. Thus the sum over n can be done, and it simply removes the constraint $n_1 + \ldots + n_m = n$. Thus

$$Z_{\Gamma_0}(a) \;=\; 1 + \sum_{m=1}^{\infty} \frac{1}{m!} \sum_{n_1,\dots,n_m \geq 1} \prod_{k=1}^{m} \frac{\tilde{\psi}(n_k)}{n_k!} \tag{2.100}$$

$$=\; 1 + \sum_{m=1}^{\infty} \frac{1}{m!} \left(\sum_{n=1}^{\infty} \frac{\tilde{\psi}(n)}{n!} \right)^m$$

$$=\; \exp\left(\sum_{n=1}^{\infty} \frac{\tilde{\psi}(n)}{n!} \right). \tag{2.101}$$

Looking at the definitions, we see that $\tilde{\psi}(n) = \psi(\{1,\dots,n\})$. ∎

2.5 The Effective Action and Connected Graphs

2.5.1 Definition and Semigroup Property

Definition 2.19. *Let $C > 0$ be a positive linear operator on \mathbb{C}^{Γ}, and V be a polynomial in the fields. Assume that either V is bounded below and $\lambda \geq 0$, or that λ is a formal parameter. The effective action for the interaction $-\lambda V$ and the propagator C is*

$$e^{G_{\text{eff}}(\psi, C, -\lambda V)} = \int d\mu_C(\phi)\, e^{-\lambda V(\phi + \psi)}. \tag{2.102}$$

Under the first assumption, the integral for the effective action is convergent for all $\lambda \geq 0$. If λ is a formal parameter, the effective action is a well-defined formal power series in λ for any polynomial V. For a nonnegative covariance, $C \geq 0$, the effective action is defined by a limiting procedure. For the graphical representation derived below, it does not make any difference whether the covariance is strictly positive or nonnegative because the values of Feynman graphs are polynomials in C.

We now discuss the relation of the effective action to the correlation functions and then motivate why it is a convenient object to study.

Shifting the integration variable and recalling the definition of the Gaussian measure, we have

$$e^{G_{\text{eff}}(\psi, C, -\lambda V)} \;=\; \int d\mu_C(\phi - \psi)\, e^{-\lambda V(\phi)}$$

$$=\; (\det(2\pi C))^{-\frac{1}{2}} \int D_{\Gamma}\psi\, e^{-\frac{1}{2}(\phi - \psi, C^{-1}(\phi - \psi))_{\Gamma} - \lambda V(\phi)}$$

$$=\; e^{-\frac{1}{2}(\psi, C^{-1}\psi)_{\Gamma}} \int d\mu_C(\phi) e^{-\lambda V(\phi) + (C^{-1}\psi, \phi)_{\Gamma}}. \tag{2.103}$$

Comparing this to (1.117), we see that apart from the explicit prefactor, this is the usual formula for the partition function with source terms, only that

the source terms are $J = C^{-1}\psi$. This modification implies that the functions are *amputated*, i.e., that in the Feynman graph representation, no propagators are assigned to the external lines of the graph. Thus we have

$$G_{\text{eff}}(\psi, C, -\lambda V) = -\frac{1}{2}(\psi, C^{-1}\psi)_\Gamma + \log Z_{\Gamma,\lambda}(C^{-1}\psi). \qquad (2.104)$$

We show in this section that the Feynman graph representation of G_{eff} contains only *connected* graphs. For this reason, $G_{\text{eff}}(\psi, C, -\lambda V)$ is called the *generating functional for the connected, amputated Green functions*. Equation (2.104) then implies that only connected (but unamputated) Feynman graphs contribute to $\log Z_{\Gamma,\lambda}(J)$, unamputated meaning that propagators are associated to external lines. Thus $\log Z_{\Gamma,\lambda}(J)$ is the generating functional for the connected Green functions.

By (1.90), $Z_{\Gamma,\lambda}(J)$, and thus also $\log Z_{\Gamma,\lambda}(J)$ and G_{eff}, determine all moments (correlation functions) of the measure $d\mu_C(\phi) \exp(-\lambda V(\phi))$. Thus if one knows G_{eff}, one can obtain all correlation functions.

Let us now consider the case where $C = C_1 + C_2$. By Theorem 1.5,

$$\begin{aligned}
e^{G_{\text{eff}}(\psi, C_1+C_2, -\lambda V)} &= \int d\mu_{C_2}(\phi_2) \int d\mu_{C_1}(\phi_1) \, e^{-\lambda V(\phi_1 + \phi_2 + \psi)} \\
&= \int d\mu_{C_2}(\phi_2) \, e^{G_{\text{eff}}(\phi_2 + \psi, C_1, -\lambda V)}. \qquad (2.105)
\end{aligned}$$

Thus, integrating out the ϕ_1 field associated to the covariance C_1 generates an effective interaction $G_{\text{eff}}(\phi_2, C_1, -\lambda V)$ for the remaining ϕ_2-field, which simply replaces V in the exponent. One can rewrite (2.105) as

$$G_{\text{eff}}(\psi, C_1 + C_2, -\lambda V) = G_{\text{eff}}(\psi, C_2, G_{\text{eff}}(\cdot, C_1, -\lambda V)); \qquad (2.106)$$

this is the semigroup property of the effective actions. If the covariance C is a sum of n terms,

$$C = C_1 + \ldots + C_n \qquad (2.107)$$

with $C_k \geq 0$ for all $k \in \{1, \ldots, n\}$, the field ϕ is a sum

$$\phi = \phi_1 + \ldots + \phi_n \qquad (2.108)$$

of n independent Gaussian fields. Successive integration over ϕ_1, \ldots, ϕ_n generates a sequence of $n + 1$ effective actions $G^{(0)}, \ldots G^{(n)}$, where $G^{(0)} = -\lambda V$ is the initial interaction and for $k \in \{1, \ldots, n\}$,

$$G^{(k)}(\psi) = G_{\text{eff}}(\psi, C_k, G^{(k-1)}) = \log \int d\mu_{C_k}(\phi_k) e^{G^{(k-1)}(\psi + \phi_k)} \qquad (2.109)$$

is obtained from $G^{(k-1)}$ by integrating over the field ϕ_k. This is the implementation of the idea of "integrating out fluctuations", originally due to Wilson. From the probabilistic point of view, one may regard the $G^{(k)}$ as random variables, and the sequence $(G^{(0)}, \ldots G^{(n)})$ is a martingale by (2.109).

Wilson's idea behind the splitting was to use this to approach the singularity in many little steps. This means that although the continuum propagator is singular, one can write it as an infinite sum of regular covariances, so that every finite partial sum is bounded and the singularity arises only in the limit. Posing a cutoff simply means retaining 'only a finite number of terms in the sum. Removal of the cutoff means taking the limit to an infinite sum. The strategy of Wilson's renormalization group is to study how the singularities build up, and how they can be prevented by adjusting the initial condition V, as the cutoff is removed. Most of modern renormalization is done that way [27, 28, 44, 45, 46, 47]. This is the subject of Chap. 3, where we consider a variant of this method (originally also due to Wilson and Wegner [14, 23]) where one writes the covariance as an integral instead of a sum.

The effective action has an expansion in the fields; more precisely, in every fixed order of the original coupling λ, G_{eff} is a polynomial in the fields. The coefficient functions G_{mr} have two interpretations: they are the (connected and amputated) Green functions for the interaction $-\lambda V$ and the propagator C_1 obtained from integration over ϕ_1, and they are also the vertex functions for the interaction of the field ϕ_2. Thus, in the integral over ϕ_2, these vertex functions are again joined by propagators to form Feynman graphs. It is the combination of these complementary interpretations that makes the effective action such a useful object. Also, one sees why it is natural that the graphs are amputated: The legs of the vertices generated by $G_{\text{eff}}(\phi_2, C_1, -\lambda V)$ are paired with propagators C_2 in the next integration step, so it would be inconsistent to associate a propagator C_1 to them in the definition of the value of the graph (this is also the reason for the convention of not counting external legs as lines of the Feynman graph: in the graphical expansions of effective actions, propagators are assigned only to the internal lines of the Feynman graph, but not to the external legs).

2.5.2 Derivation of the Graphical Representation

We shall study G_{eff} in perturbation theory, that is, by a formal expansion in λ. So we write

$$G_{\text{eff}}(\psi, C, -\lambda V) = \sum_{r=1}^{\infty} \lambda^r G_{\text{eff}}^{(r)}(\psi, C, V) \qquad (2.110)$$

with

$$G_{\text{eff}}^{(r)}(\psi, C, V) = [\frac{\partial^r}{\partial \lambda^r} \log \int d\mu_C(\phi) \, e^{\lambda V(\phi+\psi)}]_{\lambda=0}. \qquad (2.111)$$

In this section, we derive the Feynman graph representation for $G_{\text{eff}}^{(r)}(\psi, C, V)$. We shall again use Wick ordering, i.e. use $V(\phi) = \Omega_C(\tilde{V}(\phi))$. Note that when using (2.105) to integrate over ϕ_1 only, the covariance with respect to which we Wick order is in general different from the one in the measure: typically,

one Wick orders with respect to $C = C_1 + C_2$ and integrates with respect to C_1 only.

Since for any f that is n times differentiable at zero

$$(\frac{\partial}{\partial \lambda})^n f(\lambda x)\Big|_{\lambda=0} = x^n f^{(n)}(0), \tag{2.112}$$

and

$$\frac{\partial^n}{\partial \lambda_1 \ldots \partial \lambda_n} f(\lambda_1 x_1 + \ldots + \lambda_n x_n)\Big|_{\lambda_1=\ldots=\lambda_n=0} = (\prod_{k=1}^{n} x_k) f^{(n)}(0) \tag{2.113}$$

we have

$$(\frac{\partial}{\partial \lambda})^n f(\lambda x)\Big|_{\lambda=0} = \Big[\frac{\partial^n}{\partial \lambda_1 \ldots \partial \lambda_n} f(\lambda_1 x_1 + \ldots + \lambda_n x_n)\Big]_{\lambda_i=0, x_i=x \forall i}. \tag{2.114}$$

Thus

$$G_{\text{eff}}^{(r)}(\psi, C_1, V) = [\frac{\partial^r}{\partial \lambda^r} \log \int d\mu_{C_1}(\phi) \, e^{\lambda V(\phi+\psi)}]_{\lambda=0}$$

$$= \mathcal{E}^T(V, \ldots, V; C_1, \psi) \tag{2.115}$$

where V occurs r times in \mathcal{E}^T and

$$\mathcal{E}^T(V_1, \ldots, V_p; C_1, \psi) = \tag{2.116}$$

$$\Big[\frac{\partial^p}{\partial \lambda_1 \ldots \partial \lambda_p} \log \int d\mu_{C_1}(\phi) \, e^{\lambda_1 V_1(\phi+\psi)+\ldots+\lambda_p V_p(\phi+\psi)}\Big]_{\lambda_1=\ldots=\lambda_p=0}.$$

We may regard $\lambda_1, \ldots, \lambda_p$ as formal variables since they are only used to do combinatorics. Note that $\mathcal{E}^T(V_1, \ldots, V_p; C_1, \psi)$ is linear in every V_k since it is the p'th derivative applied to V_1, \ldots, V_p, and a p'th derivative is a multilinear map. We assume that every V_k is a polynomial, i.e.

$$V_k(\phi) = \sum_{m_k=1}^{M_k} \int_{\Gamma^{m_k}} \prod_{i=1}^{m_k} dx_{k,i} \, v_k^{(m_k)}(x_{k,1}, \ldots, x_{k,m_k}) \, \Omega_C \Big(\prod_{i=1}^{m_k} \phi(x_{k,i})\Big). \tag{2.117}$$

This assumption is natural because at the end, the effective action will appear as exactly such a polynomial as well. We now calculate \mathcal{E}^T in a similar way as we calculated $I(V_1, \ldots, V_p, \psi)$ in Sect. 2.3.

We use the linearity of Ω_C to write

$$V_k(\phi) = \sum_{m_k=1}^{M_k} \int_{\Gamma^{m_k}} dx_{k,1} \ldots dx_{k,m_k} v_k^{(m_k)}(x_{k,1}, \ldots, x_{k,m_k})$$

$$\Big[i^{-m_k} \frac{\partial^{m_k}}{\partial \alpha_{k,1} \ldots \partial \alpha_{k,m_k}} \Omega_C(e^{i(\alpha^{(k)},\phi)})\Big]_{\alpha=0} \tag{2.118}$$

with

$$\alpha^{(k)}(x) = \sum_{i=1}^{m_k} \alpha_{k,i} \delta_\Gamma(x, x_{k,i}). \tag{2.119}$$

The important point is now that $\mathcal{E}^T(V_1, \ldots, V_p; C_1, \psi)$ is linear in every V_k, so we may take out all sums *and* the derivatives with respect to the α variables. Denoting again

$$B_0 = \{(k, i_k) : k \in \{1, \ldots, p\}, i_k \in \{1, \ldots, m_k\}\} \tag{2.120}$$

and $X_k = (x_{k,1}, \ldots, x_{k,m_k})$, we get

$$\mathcal{E}^T(V_1, \ldots, V_p; C_1, \psi) \tag{2.121}$$

$$= \sum_{m_1, \ldots m_p} \int \prod_{b \in B_0} dx_b \prod_{k=1}^p v_k^{(m_k)}(X_k) \left[\prod_{b \in B_0} \frac{\partial}{\partial \alpha_b} \mathbb{E}(\alpha, C, C_1, \psi) \right]_{\alpha=0}$$

with

$$\mathbb{E}(\alpha, C, C_1, \psi) = \mathcal{E}^T(\Omega_C(e^{i(\alpha^{(1)}, \cdot)}), \ldots, \Omega_C(e^{i(\alpha^{(p)}, \cdot)}); C_1, \psi). \tag{2.122}$$

Let

$$Z' = \int d\mu_{C_1}(\phi) \exp\left(\sum_{k=1}^p \lambda_k e^{i\Omega_C[(\alpha^{(k)}, \phi+\psi)]}\right) \tag{2.123}$$

then

$$\mathbb{E}(\alpha, C, C_1, \psi) = \left[\frac{\partial^p}{\partial\lambda_1 \ldots \partial\lambda_p} \log Z'\right]_{\lambda=0}. \tag{2.124}$$

The integral for Z' converges for all $(\lambda_1, \ldots, \lambda_p) \in \mathbb{C}^p$ because $|e^{i(\alpha, \phi)}| = 1$. So, in Z', we may regard the λ's again as complex, not just formal variables. This would not have been the case if we had not put an i in the exponent.

However, we are only interested in $\frac{\partial^p}{\partial\lambda_1 \ldots \partial\lambda_p} \log Z$ at $\lambda = 0$, so we need only a polynomial of degree 1 in every λ_i. Thus

$$\mathbb{E}(\alpha, C, C_1, \psi) = \left[\frac{\partial^p}{\partial\lambda_1 \ldots \partial\lambda_p} \log Z\right]_{\lambda=0} \tag{2.125}$$

where

$$Z = \int d\mu_{C_1}(\phi) \prod_{k=1}^p \left(1 + \lambda_k \Omega_C(e^{i(\alpha^{(k)}, \phi+\psi)})\right)$$

$$= \sum_{M \subset \{1, \ldots, p\}} \prod_{r \in M} \lambda_r \int d\mu_{C_1}(\phi) \prod_{k \in M} \Omega_C(e^{i(\alpha^{(k)}, \phi+\psi)}). \tag{2.126}$$

Inserting the definition of Ω_C and doing the Gaussian integral, we get

$$Z = \sum_{M \subset \{1, \ldots, p\}} \left(\prod_{r \in M} \lambda_r e^{i(\alpha^{(r)}, \psi) + \frac{1}{2}(\alpha^{(r)}, C\alpha^{(r)})}\right) e^{-\frac{1}{2} \sum_{k, l \in M} (\alpha^{(k)}, C_1\alpha^{(l)})}. \tag{2.127}$$

Collecting the diagonal terms and using $C = C_1 + C_2$, we get

$$Z = \sum_{M \subset \{1,\ldots,p\}} \prod_{r \in M} \zeta_r \prod_{\substack{k,l \in M \\ k < l}} e^{-(\alpha^{(k)}, C_1 \alpha^{(l)})} \qquad (2.128)$$

with

$$\zeta_r = \lambda_r \, e^{i(\alpha^{(r)}, \psi) + \frac{1}{2}(\alpha^{(r)}, C_2 \alpha^{(r)})}. \qquad (2.129)$$

We now employ the standard trick of writing

$$e^{-(\alpha^{(k)}, C\alpha^{(l)})} = 1 + b_{kl} \qquad (2.130)$$

with $b_{kl} = e^{-(\alpha^{(k)}, C\alpha^{(l)})} - 1$. With this,

$$Z = \sum_{M \subset \{1,\ldots,p\}} \prod_{k \in M} \zeta_k \prod_{\substack{k < l \\ k,l \in M}} (1 + b_{kl}). \qquad (2.131)$$

Using Lemma 2.15, we get

$$Z = 1 + \sum_{\substack{M \subset \{1,\ldots,p\} \\ M \neq \emptyset}} \prod_{k \in M} \zeta_k \sum_{m=1}^{|M|} \frac{1}{m!} \sum_{\substack{N_1,\ldots,N_m \neq \emptyset \\ N_1 \cup \ldots \cup N_k = M}} \prod_{l=1}^{m} \left(\sum_{G \in \mathcal{G}_c(N_l)} \prod_{(i,j) \in G} b_{ij} \right). \qquad (2.132)$$

It may seem at this point that we are just redoing all calculations that led to the formula for the logarithm of a polymer partition function. That is true, but it is exactly here that we cannot proceed as before because the b_{ij} and ζ_k all depend on the α and ψ variables, so they are certainly not going to produce permutation invariant objects. But we can still rewrite Z as a polymer partition function: using $\prod_{k \in M} \zeta_k = \prod_{l=1}^{m} (\prod_{r \in N_l} \zeta_r)$ and denoting

$$a(N_k) = \prod_{r \in N_k} \zeta_k \sum_{G \in \mathcal{G}_c(N_k)} \prod_{(i,j) \in G} b_{ij}, \qquad (2.133)$$

we have

$$Z = 1 + \sum_{m=1}^{p} \frac{1}{m!} \sum_{\substack{N_1,\ldots,N_m \neq \emptyset \\ N_1 \cup \ldots \cup N_m \subset \{1,\ldots,p\}}} \prod_{l=1}^{m} a(N_l). \qquad (2.134)$$

Here, we have joined the sum over $M \subset \{1,\ldots,p\}$ and the one over N_1,\ldots,N_m with the constraint $N_1 \cup \ldots \cup N_m = M$ into a single sum with the constraint $N_1 \cup \ldots \cup N_m \subset \{1,\ldots,p\}$.

Note that by definition (2.129) of the ζ_k, the λ-dependence of $a(N)$ is

$$a(N) \sim \prod_{r \in N} \lambda_r. \qquad (2.135)$$

We now recognize that Z is the partition function of the following polymer system: Γ_0 is the set of nonempty subsets of $\{1, \ldots, p\}$,

$$N \subset \{1, \ldots, p\} \quad \text{is a polymer} \quad \Leftrightarrow \quad N \neq \emptyset$$
$$N \sim N' \quad \Leftrightarrow \quad N \cap N' = \emptyset. \tag{2.136}$$

By Theorem 2.17,

$$\log Z = \sum_{n=1}^{\infty} \frac{1}{n!} \sum_{N_1, \ldots, N_n \subset \{1, \ldots, p\}} U_c(N_1, \ldots, N_n) \prod_{l=1}^{n} a(N_l). \tag{2.137}$$

At this point a big simplification arises: only $n = 1$ contributes to (2.125), because of the following. For $n \geq 2$, $U_c(N_1, \ldots, N_n) = 0$ unless there is a $k \in \{1, \ldots, n\}$ such that N_1 and N_k are incompatible, i.e. $N_1 \cap N_k \neq \emptyset$. But by (2.135), $a(N_1)a(N_k)$ contains $(\prod_{r \in N_1} \lambda_r)(\prod_{r \in N_k} \lambda_r)$, in which λ_r appears quadratically for every $r \in N_1 \cap N_k$. Thus, taking only one derivative $\frac{\partial}{\partial \lambda_r}$ and evaluating at $\lambda_r = 0$ gives zero. So only the term $n = 1$ contributes, and

$$\left[\frac{\partial^p}{\partial \lambda_1 \ldots \partial \lambda_p} \log Z \right]_{\lambda=0} = \left[\frac{\partial^p}{\partial \lambda_1 \ldots \partial \lambda_p} \sum_{N_1 \subset \{1, \ldots, p\}} a(N_1) \right]_{\lambda=0} \tag{2.138}$$

since $U_c(N_1) = 1$. Since $a(N_1)$ contains $\prod_{r \in N_1} \lambda_r$, and no additional factors of λ_s, only $N_1 = \{1, \ldots, p\}$ contributes to (2.138), and we obtain

$$\mathbb{E}(\alpha, C, C_1, \psi) = \left(\prod_{i=1}^{p} \lambda_i \right)^{-1} a(\{1, \ldots, p\}) \tag{2.139}$$

$$= \prod_{k=1}^{p} \left(e^{i(\alpha^{(k)}, \psi) + \frac{1}{2}(\alpha^{(k)}, C_2 \alpha^{(k)})} \right) \sum_{G \in \mathcal{G}_c(\{1, \ldots, p\})} \prod_{(i,j) \in G} \left(e^{-(\alpha^{(i)}, C_1 \alpha^{(j)})} - 1 \right).$$

We have now again reached the point where we only have to use Lemma 2.6 and differentiate, to complete the derivation of the representation by Feynman graphs. Before doing that, let us compare (2.139) to the result we had for the analogous object in the equation (2.53) for $J(X_1, \ldots, X_p; \psi)$. There we had $C_2 = 0$, so $C = C_1$, and by the first equality in Lemma 2.15,

$$e^{i \sum_{k=1}^{p} (\alpha^{(k)}, \psi) - \sum_{k < l} (\alpha^{(k)}, C_1 \alpha^{(l)})}$$

$$= \prod_{k=1}^{p} e^{i(\alpha^{(k)}, \psi)} \prod_{k < l} (1 + e^{-(\alpha^{(k)}, C_1 \alpha^{(l)})} - 1) \tag{2.140}$$

$$= \prod_{k=1}^{p} e^{i(\alpha^{(k)}, \psi)} \sum_{G \in \mathcal{G}(\{1, \ldots, p\})} \prod_{(i,j) \in G} (e^{-(\alpha^{(k)}, C_1 \alpha^{(l)})} - 1).$$

Comparing this to (2.139) (with $C_2 = 0$), we see that the only difference is that in (2.140), the sum over G runs only over connected graphs instead of all graphs. Since

$$e^{-(\alpha^{(k)}, C\alpha^{(l)})} - 1 = -(\alpha^{(k)}, C\alpha^{(l)}) + O(\alpha^4), \tag{2.141}$$

the vertices k and l *must* be joined by a line whenever a line (k, l) appears in G because we differentiate and evaluate at $\alpha = 0$. It should now be obvious that only connected Feynman graphs contribute to $\mathcal{E}^T(V_1, \ldots, V_p; C, \psi)$, where we define:

Definition 2.20. *Let* $B \subset B_0(p, m_1, \ldots, m_p)$. *A Feynman graph* $L \in \mathcal{L}(B)$ *is connected iff*

$$\begin{aligned}
\mathcal{T}(L) &= \{(k, l) : 1 \leq k < l \leq p, \text{ there are } i \in \{1, \ldots, m_k\} \\
&\quad \text{and } j \in \{1, \ldots, m_l\} \text{ such that } ((k, i), (l, j)) \in L\} \tag{2.142}
\end{aligned}$$

satisfies $\mathcal{T}(L) \in \mathcal{G}_c(\{1, \ldots, p\})$. *We denote the set of all connected* $L \in \mathcal{L}(B)$ *by* $\mathcal{L}_c(B)$.

This definition corresponds to the intuition one has about connected graphs; it also implies the existence of a *spanning tree* for any connected Feynman graph.

Remark 2.21. Note also that there are no additional combinatorial factors. The effect of taking the logarithm on the graphical sum is only that the disconnected graphs are deleted, nothing else. This is because only the term with $n = 1$ in the sum (2.137) contributes to $\mathbb{E}(\alpha, C, C_1, \psi)$ (the terms with $n \geq 2$ would give extra signs and combinatorial factors by definition of the higher Ursell functions U_c). If one is used to writing the expansion in terms of topological Feynman graphs and the associated symmetry factors, this means that taking the logarithm does not change the symmetry factors; its only effect is to remove the disconnected Feynman graphs.

The procedure to derive the Feynman graph representation is now identical to the one used for $I(V_1, \ldots, V_p; \psi)$. The case $C_1 = C, C_2 = 0$ has just been discussed. We briefly discuss the more important case $C = C_1 + C_2$, with $C_2 \neq 0$. It is more important because we have assumed that the starting interaction V is Wick ordered. To iterate the semigroup equation for the effective action (this generates the renormalization group), we need to express the result in Wick ordered form as well. We only have to Wick order the function in (2.139) that depends on ψ with respect to C_2. This gives

$$e^{i(\sum_{k=1}^{P} \alpha^{(k)}, \psi)} e^{\frac{1}{2} \sum_{k=1}^{P} (\alpha^{(k)}, C_2\alpha^{(k)})} = \Omega_{C_2} \left(e^{i \sum_{k=1}^{P} (\alpha^{(k)}, \psi)} \right) e^{-\sum_{k<l} (\alpha^{(k)}, C_2\alpha^{(l)})}. \tag{2.143}$$

Thus in that case

$$\mathbb{E}(\alpha, C, C_1, \psi) = \Omega_{C_2}\left(e^{i\sum_{k=1}^{p}(\alpha^{(k)}, \psi)}\right) e^{\sum_{k<l}(i\alpha^{(k)}, C_2 i\alpha^{(l)})}$$

$$\sum_{G \in \mathcal{G}_c(\{1,\dots,p\})} \prod_{(k,l) \in G} \left(e^{(i\alpha^{(k)}, C_1 i\alpha^{(l)})} - 1\right). \quad (2.144)$$

When one now does the α-derivatives to get the sum over Feynman graphs, the graphs can have either C_1 or C_2 on their lines. Usually the C_1 lines (corresponding to the integrated fields ϕ_1) are called the *hard lines* and the C_2-lines (coming from writing the result in terms of Wick ordered polynomials) are called the *soft lines* or *Wick lines*. It is obvious from (2.144) and the above discussion of connectedness that only those combinations of hard and soft lines occur for which the Feynman graph is still connected by the hard lines alone.

2.5.3 Result and Discussion

The resulting representation of the effective action as a sum over values of Feynman graphs is as follows. Recall that by (2.110) and (2.115), it suffices to represent $\mathcal{E}^T(V_1, \dots, V_p, C; \psi)$ as a sum over values of Feynman graphs. We give both the expansion in terms of ordinary and in terms of Wick ordered monomials.

Theorem 2.22. Let V_1, \dots, V_p be given by (2.117), where $C = C_1$ and $C_2 = 0$, and let $\bar{m} = \sum m_i - 2(p-1)$. Then

$$\mathcal{E}^T(V_1, \dots, V_p, C; \psi) = \sum_{\substack{m_1, \dots, m_p \\ 1 \le m_i \le M_i}} \sum_{m=0}^{\bar{m}} \int \prod_{k=1}^{m} d\zeta_k \, \psi(\zeta_k) \kappa(C, \underline{\zeta}) \quad (2.145)$$

where $\underline{\zeta} = (\zeta_1, \dots, \zeta_m)$ and

$$\kappa(C, \underline{\zeta}) = \frac{1}{m!} \sum_{\substack{b_1, \dots, b_m \in B_0 \\ \text{all distinct}}} \sum_{L \in \mathcal{L}_C(B_0 \setminus \{b_1, \dots, b_m\})} Val(G_L(C, \underline{\zeta})). \quad (2.146)$$

Here, the Feynman graph G_L is identified with its set of lines L, and

$$Val(G_L(C, \underline{\zeta})) = \int \prod_{b \in B_0 \setminus E} dx_b \prod_{k=1}^{p} v_k(X_k) \prod_{l=(b,b') \in L} C(x_b, x_{b'}) \quad (2.147)$$

where $x_{b_l} = \zeta_l$ for all $l \in \{1, \dots, m\}$ and $E = \{b_1, \dots, b_m\}$. Both $Val(G_L)$ and κ depend on v_1, \dots, v_p.

We shall refer to this more briefly as

$$\kappa(C, \zeta_1, \ldots, \zeta_m) = \sum_G Val(G) \tag{2.148}$$

where the sum is over all connected Feynman graphs with vertices v_1, \ldots, v_p (v_k has m_k external legs), m external legs (the sum over all b_1, \ldots, b_m is just the sum over all choices of external legs), and a set of internal lines $L \in \mathcal{L}_C$. The restriction on the sum over m is there because the Feynman graph has to be connected, so it must at least be a connected tree. A tree with p vertices has $p - 1$ internal lines. Every internal line binds two legs $b, b' \in B_0$ of the vertices it connects. The number of available lines is $|B_0| = m_1 + \ldots + m_p$. By these considerations the total number of external legs is at most that of a connected tree:

$$m = |B_0| - 2|L| \le |B_0| - 2(p-1) = \sum_{i=1}^{p} m_i - 2(p-1). \tag{2.149}$$

Theorem 2.23. *Let V_1, \ldots, V_p be given by (2.117), where $C = C_1 + C_2$, and let $\bar{m} = \sum m_i - 2(p-1)$. Then*

$$\mathcal{E}^T(V_1, \ldots, V_p, C; \psi) \tag{2.150}$$

$$= \sum_{m_1, \ldots, m_p} \sum_{m=0}^{\bar{m}} \int d\zeta_1 \ldots d\zeta_m \Omega_{C_2}(\psi(\zeta_1) \ldots \psi(\zeta_m)) \mathcal{V}_m(\underline{\zeta})$$

with

$$\mathcal{V}_m(\underline{\zeta}) = \frac{1}{m!} \sum_{\substack{b_1, \ldots, b_m \in B_0 \\ \text{all distinct}}} \sum_{\substack{L \in \mathcal{L}_C(B_0 \backslash \\ \{b_1, \ldots, b_m\})}} \sum_{h: L \to \{1, 2\}} ValG_L^{(h)}(C_1, C_2, \underline{\zeta}). \tag{2.151}$$

The value of the labelled graph $G_L^{(h)}$ is

$$Val(G_L^{(h)})(C_1, C_2, \underline{\zeta}) = \int \prod_{b \in B_0 \backslash E} dx_b \prod_{k=1}^{p} v_k(x_{k,1}, \ldots, x_{k,m_k})$$

$$\prod_{l=(b,b') \in L} C_{h(l)}(x_b, x_{b'}) \tag{2.152}$$

if $G_L^{(h)}$ is connected by lines with $h(l) = 1$, and $Val(G_L^{(h)})(C_1, C_2, \underline{\zeta}) = 0$ otherwise. Here $E = \{b_1, \ldots, b_m\}$ and $x_{b_l} = \zeta_l$ $\forall l \in \{1, \ldots, m\}$, and the labelled graph $G_L^{(h)}$ is defined as the pair (L, h) where $L \in \mathcal{L}_c(B_0)$ is the set of lines and $h : L \to \{1, 2\}$ is the map that determines which lines get a factor C_1 and which ones get a factor C_2.

The graph must be connected by C_1 ("hard") lines because those lines came from factors $\exp((\alpha^{(k)}, C_1\alpha^{(l)})) - 1 = O(\alpha^2)$, whereas the C_2 ("soft") lines came from $\exp((\alpha^{(k)}, C_2\alpha^{(l)})) = 1 + O(\alpha^2)$ lines.

For iterating the renormalization map that relates effective actions at different scales, the representation given in Theorem 2.23 is more important because the result is written in the same form as the input, that is, (2.150) is an expansion in Wick monomials in the same way as (2.117). The covariance with respect to which Wick ordering is done also matches the semigroup property of the flow of effective actions: Wick ordering is always done with respect to the unintegrated covariance. Once the fields corresponding to C_1 are integrated over, one has to expand $G_{\text{eff}}(\phi_2, C_1, -\lambda V)$ in Wick ordered monomials with respect to the covariance C_2 of the fields ϕ_2 that are not yet integrated. If C is a sum of n terms, the sum of all unintegrated covariances enters into Wick ordering at every step in the sequence.

Remark 2.24. G_{eff} is the generating functional for the connected, amputated Green functions. This means the following. G_{eff} is a formal power series in the coupling λ. The coefficient of λ^r is a polynomial in the fields whose coefficients are given by a sum over *connected* Feynman graphs, as proven above. The graphs are called amputated because by the graphical rules, there are no propagators associated to the external legs. It was discussed above that this is the natural convention for calculating effective actions.

Remark 2.25. It is not essential for the results derived here that we are dealing with bosons. The generalization to fermions is given in Appendix B.6.

2.6 Graphical Representations: Conclusions

The above representations by sums over values of Feynman graphs have many applications. We shall discuss only the most basic ones. We can summarize the calculations of the last section as follows. As a formal power series in λ,

$$\mathcal{G}(\phi, C_1, C_2, V) = \log \int d\mu_{C_1}(\psi) \, e^{-\lambda \Omega_C(V(\phi+\psi))} \tag{2.153}$$

$$= \sum_{r=1}^{\infty} \frac{(-\lambda)^r}{r!} \sum_{m=0}^{\bar{m}(r)} \int \prod_{k=1}^{m} dx_k \, \Omega_{C_2}\Big(\prod_{k=1}^{m} \phi(x_k)\Big) G_{m,r}(x_1, \ldots, x_m)$$

with

$$G_{m,r}(x_1, \ldots, x_m) = \sum_{\Gamma} Val(\Gamma) \tag{2.154}$$

where the sum is over all connected, amputated Feynman graphs with m external lines and r interaction vertices, and with hard and soft lines.

One result is already built into the above formula: for a given r, m cannot exceed a certain $\bar{m}(r)$, because of the following. If the initial interaction V

is of degree m_V in ϕ, then the maximal number of legs of vertices that can occur in r'th order is $|B_0| = m_V r$. By the formula

$$|E| = |B_0| - 2|L| \qquad (2.155)$$

which holds because the graphs contributing to $G_{m,r}$ are connected (and which we have already used in (2.149)), we have

$$\bar{m}(r) \leq m_V r - 2(r-1). \qquad (2.156)$$

So, e.g. for $V(\phi) = \int_\Gamma \phi(x)^4 dx$, we have

$$\bar{m}(r) \leq 4r - 2(r-1) = 2(r+1). \qquad (2.157)$$

Also, the sum over r starts at $r = 1$, since for $\lambda = 0$, the logarithm gives zero because $\int d\mu_C(\phi) = 1$.

The third observation is that $G_{m,r}$ is invariant under any transformation that leaves $\int_\Gamma dx$, $C(x, x')$, and all $v_k(x_1, \ldots, x_{m_k})$ invariant. This symmetry can also be deduced from the representation of \mathcal{G} by a Gaussian integral, but it also holds termwise in the sum (2.154), i.e. separately for the value of every graph.

In particular, in our lattice field theory, C is invariant under translations,

$$C(x + a, y + b) = C(x, y) \qquad \forall a \in \Gamma. \qquad (2.158)$$

If the same holds for all v_k, i.e. if

$$v_k(x_1 + a, \ldots, x_{m_k} + a) = v_k(x_1, \ldots, x_{m_k}) \qquad \forall a \in \Gamma \qquad (2.159)$$

then $Val\, G_L(C, \xi_1, \ldots, \xi_m)$ (which depends on v_1, \ldots, v_p) has this property as well:

$$Val\, G_L(C, \xi_1 + a, \ldots, \xi_m + a) = Val\, G_L(C, \xi_1, \ldots, \xi_m) \qquad (2.160)$$

for all $a \in \Gamma$.

Using Fourier transformation, we can rewrite \mathcal{G} as a polynomial in the $\hat{\phi}(p)$. Noting that the Wick ordering formula for $\Omega_C(\hat{\phi}(p_1) \ldots \hat{\phi}(p_m))$ follows directly from the formula for the generating functional

$$\Omega_C(e^{i(\alpha,\phi)r}) = \Omega_C(e^{i(\hat{\alpha},\hat{\phi})r\bullet}) = e^{i(\hat{\alpha},\hat{\phi})r\bullet + \frac{1}{2}(\hat{\alpha},\hat{C}\hat{\alpha})r\bullet} \qquad (2.161)$$

we have

$$\mathcal{G}(\phi, C_1, C_2, V) \qquad (2.162)$$

$$= \sum_{r=0}^{\infty} \frac{(-\lambda)^r}{r!} \sum_{m=0}^{\bar{m}(r)} \int \prod_{k=1}^{m} dp_k\, \Omega_{C_2}\left(\prod_{k=1}^{m} \hat{\phi}(p_k) \right) \hat{G}_{m,r}(p_1, \ldots, p_m)$$

with

$$\hat{G}_{m,r}(p_1,\ldots,p_m) = \int \prod_{k=1}^{m} \mathrm{d}x_k \; e^{ip_k x_k} G_{m,r}(x_1,\ldots,x_m). \tag{2.163}$$

For any translation invariant function $F(x_1,\ldots,x_m)$,

$$
\begin{aligned}
\hat{F}(p_1,\ldots,p_m) &= \int \prod_{k=1}^{m} \mathrm{d}x_k \; e^{ip_k x_k} F(x_1,\ldots,x_m) \\
&= \int \prod_{k=1}^{m} \mathrm{d}x_k \; e^{ip_k x_k} F(0, x_2-x_1,\ldots,x_m-x_1) \\
&= \int_{\Gamma} \mathrm{d}x_1 e^{ix_1(p_1+\ldots+p_m)} \tilde{F}(p_2,\ldots,p_m) \\
&= \delta_{\Gamma\bullet}(p_1+\ldots+p_m, 0)\, \tilde{F}(p_2,\ldots,p_m) \tag{2.164}
\end{aligned}
$$

with

$$\tilde{F}(p_2,\ldots,p_m) = \int \prod_{k=1}^{m} \mathrm{d}\xi_k \; e^{ip_k \xi_k} F(0, \xi_2,\ldots,\xi_m). \tag{2.165}$$

If F is permutation invariant (as in the case for v_k and $G_{m,r}$) then \tilde{F} is also permutation invariant in the sense that

$$\tilde{F}(p_{\pi(2)},\ldots,p_{\pi(m)}) = \tilde{F}(p_2,\ldots,p_m) \qquad \forall \pi \in \mathcal{S}_m \tag{2.166}$$

where p_1 is defined as $p_1 = -\sum_{l=2}^{m} p_l$ whenever it appears as $p_{\pi(k)}$ on the left hand side. Thus

$$\hat{G}_{m,r}(p_1,\ldots,p_m) = L^d \delta_{p_1+\ldots+p_m, 0}\, \tilde{G}_{m,r}(p_2,\ldots,p_m). \tag{2.167}$$

We shall show in the following that if $C(x,y)$ decays fast enough for large $d_\Gamma(x,y)$, and if an analogous statement holds for the v_k, then $\tilde{G}_{m,r}$ converges to a smooth function as $L \to \infty$. Here $d_\Gamma(x,y)$ is the distance on the torus.

We shall actually show this for the more general graphs G_L encountered in $\mathcal{E}^T(V_1,\ldots,V_r; \phi)$. Let $\hat{V}(G_L^{(h)})$ be the Fourier transform of $Val(G_L^{(h)})$, then, setting $p_{b_l} = q_l$ and relabelling the integration variables, we have

$$\hat{V}(G_L^{(h)})(q_1,\ldots,q_m) \tag{2.168}$$

$$= \int \prod_{b\in B_0} \mathrm{d}x_b \prod_{b\in E} e^{ip_b x_b} \prod_{k=1}^{r} v_k(X_k) \prod_{l=(b,b')\in L} C_{h(l)}(x_b, x_{b'}).$$

Now insert

$$C_{h(l)}(x_b, x_{b'}) = \int \mathrm{d}p_l \; e^{ip_l(x_b-x_{b'})} \tilde{C}_{h(l)}(p_l), \tag{2.169}$$

then

$$\hat{V}(G_L^{(h)})(q_1,\ldots,q_m) = \int \prod_{b \in B_0} dx_b \int \prod (dp_l\, \tilde{C}_{h(l)}(p_l))$$

$$\prod_{k=1}^{r} \left(v_k(X_k) \prod_{i=1}^{m_k} e^{ix_{(k,i)} \sum_{l \in LUE} \varepsilon_{(k,i)}^{(l)} p_l} \right) \qquad (2.170)$$

where $\varepsilon_{(k,i)} = 1$ if $l = (b,(k,i))$ for some $b \prec (k,i)$ or $l = b \in E$, $\varepsilon_{(k,i)} = -1$ if $l = ((k,i),b')$ for some $b' \succ (k,i)$, and $\varepsilon_{(k,i)}^{(l)} = 0$ otherwise. $\varepsilon_b^{(l)}$ is related to the incidence matrix of the graph. This is just a formal way of saying that $e^{ix_{k,j}p_l}$ occurs if l joins leg number j of vertex k to a leg of vertex number k', with $k' > k$, etc. We can now do the x-integrals and get

$$\hat{V}(G_L^{(h)})(q_1,\ldots,q_m) = \int \prod_{l \in L} dp_l\, \tilde{C}_{h(l)}(p_l) \prod_{k=1}^{r} \hat{v}_k(P_k) \qquad (2.171)$$

with $P_k = (p_1^{(k)},\ldots,p_{m_k}^{(k0)})$, the $p_i^{(k)}$ given as above,

$$p_i^{(k)} = \sum_{l \in L} \varepsilon_{(k,i)}^{(l)} p_l + \sum_{b \in E} \varepsilon_{(k,i)}^{(b)} p_b \qquad (2.172)$$

(note that by definition, only one term in this sum is nonzero). Less formally, but as precisely, one may say that the momentum p_l that "flows" in line $l = (b,b')$ is the momentum exiting from leg b (hence $+p_i^{(k)}$ if $b = (k,i)$) of vertex k, and it is the momentum entering leg b' of vertex k' (hence $-p_i^{(k)}$ exits from that leg of vertex k'). Thus the signs are due to the convention that every momentum is counted as exiting the vertex.

Theorem 2.26. *Let $C_h(x,y)$ be defined as in (2.169), with Fourier transform $\tilde{C}_{h(l)} \in C^s(\mathbb{R}^d/\frac{2\pi}{\varepsilon}\mathbb{Z}^d, \mathbb{C})$, and assume that for all vertices $k \in \{1,\ldots,r\}$*

$$\hat{v}_k(p_1^{(k)},\ldots,p_m^{(k)}) = L^d \delta_{p_1^{(k)}+\ldots+p_m^{(k)},0} \tilde{v}_k(p_2^{(k)},\ldots,p_{m_k}^{(k)}), \qquad (2.173)$$

with $\tilde{v}_k \in C^s((\mathbb{R}^d/\frac{2\pi}{\varepsilon}\mathbb{Z}^d)^{m_k-1}, \mathbb{C})$. Then, for all $E \subset B_0$ (with $|E| = m$) and all $\mathbb{L} \in \mathcal{L}_C(B_0 \setminus E)$: If $m = 0$ $(E = \emptyset)$, then

$$\hat{V}(G_\mathbb{L}^{(h)}) = L^d g_L^{(h)}(\mathbb{L}, C_1, C_2, v_1,\ldots,v_p) \qquad (2.174)$$

and $\lim_{L\to\infty} g_L^{(h)} = g^{(h)}$ exists. If $m > 0$ $(E \neq \emptyset)$, then

$$\hat{V}(G_\mathbb{L}^{(h)})(q_1,\ldots,q_m) = L^d \delta_{q_1+\ldots+q_m,0}\, \tilde{\gamma}_{mr}^{(L,h)}(q_2,\ldots,q_m) \qquad (2.175)$$

with $\tilde{\gamma}_{mr}^{(L,h)} \in C^s((\mathbb{R}^d/2\pi\mathbb{Z})^{m-1}, \mathbb{C})$, and $\tilde{\gamma}_{mr}^{(L,h)}$ and its derivatives up to order s converge to functions $\tilde{\gamma}_{mr}^{(h)} \in C^s((\mathbb{R}^d/2\pi\mathbb{Z})^{m-1}, \mathbb{C})$ as $L \to \infty$.

Remark 2.27. 1. If $\tilde{C}_{h(l)}$ and the \tilde{v}_k are analytic in a strip around the real axis, $\tilde{\gamma}_{mr}^{(L,h)}$ is also analytic in every q_l in the same strip.

2. For the covariance given by $\tilde{C}(p) = (m_\phi{}^2 + D_\varepsilon(p))^{-1}$, all functions depend on ε and they diverge as $\varepsilon \to 0$. They also depend on m_ϕ, and it is essential in the following proof that $m_\phi{}^2 > 0$.

Proof. We do an induction in the number of vertices r of the graph. Recall that

$$\hat{V}(G_L^{(h)})(q_1, \ldots, q_m)$$
$$= \int \prod_{l \in L} dp_l \, \tilde{C}_{h(l)}(p_l) \prod_{k=1}^{r} L^d \delta_{p_1^{(k)} + \ldots + p_m^{(k)}, 0} \tilde{v}_k(P_k), \qquad (2.176)$$

where, as before, $P_k = (p_1^{(k)}, \ldots, p_{m_k}^{(k)})$.

If $r = 1$, all legs of the vertex v_1 are external legs of the graph, and $\hat{V}(G_\emptyset^{(h)})(q_1, \ldots, q_{m_1}) = \hat{v}(q_1, \ldots, q_{m_1})$, so the claim holds by the hypotheses of the theorem. Let $r \geq 2$, and assume the statement to be proven for all $r' < r$. There is a vertex k of $G_L^{(h)}$ such that the subgraph G', obtained by deleting the vertex k and all lines that end at k, is still connected. Without loss of generality, we may assume that $k = 1$. Let \mathbb{L}_1 be the set of lines $l \in \mathbb{L}$ joining to this vertex. The graph is connected, so $\mathbb{L}_1 \neq \emptyset$. See also Fig. 2.7.

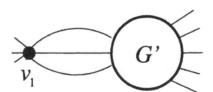

Fig. 2.7. The decomposition of G in terms of the vertex v_1 and the subgraph G'

Then

$$\hat{V}(G_L^{(h)})(q_1, \ldots, q_m) = \int \prod_{l \in \mathbb{L}_1} dp_l \, \tilde{C}_{h(l)} \, \hat{v}(P_1) \, \hat{V}(G'^{(h)}_{L \setminus L_1})(P') \qquad (2.177)$$

where $P'_b = -(P_1)_a$ if $(a, b) \in \mathbb{L}_1$. Denote $P' = (p'_1, \ldots, p'_{m'})$. The number of external legs m' of G' must of course be at least one since otherwise G' could not be joined to vertex 1. The inductive hypothesis applies to G', so

$$\hat{V}(G'^{(h)}_{L \setminus L_1})(P') = L^d \delta_{p'_1 + \ldots + p'_{m'}, 0} \, \tilde{\gamma}'_{m', r-1}(p'_2, \ldots, p'_{m'}). \qquad (2.178)$$

By our hypotheses,

$$\hat{v}(p_1^{(1)}, \ldots, p_{m_1}^{(1)}) = L^d \delta_{p_1^{(1)} + \ldots + p_{m_1}^{(1)}, 0} \, \tilde{v}(p_2^{(1)}, \ldots, p_{m_1}^{(1)}) \qquad (2.179)$$

where $\tilde{v} \in C^s$, and $\tilde{\gamma}'$ converges to a C^s function as $L \to \infty$.

Suppose first, that (as drawn Fig. 2.7) some legs of v_1 and some external legs of G' are external legs of G. Then $(p'_1, \ldots, p'_{m'})$ contains those q_m that enter G', and $(p_1^{(1)}, \ldots, p_{m_1}^{(1)})$ contains those q_m that enter vertex 1, and therefore the product of the two δ's is

$$L^d \delta_{p_1^{(1)} + \ldots + p_{m_1}^{(1)}, 0} \, L^d \delta_{p'_1 + \ldots + p'_{m'}, 0} = L^d \, \delta_{q_1 + \ldots + q_m, 0} \, L^d \, \delta_{p_1^{(1)} + \ldots + p_{m_1}^{(1)}, 0}. \quad (2.180)$$

If no legs of vertex 1 are external legs of the graph G, then $(p'_1, \ldots, p'_{m'}) = -(p_1^{(1)}, \ldots, p_{m_1}^{(1)})$, so (2.180) still holds. Since $p_1^{(1)} + \ldots + p_{m_1}^{(1)} = \sum_{l \in \mathbb{L}_1} p_l + Q$ where $Q = \sum_m q_m$, the sum running only over those external moments for which q_m is associated to an external leg of vertex 1 ($Q = 0$ if vertex 1 joins only to internal lines).

$$\hat{V}(G_{\mathbf{L}}^{(h)})(q_1, \ldots, q_m) \quad (2.181)$$

$$= L^d \delta_{q_1 + \ldots + q_m, 0} \int \prod_{l \in \mathbb{L}_1} dp_l \, \tilde{C}_{h(l)}(p_l) \, L^d \delta_{\sum_{l \in \mathbb{L}_1} p_l + Q, 0} \, \tilde{\gamma}'_{m', r-1}(\tilde{P}').$$

If $m = 0$, $\delta_{q_1 + \ldots + q_m, 0}$ is replaced by 1 and Q by 0 in this formula. Remove the summation $\int_{\Gamma^\bullet} dp_{l_1} = L^{-d} \sum_{p_{l_1} \in \Gamma^\bullet}$ of one line $l_1 \in \mathbb{L}_1$ by the delta function, then

$$\hat{V}(G_{\mathbf{L}}^{(h)})(q_1, \ldots, q_m) = L^d \delta_{q_1 + \ldots + q_m, 0} \, W(q_2, \ldots, q_m) \quad (2.182)$$

where

$$W(q_2, \ldots, q_m) = \int \Big(\prod_{\substack{l \in \mathbb{L}_1 \\ l \neq l_1}} dp_l \, \tilde{C}_{h(l)}(p_l) \Big) \tilde{C}_{h(l)}\Big(- \sum_{\substack{l \in \mathbb{L}_1 \\ l \neq l_1}} p_l - Q \Big) \tilde{\gamma}'_{m', r-1}(\tilde{P}')$$

$$(2.183)$$

with $q_1 = -\sum_{k=2}^{m} q_k$. This is the approximation, by a Riemann sum, of an integral over the bounded region $[-\pi/\varepsilon, \pi/\varepsilon]^{d(|\mathbb{L}_1| - 1)}$. The integrand is bounded because $\tilde{C}_{h(l)} \in C^s(\mathbb{R}^d / \frac{2\pi}{\varepsilon} \mathbb{Z}^d, \mathbb{C})$ for all l and because $\mathbb{R}^d / \frac{2\pi}{\varepsilon} \mathbb{Z}^d$ is compact, and the same holds for $\tilde{\gamma}'$ by the inductive hypothesis. By the dominated convergence theorem, the limit $L \to \infty$ is finite and (by repeating this argument after taking up to s derivatives) C^s in q_1, \ldots, q_m. ∎

At this point, we have reached the stage that we know the preliminaries for reading at least part of the current literature on field theory without too much trouble, for instance [26, 27, 28, 44, 45, 46, 47, 58, 59]. In the next chapter, we introduce a renormalization group differential equation that allows us to analyze the renormalization problem in a very efficient way.

3. The Renormalization Group

We now turn to the problem of renormalizing our field theory to all orders in perturbation theory. To this end, we shall define a flow of effective actions, by defining an effective action that depends on a scale parameter Λ. The differential equation obtained by applying $\partial/\partial\Lambda$ to the functional integral for the effective action is the renormalization group (RG) equation. The graphical expansion derived in Chap. 2 provides an explicit solution of the RG equation. However, the differential equation itself is very useful for deriving bounds for the solutions without using the details of the graphical representation. We shall learn more about the connection between the two approaches as we develop the theory.

The main ideas for the renormalizability proofs using the RG differential equation are due to Polchinski [29]. His proof, which uses a non-Wick–ordered form of the equation, was further simplified in [31] and then extended to gauge theories by Keller and Kopper [32]; a brief but very readable review of that work appeared in [34]. Here, we shall use the Wick–ordered form of the differential equation which was introduced soon after Polchinski's work by Wieczerkowski [30]. Using this form of the equation makes the inductive proofs still simpler.

The renormalization group is in truth not a group, but a semigroup, namely the semigroup of effective actions introduced in Sect. 2.5. It is only a semigroup because Gaussian integrals are defined only for nonnegative covariances. Thus integrating over fields with covariance $C \geq 0$, which is a convolution with the Gaussian measure $d\mu_C$, cannot be undone by another convolution with the measure associated to $-C$ because the latter does not exist. If a ϕ^4 term is present, one may argue that the integral may still be convergent for C with negative eigenvalues, but then one has to prove that the sign of the ϕ^4 coupling does not change when integration is done, which is not obvious. When considering formal power series expansions of the effective action, the argument that the ϕ^4-term controls the growth of the quadratic part does not apply anyway).

3.1 A Cutoff in Momentum Space

To motivate the choice of flow parameter, recall that the cause of the divergences was the singularity of $C(x,y) = \lim_{\varepsilon \to 0} C_{\Gamma_\varepsilon, L}(x,y)$ at $x = y$. This singularity is due to the slow decay of the propagator in momentum space:

$$\hat{C}_\varepsilon(p) = \frac{1}{m_\phi^2 + D_\varepsilon(p)} \xrightarrow{\varepsilon \to 0} \frac{1}{p^2 + m_\phi^2} = \hat{C}(p). \tag{3.1}$$

In particular, $\hat{C} \notin L^1(\mathbb{R}^d)$ and thus $C(x,x)$ is divergent. In principle, we could now study how the singularity builds up as $\varepsilon \to 0$. This is, however, messy, because the theory on a lattice has no rotation symmetry. Instead, we shall take a continuum limit at a fixed momentum space cutoff $\Lambda_0 \gg 1$, which is possible on the level of perturbation theory as a formal power series, and then renormalize directly in the continuum. As mentioned, the main motivation for introducing a lattice in the present setting was to give clean definitions of Gaussian integrals and objects like $\frac{\delta}{\delta\phi}$ without going through easy but lengthy (and in the end, irrelevant) functional analytical details of defining functional derivatives in the continuum.

Let $K \in C^\infty(\mathbb{R}_0^+, [0,1])$ obey

$$K(x) = \begin{cases} 0 & \text{if } x \geq 4 \\ 1 & \text{if } x \leq 1, \end{cases} \tag{3.2}$$

and $K'(x) < 0$ for $x \in (1,4)$. For convenience, we may choose K such that

$$\|K'\|_\infty \leq 1. \tag{3.3}$$

Let

$$C_{0,\Lambda_0}^{(\Gamma)}(x,y) = \int_{\Gamma^*} dp \, \frac{e^{ip(x-y)}}{m_\phi^2 + D_\varepsilon(p)} K\left(\frac{p^2}{\Lambda_0^2}\right). \tag{3.4}$$

The function $K(p^2/\Lambda_0^2)$ cuts off large values of p because, by definition of K, it vanishes whenever $|p| > 2\Lambda_0$. In the thermodynamic limit we get

$$C_{0,\Lambda_0}^{(\Gamma_\varepsilon,\infty)}(x,y) = \int_{[-\frac{\pi}{\varepsilon},\frac{\pi}{\varepsilon}]} \frac{d^d p}{(2\pi)^d} \frac{e^{ip(x-y)}}{m_\phi^2 + D_\varepsilon(p)} K\left(\frac{p^2}{\Lambda_0^2}\right) \tag{3.5}$$

which converges for all x, y, in particular for $x = y$ by the dominated convergence theorem because the integrand converges pointwise and is bounded by the function $\frac{1}{m_\phi^2}\Theta(\Lambda_0 - |p|)$ which is integrable. Thus

$$C_{0,\Lambda_0} = \lim_{\varepsilon \to 0} \lim_{L \to \infty} C_{0,\Lambda_0}^{\Gamma_\varepsilon, L} \tag{3.6}$$

exists and is a bounded function. For $x = y$,

$$C_{0,\Lambda_0}(x,x) \;=\; \frac{1}{(2\pi)^d} \int d^d p \, \frac{K(\frac{p^2}{\Lambda_0^2})}{p^2 + m_\phi^2}$$

$$= \; A_d \int_0^\infty \frac{r^{d-1}}{r^2 + m_\phi^2} K(\frac{r^2}{\Lambda_0^2}) dr \qquad (3.7)$$

with $A_d = \frac{|S^{d-1}|}{(2\pi)^d}$. Thus

$$C_{0,\Lambda_0}(x,x) \;=\; A_d \left(\int_0^1 \frac{r^{d-1}}{r^2 + m_\phi^2} dr + \int_1^\infty \frac{r^{d-1}}{r^2 + m_\phi^2} K(\frac{r^2}{\Lambda_0^2}) dr \right)$$

$$\leq \; A_d \left(\nu(m_\phi) + \int_1^{2\Lambda_0} r^{d-3} dr \right)$$

$$\leq \; A_d \nu(m_\phi) + A_d \begin{cases} \log 2\Lambda_0 & d = 2 \\ \frac{1}{d-2}(2\Lambda_0)^{d-2} & d \geq 3. \end{cases} \qquad (3.8)$$

This bound gives the correct behaviour for large Λ_0. The function $\nu(m_\phi)$ is uniformly bounded in m_ϕ for all $d \geq 3$, but not for $d = 2$ (however, $m_\phi > 0$ is fixed).

The continuum propagator $C_{0,\Lambda_0}(x,y)$ is rotation invariant because a rotation $R(x-y)$ can be undone by a change in the integration variable from p to $R^{-1}p$ since p^2 and the measure $d^d p$ are rotation invariant.

Before taking the continuum limit $\varepsilon \to 0$ of the effective action at fixed Λ_0, we derive the RG equation.

3.2 The Semigroup Structure of Renormalization

Let $0 < \Lambda_1 < \Lambda_0$ and $\Lambda \in [\Lambda_1, \Lambda_0]$. Let $C_{\Lambda,\Lambda_0}^{(\Gamma_\varepsilon, L)}(x,y)$ be the covariance

$$C_{\Lambda,\Lambda_0}^{(\Gamma)}(x,y) \;=\; C_{0,\Lambda_0}^{(\Gamma)}(x,y) - C_{0,\Lambda}^{(\Gamma)}(x,y)$$

$$= \; \int_{\Gamma^*} dp \, \frac{e^{ip(x-y)}}{m_\phi^2 + D_\varepsilon(p)} \, (K(\frac{p^2}{\Lambda_0^2}) - K(\frac{p^2}{\Lambda^2})). \qquad (3.9)$$

Its continuum limit is the inverse Fourier transform of

$$\hat{C}_{\Lambda,\Lambda_0}(p) = \frac{1}{p^2 + m_\phi^2}(K(\frac{p^2}{\Lambda_0^2}) - K(\frac{p^2}{\Lambda^2})). \qquad (3.10)$$

Note that

$$\hat{C}_{\Lambda,\Lambda_0}(p) \geq 0 \quad \forall p \in \mathbb{R}^d \qquad (3.11)$$

because $\Lambda \leq \Lambda_0$, hence $\frac{p^2}{\Lambda_0^2} \leq \frac{p^2}{\Lambda^2}$, and because K is a decreasing function.

For $\Lambda = \Lambda_0$, $\hat{C}_{\Lambda,\Lambda_0}(p) = 0$ for all p. Λ plays the role of an infrared cutoff, or, alternatively, that of a varying energy scale. We define the effective action for a given initial interaction $-\lambda V^{(\Lambda_0)}$ as

$$e^{\mathcal{G}(\phi, C_{\Lambda,\Lambda_0}^{(\Gamma)}, -\lambda V^{(\Lambda_0)})} = \int d\mu_{C_{\Lambda,\Lambda_0}^{(\Gamma)}}(\psi) e^{-\lambda V^{(\Lambda_0)}(\psi+\phi)}. \tag{3.12}$$

Here, we shall allow V to depend on Λ_0, since we have already seen in low orders that otherwise the limit $\Lambda_0 \to \infty$, which is our main interest, will not exist. V does not depend on Λ.

We now come to one of the main points of renormalization using the effective action, namely the semigroup property. It is this property that allows us to separate large and small energy scales in a conceptually clear and technically convenient way.

Let $V^{(\Lambda_0)}$ be a polynomial in the fields with C^∞ kernels (coefficients) $V_{mr}^{(\Lambda_0)}$. As shown in Chap. 2, \mathcal{G} has the formal power series expansion in λ

$$\mathcal{G}(\phi, C_{\Lambda,\Lambda_0}^{(\Gamma)}, -\lambda V^{(\Lambda_0)})$$

$$= \sum_{r=1}^{\infty} \lambda^r \sum_{m=0}^{\bar{m}(r)} \int_{(\Gamma^*)^m} \prod_{k=1}^{m} dp_k \, \delta_{\Gamma^*}(p_1 + \ldots + p_m, 0) \tag{3.13}$$

$$G_{m,r}^{\Gamma,\Lambda,\Lambda_0}(p_2, \ldots, p_m) \, \Omega_{C_0,\Lambda}(\hat{\phi}(p_1) \ldots \hat{\phi}(p_m)).$$

The cutoff $\Lambda_0 < \infty$ removes the short-distance singularity of $C_{\Lambda,\Lambda_0} = \lim_{\varepsilon \to 0} C_{\Lambda,\Lambda_0}^\Gamma$, that is, $\hat{C}_{\Lambda,\Lambda_0}$ is a C^∞ function of p, uniformly in ε (but not in Λ_0, which is kept fixed at the moment). By Theorem 2.26, $G_{m,r}^{\Gamma,\Lambda,\Lambda_0}$ converges uniformly to a C^∞ function $G_{m,r}^{\Lambda,\Lambda_0}$ as $L \to \infty$, *and as $\varepsilon \to 0$*. The coefficients $G_{m,r}^{\Lambda,\Lambda_0}$ have two complementary interpretations:

(i) $G_{m,r}^{\Lambda,\Lambda_0}$ is the order r (in λ) connected, and amputated m-point function with infrared cutoff Λ and ultraviolett cutoff $2\Lambda_0$. This follows from the graphical representation derived in Chap. 2. The ultraviolet cutoff is $2\Lambda_0$ because values $|p| > 2\Lambda_0$ are forbidden in all loop integrals. Similarly, the propagators vanish when $|p| < \Lambda$, so there is an infrared cutoff given by Λ.

(ii) The one-parameter family of \mathcal{G} obtained by varying Λ obeys the semigroup property: for any $\Lambda < \Lambda'$

$$\hat{C}_{\Lambda,\Lambda_0}(p) = \frac{1}{p^2 + m_\phi^2} \left(K(\frac{p^2}{\Lambda'^2}) - K(\frac{p^2}{\Lambda^2}) + K(\frac{p^2}{\Lambda_0^2}) - K(\frac{p^2}{\Lambda'^2}) \right)$$

$$= \hat{C}_{\Lambda,\Lambda'}(p) + \hat{C}_{\Lambda',\Lambda_0}(p) \tag{3.14}$$

and $C_{\Lambda,\Lambda'} \geq 0$, $C_{\Lambda',\Lambda_0} \geq 0$ because $\Lambda < \Lambda'$. So, by the covariance splitting formula,

$$\mathcal{G}(\phi, C_{\Lambda,\Lambda_0}, -\lambda V^{(\Lambda_0)}) = \mathcal{G}(\phi, C_{\Lambda,\Lambda'}, \mathcal{G}(\cdot, C_{\Lambda',\Lambda_0}, -\lambda V^{(\Lambda_0)})) \tag{3.15}$$

holds (for details, see (2.105)). This identity is the semigroup property. Together with the graphical representation, it implies that the coefficient functions $G_{m,r}^{\Lambda',\Lambda_0}$ also appear as effective interaction vertices for a field theory with ultraviolet cutoff Λ' (and an infrared cutoff $\Lambda < \Lambda'$). Note that the semigroup property also shows that it is natural that amputated Green functions appear as effective vertices: in the graphical representation of the $G_{m,r}^{\Lambda,\Lambda'}$ in terms of the $G_{m,r}^{\Lambda',\Lambda_0}$ and the propagator $C_{\Lambda,\Lambda'}$, pairs of legs of the vertices defined by the $G_{m,r}^{\Lambda',\Lambda_0}$ are joined to lines associated to propagators $C_{\Lambda,\Lambda'}$. Thus it is natural that no propagators C_{Λ',Λ_0} are associated to these legs of vertices. Note also that the Wick ordering with respect to $C_{\Lambda,\Lambda'}$, as implied by the representation given in Theorem 2.23, is also natural in this context: Wick ordering is always done with respect to the covariance of the fields that have not yet been integrated over.

For $\Lambda = \Lambda_0$, $C_{\Lambda_0,\Lambda_0} = 0$, and the effective action indeed consists only of the vertices originally present in V. For $\Lambda = 0$, all fields have been integrated over, thus one has a theory without infrared cutoff, only with the ultraviolet cutoff Λ_0.

This has the following significance for the problem of perturbative renormalization, which we may now formulate. We say that the theory is perturbatively renormalizable if we can adjust a finite number of parameters in $V^{(\Lambda_0)}$ such that for all m and r, the continuum Green functions $G_{m,r}^{0,\Lambda_0}$ have finite limits as $\Lambda_0 \to \infty$. This is physically sensible because the $G_{m,r}^{0,\Lambda_0}$ can be used to construct the S-matrix and hence to predict the outcome of experiments. If only a finite number of parameters has to be fixed, the theory remains predictive. Moreover, the details of the interaction at very high energies $\Lambda_0 \to \infty$ cannot be observed, so changing $V^{(\Lambda_0)}$ is allowed.

Remark 3.1. The importance of the semigroup property is that it allows us to separate low and high energy scales in a clear way: Suppose we know that $G_{m,r}^{\Lambda_1,\Lambda_0}$ has a finite limit as $\Lambda_0 \to \infty$. Then the same holds for $G_{m,r}^{0,\Lambda_0}$ because

$$\mathcal{G}(\phi, C_{0,\Lambda_0}, -\lambda V^{(\Lambda_0)}) = \mathcal{G}(\phi, C_{0,\Lambda'}, \mathcal{G}(\cdot, C_{\Lambda_1,\Lambda_0}, -\lambda V^{(\Lambda_0)})). \tag{3.16}$$

In other words, once we know that the effective vertices $\lim_{\Lambda_0 \to \infty} G_{m,r}^{\Lambda_1,\Lambda_0}$ for the fields with energy scales below Λ_1 are finite, the same holds for $G_{m,r}^{0,\Lambda_0}$ simply because they are given by finite-dimensional Feynman integrals over $|p_i| \leq 2\Lambda_1$ and with bounded integrands. In fact, once the limit $\Lambda_0 \to \infty$ has been taken in $G_{m,r}^{\Lambda_1,\Lambda_0}$, any reference to Λ_0 is removed, so the limiting effective action $G_{m,r}^{\Lambda_1,\Lambda_0} = \lim_{\Lambda_0 \to \infty} G_{m,r}^{\Lambda_1,\Lambda_0}$ contains all the information about the theory at energies below Λ_1. In more physical jargon, the effective action represents all the data that are accessible to us at low energies.

We show finiteness of $G_{m,r}^{\Lambda_1,\Lambda_0}$ for fixed $\Lambda_1 > 0$ by use of the RG differential equation.

3.3 The Renormalization Group Equation

3.3.1 The Functional Form

The RG equation is obtained by taking a derivative of $\mathcal{G}(\phi, C_{\Lambda,\Lambda_0}, -\lambda V^{(\Lambda_0)})$ with respect to Λ. To do this efficiently, we introduce sources $J = (J(x))_{x \in \Gamma}$ and write

$$
\begin{aligned}
e^{\mathcal{G}(\phi, C_{\Lambda\Lambda_0}, -\lambda V)} &= \int d\mu_{C_{\Lambda\Lambda_0}}(\chi)\, e^{-\lambda V(\chi+\phi)} \\
&= \left[e^{-\lambda V(\frac{\delta}{\delta J})} \int d\mu_{C_{\Lambda\Lambda_0}}(\chi)\, e^{+(J,\chi+\phi)} \right]_{J=0} \quad (3.17) \\
&= \left[e^{-\lambda V(\frac{\delta}{\delta J})}\, e^{(J,\phi)}\, e^{\frac{1}{2}(J, C_{\Lambda\Lambda_0} J)} \right]_{J=0}.
\end{aligned}
$$

This identity holds in a formal power series expansion in λ because V is a polynomial, so in every order in λ, the differential operator is just a polynomial in $\frac{\delta}{\delta J}$. Differentiating with respect to Λ, we get

$$
\frac{\partial}{\partial \Lambda} e^{\mathcal{G}(\phi, C_{\Lambda\Lambda_0}, -\lambda V)} = \left[e^{-\lambda V(\frac{\delta}{\delta J})}\, e^{(J,\phi)}\, e^{\frac{1}{2}(J, C_{\Lambda\Lambda_0} J)} \frac{1}{2}\left(J, \frac{\partial C_{\Lambda\Lambda_0}}{\partial \Lambda} J\right) \right]_{J=0}.
$$
(3.18)

Rewriting

$$
\frac{1}{2}\left(J, \frac{\partial C_{\Lambda\Lambda_0}}{\partial \Lambda} J\right) e^{(J,\phi)} = \frac{1}{2}\left(\frac{\delta}{\delta \phi}, \frac{\partial C_{\Lambda\Lambda_0}}{\partial \Lambda} \frac{\delta}{\delta \phi}\right) e^{(J,\phi)},
\tag{3.19}
$$

and taking out the ϕ-derivatives, we get

$$
\frac{\partial}{\partial \Lambda} e^{\mathcal{G}(\phi, C_{\Lambda\Lambda_0}, V)} = \frac{1}{2}\left(\frac{\delta}{\delta \phi}, \frac{\partial C_{\Lambda\Lambda_0}}{\partial \Lambda} \frac{\delta}{\delta \phi}\right) e^{\mathcal{G}(\phi, C_{\Lambda\Lambda_0}, V)}.
\tag{3.20}
$$

This is a heat equation for $e^{\mathcal{G}}$:

$$
\frac{\partial}{\partial \Lambda} e^{\mathcal{G}_\Lambda(\phi)} = \frac{\partial \Delta_{\Lambda,\Lambda_0}}{\partial \Lambda} e^{\mathcal{G}_\Lambda(\phi)}
\tag{3.21}
$$

with

$$
\Delta_{\Lambda,\Lambda_0} = \frac{1}{2}\left(\frac{\delta}{\delta \phi}, C_{\Lambda\Lambda_0} \frac{\delta}{\delta \phi}\right)
\tag{3.22}
$$

the Laplacian associated to C_{Λ,Λ_0} in field space. In the formal continuum limit, $\Delta_{\Lambda,\Lambda_0}$ becomes a functional Laplacian. For $\varepsilon > 0$ and $L < \infty$, the number of field variables is finite, and therefore $\Delta_{\Lambda,\Lambda_0}$ is an ordinary Laplacian in finitely many (but very many) variables $(\phi(x))_{x \in \Gamma_{\varepsilon,L}}$. Doing the derivatives, we get

$$
\frac{\partial}{\partial \Lambda} \mathcal{G} = \frac{\partial \Delta_{\Lambda,\Lambda_0}}{\partial \Lambda} \mathcal{G} + \frac{1}{2}\left(\frac{\delta \mathcal{G}}{\delta \phi}, \frac{\partial C_{\Lambda\Lambda_0}}{\partial \Lambda} \frac{\delta \mathcal{G}}{\delta \phi}\right).
\tag{3.23}
$$

Remark 3.2. It is obvious that the form of this equation does not depend on the nature of the parameter with respect to which we differentiate. That is, as long as only the covariance, but not the the initial interaction, depends on a parameter t, application of $\frac{\partial}{\partial t}$ will lead to an equation of the same form as (3.23).

Remark 3.3.

$$e^{\mathcal{G}(\phi, C_{\Lambda, \Lambda_0}, -\lambda V)} = e^{\Delta_{C_{\Lambda, \Lambda_0}}} e^{-\lambda V(\phi)}. \tag{3.24}$$

Proof. The right hand side of (3.24) also satisfies the heat equation (3.21) because $\Delta_{C_{\Lambda, \Lambda_0}}$ and its derivative with respect to Λ commute. Since $\Delta_{C_{\Lambda_0, \Lambda_0}} = 0$, both functions coincide at $\Lambda = \Lambda_0$. By the uniqueness of the solution to the heat equation, they must be equal for all Λ. ∎

We know already that the solutions to this differential equation are given by (3.13), but the differential equation itself will be very useful for proving bounds for the $G_{m,r}^{\Lambda, \Lambda_0}$.

3.3.2 The Component Form

We now derive the component form of the RG equation, that is, the corresponding equation for the $G_{m,r}^{\Gamma, \Lambda, \Lambda_0}$, simply by inserting the expansion (3.13) into (3.23). We begin by motivating how the flow equation will look like in terms of the functions $G_{m,r}^{\Gamma, \Lambda, \Lambda_0}$. If we denote the associated polynomial graphically by the vertex shown in Fig. 3.1, where every leg carries a field variable $\phi(x)$, then the two terms on the right hand side of (3.23) can be visualized as in Fig. 3.2, where the slash on the internal line indicates that it carries propagator $\partial_\Lambda C_{\Lambda, \Lambda_0}$.

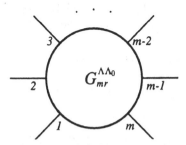

Fig. 3.1. The vertex associated to $G_{mr}^{\Lambda, \Lambda_0}$

At this point, there are two different ways to proceed: one can expand in the ordinary monomial basis $\hat{\phi}(p_1) \ldots \hat{\phi}(p_m)$, as done in Polchinski's original work [29] and in [31], or one can expand in Wick ordered monomials, as done in [30]. We follow the second route and use (3.13). This removes the linear

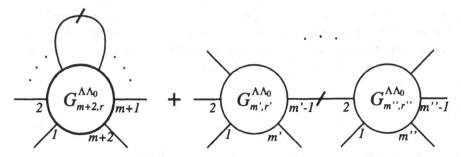

Fig. 3.2. The right hand side of the non-Wick-ordered RG equation

term, i.e. the one containing the self-contraction, and thereby simplifies the inductive proofs (as we shall see).

The details of the calculation are similar to those of the derivation of the Feynman graph expansion, only a little simpler. We show why the term in (3.23) that is linear in \mathcal{G} is removed, and defer the details of rearrangement of $\Omega_C(\prod \hat{\phi}(p_i))\Omega_C(\prod \hat{\phi}(q_i))$ occuring in the quadratic term to the appendix, where we also give the formulas in position space.

It is evident from (3.13) that a derivative $\frac{\partial}{\partial \Lambda}$ will produce two terms, one where $\frac{\partial \mathcal{G}}{\partial \Lambda}$ appears, and one where the Wick ordered monomial gets differentiated. This second term cancels the first term in (3.23): by Theorem 2.4, and because $\partial_\Lambda C_{0,\Lambda} = -\partial_\Lambda C_{\Lambda,\Lambda_0}$ implies $\partial_\Lambda \Delta_{0,\Lambda} = -\partial_\Lambda \Delta_{\Lambda,\Lambda_0}$,

$$
\frac{\partial}{\partial \Lambda} \, \Omega_{C_{0,\Lambda}}(\hat{\phi}(p_1) \dots \hat{\phi}(p_m))
$$
$$
= -\partial_\Lambda \Delta_{0,\Lambda} \Omega_{C_{0,\Lambda}}(\hat{\phi}(p_1) \dots \hat{\phi}(p_m)) \tag{3.25}
$$
$$
= \partial_\Lambda \Delta_{\Lambda,\Lambda_0} \Omega_{C_{0,\Lambda}}(\hat{\phi}(p_1) \dots \hat{\phi}(p_m)).
$$

When multiplied with $\delta_{\Gamma^\bullet}(p_1 + \dots + p_m, 0) G^{\Lambda,\Lambda_0}_{m,r}(p_2, \dots, p_m)$ and integrated over p_1, \dots, p_m, this gives exactly the same as the first term on the right hand side of (3.23).

Thus, the flow equation (3.23) now reads

$$
\sum_{r=1}^{\infty} \lambda^r \sum_{m=0}^{\bar{m}(r)} \int \prod_{j=1}^{m} dk_j \, \delta_{\Gamma^\bullet}(k_1 + \dots + k_m, 0) \, \Omega_{C_{0,\Lambda}}(\hat{\phi}(k_1), \dots, \hat{\phi}(k_m))
$$
$$
\cdot \, \partial_\Lambda G^{\Gamma\Lambda\Lambda_0}_{mr}(k_2, \dots, k_m) = \frac{1}{2}(\frac{\delta \mathcal{G}}{\delta \hat{\phi}}, \partial_\Lambda C^{(\Gamma)}_{\Lambda,\Lambda_0} \frac{\delta \mathcal{G}}{\delta \hat{\phi}}). \tag{3.26}
$$

If we rewrite the right hand side in the same form as the left hand side, we can compare coefficients of $\Omega_{C_{0,\Lambda}}(\hat{\phi}(k_1), \dots, \hat{\phi}(k_m))$ to get the component form of the RG equation. When comparing coefficients, we have to remember that $\Omega_{C_{0,\Lambda}}(\hat{\phi}(k_1), \dots, \hat{\phi}(k_m))$ is invariant under permutations of $\{1, \dots, m\}$, so only the symmetric part of the coefficients is nonzero. This means that we have to symmetrize the coefficients. For a function F of (k_2, \dots, k_m), let

$$(\mathbb{S}_m F)(k_2, \ldots, k_m) = \frac{1}{m!} \sum_{\pi \in S_m} F(k_{\pi(2)}, \ldots, k_{\pi(m)}) \qquad (3.27)$$

where $k_1 = -\sum_{j=2}^{m} k_j$ wherever it appears as $k_{\pi(l)}$ for some $l \in \{2, \ldots, m\}$.

In Appendix A.4, it is shown that, as a formal power series in λ,

$$\left(\frac{\delta \mathcal{G}}{\delta \hat{\phi}}, \partial_\Lambda \hat{C}_{\Lambda, \Lambda_0}^{(\Gamma)} \frac{\delta \mathcal{G}}{\delta \hat{\phi}}\right) \Gamma_* = \sum_{r=2}^{\infty} \lambda^r \mathcal{Q}_r^{(\Gamma, \Lambda, \Lambda_0)}(\phi) \qquad (3.28)$$

where

$$\mathcal{Q}_r^{(\Gamma, \Lambda, \Lambda_0)}(\phi) = \sum_{m=0}^{\tilde{m}(r)} \int \prod_{i=1}^{m} dk_i \, \delta_{\Gamma^*} \left(\sum_{i=1}^{m} k_i, 0\right) Q_{mr}^{\Gamma \Lambda \Lambda_0}(\underline{k}) \Omega_{C_0, \Lambda} \left(\prod_{i=1}^{m} \hat{\phi}(k_i)\right) \qquad (3.29)$$

with

$$\underline{k} = (k_2, \ldots, k_m) \in \left(\mathbb{R}^d\right)^{m-1} \qquad (3.30)$$

and

$$Q_{mr}^{\Gamma \Lambda \Lambda_0}(\underline{k}) = \int d\kappa_{mr} \int d\omega_\Lambda(\underline{p}) \, \partial_\Lambda C_{\Lambda \Lambda_0}(p_0) \, G_{m'r'}^{\Lambda \Lambda_0}(\underline{p}, \underline{k}') \, G_{m''r''}^{\Lambda \Lambda_0}(-\underline{p}, \underline{k}''). \qquad (3.31)$$

In this expression, the integral $\int d\kappa_{mr}$ is written for a weighted sum

$$\int d\kappa_{mr}(m', m'', r', r'', l) \, F(m', m'', r', r'', l) \qquad (3.32)$$

$$= \sum_{\substack{r'+r''=r \\ r' \geq 1, r'' \geq 1}} \sum_{(m', m'', l) \in \mathcal{M}_{r'r''m}} \kappa_{m'm''l} \, F(m', m'', r', r'', l)$$

where

$$\mathcal{M}_{r'r''m} = \{(m', m'', l) : l \geq 0, m' \in \{1, \ldots, \tilde{m}(r')\},$$
$$m'' \in \{1, \ldots, \tilde{m}(r'')\}, m' + m'' = m + 2l + 2\} \qquad (3.33)$$

and where the weight $\kappa_{m'm''l}$ is the combinatorial factor

$$\kappa_{m'm''l} = m' \, m'' \, l! \binom{m'-1}{l} \binom{m''-1}{l}. \qquad (3.34)$$

Moreover, we use the notations

$$\underline{p} = (p_1, \ldots, p_l) \in \left(\mathbb{R}^d\right)^l$$
$$\underline{k}' = (-\sum_{s=2}^{m} k_s, k_2, \ldots, k_{m'-1-l}) \qquad (3.35)$$
$$\underline{k}'' = (k_{m'-l}, \ldots, k_m)$$

and p_0 is given by

$$p_0 = -\sum_{j=1}^{l} p_j - \sum_{j=1}^{m'-l-1} k_j = -\sum_{j=1}^{l} p_j + \sum_{j=2}^{m} k_j - \sum_{j=1}^{m'-l-1} k_j, . \tag{3.36}$$

Finally,

$$\int d\omega_\Lambda(\underline{p}) = \int \prod_{i=1}^{l} \frac{d^d p_i}{(2\pi)^d} C_{0\Lambda}(p_i). \tag{3.37}$$

$\underline{p}, \underline{k}', \underline{k}''$, and p_0 depend, of course, on $(m, r, m', m'', r', r'', l)$. We do not denote this dependence explicitly because we are soon going to take sup norms where the dependence of the G_{mr} on $\underline{p}, \underline{k}', \underline{k}''$, and p_0 drops out.

We have omitted the caret and the superscript Γ on C since we shall only use momentum space in the following, and because we shall take the continuum limit soon.

With all this, the component form of the RG equation is

$$\partial_\Lambda G_{mr}^{\Gamma,\Lambda,\Lambda_0}(k_2, \ldots, k_m) = \frac{1}{2} \mathbb{S}_m \, Q_{mr}^{\Gamma,\Lambda,\Lambda_0}(k_2, \ldots, k_m), \tag{3.38}$$

with $Q_{mr}^{\Gamma,\Lambda,\Lambda_0}$ given by (3.31) and \mathbb{S} given by (3.27).

Contrary to what one might believe at a first glance, the details of this equation and of its derivation are quite simple to understand, and it has a simple and convenient structure, which we explain in some detail in the next section. Also, we relate it to the Feynman graph representations derived in Chap. 2, which, as mentioned, provide explicit solutions to this equation.

3.4 The Structure of the RG equation

3.4.1 The Graphical Representation

Visualizing $G_{mr}^{\Gamma\Lambda\Lambda_0}$ as in Fig. 3.1, every term in the sum over r', r'', m', m'', and l has the graphical representation shown in Fig. 3.3.

In other words, it is the value of a graph with only two vertices, with m' and m'' legs, such that the number of external legs is m. The graph has $l \geq 0$ loops. The sum over m' and m'' starts at 1 because otherwise $\delta/\delta\hat{\phi}(p)$ in (3.23) would have had nothing to act on. Thus $G_{0,r}$ does not appear on the right hand side of the flow equation. The graph is connected for all $l \geq 0$ because there is always the line carrying $\partial_\Lambda C_{\Lambda,\Lambda_0}(p_0)$ connecting the two vertices. If $l = 0$ it is a tree graph. If $l = 1$ it is a one-loop graph, usually called a bubble graph. Retaining only this term in the sum over l and integrating the RG equation gives the so-called ladder approximation, done in Sect. 3.9.

A brief description of the procedure used in Appendix A.4 to derive the right hand side of the RG equation is as follows. The functional Laplacian generates the tree graph shown in Fig. 3.2, where both vertices correspond to

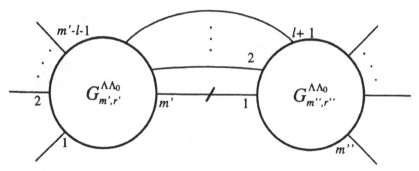

Fig. 3.3. The right hand side of the Wick ordered RG equation

coefficients of Wick-ordered monomials. There are m' legs of the first vertex on which the first ϕ-derivative can act, and m'' legs of the second vertex on which the second derivative can act. By the permutational symmetry of the G_{mr}, we may permute and thus get a factor $m'm''$ (this argument involving the permutational symmetry will be applied also in the following to derive the combinatorial factor by counting the number of graphs). We now have to reorder the product of two Wick ordered monomials; this gives a sum over Wick monomials, indexed by l. In every term, all possible self-contractions of the new vertex associated to the entire graph are subtracted, hence appear as soft lines. Since both vertices already correspond to Wick-ordered monomials, all soft lines must join vertex 1 to vertex 2. If $l \geq 0$ is the number of Wick lines, there are $\binom{m'-1}{l}$ possibilities to choose a subset of l legs of vertex 1, and $\binom{m''-1}{l}$ possibilities to choose a subset of l legs of vertex 2, to form pairs. Given these legs, there are $l!$ ways of pairing them together. This explains the combinatorial factor. The remaining m legs are the external legs of the graph. Counting legs and internal lines gives the constraint $m' + m'' = m + 2l + 2$.

In Fig. 3.4, we show how the flow equation looks for small values of m and even interactions.

$$\frac{d}{d\Lambda}\,\bigcirc = -\bigcirc\!\!\!-\!\!\!\bigcirc + \overset{\bigcirc}{\underset{}{\bigcirc}} + \bigcirc\!\!\!-\!\!\!\bigcirc + \ldots$$

$$\frac{d}{d\Lambda}\,\diagup\!\!\!\bigcirc\!\!\!\diagdown = \diagup\!\!\!\bigcirc\!\!\!\diagdown^{\!\!\circ} + \diagup\!\!\!\bigcirc\!\!\!-\!\!\!\bigcirc\!\!\!\diagdown + \overset{\bigcirc}{\diagup\!\!\!\bigcirc\!\!\!\diagdown} + \ldots$$

Fig. 3.4. The RG equation for $m = 2$ and $m = 4$

3.4.2 The Relation to the Feynman Graph Expansion

The RG equation is a differential equation for the kernels of the effective action, for which we derived a representation in terms of values of connected Feynman graphs in Chap. 2. Thus the sum over values of connected Feynman graphs is a solution to the RG equation. It is a consequence of the Feynman graph representation that

$$\tilde{m}(r) \leq \max_{r'+r''=r} (\bar{m}(r') + \bar{m}(r'') - 2). \tag{3.39}$$

Conversely, using the RG equation, one can derive both (3.39) and give an inductive proof in r that the Feynman graph representation holds by transforming the RG equation to its equivalent integral equation by an integration over Λ,

$$G_{mr}^{\Lambda,\Lambda_0} = G_{mr}^{\Lambda_0,\Lambda_0} + \int_{\Lambda}^{\Lambda_0} ds\ Q_{mr}^{s\Lambda_0}, \tag{3.40}$$

inserting the definition of Q, and then iterating this integral equation in the way familiar from generating series solutions to integral equations. We shall not go into the details of this here, but only note that when regarded in that way, it is almost obvious that only connected Feynman graphs contribute to the sum because the graph in Fig. 3.3 is connected, and the iteration thus produces only connected graphs.

When renormalizing field theory using Feynman graphs, one often starts by proving that graphs without two–legged subgraphs, or for $d = 4$, graphs without two– and four–legged subgraphs are finite without renormalization. We shall do the same using the RG equation because there is a truncation of the RG equation which produces these partial sums. To get the sum over all graphs without two–legged subgraphs, one omits all terms with $m' = 2$ or $m'' = 2$ in the sum $\int d\kappa_{mr}$ in (3.31). Obviously, then, no two–legged subgraphs can appear, since the only way they could have been produced in the iteration has been removed. To get the sum over all graphs without nontrivial four–legged subgraphs (i.e. fourlegged subgraphs with more than one vertex), one omits all terms with $m' = 4$ and $r' > 1$, and all those with $m'' = 4$ and $r'' > 1$.

3.4.3 The Continuum Limit at Fixed Λ_0

Thus, the functional form gets translated into a system of differential equations, with the following structure. The system is labelled by (m, r) where, as shown in the representation by Feynman graphs, $m \leq \bar{m}(r)$ for fixed r. On the right side $m' \geq 1$ and $m'' \geq 1$, and $r' < r$ and $r'' < r$, since $r' + r'' = r$ and $r' \geq 1, r'' \geq 1$. Thus the $m = 0$ term never appears on the RHS. Since we are interested in the $m \geq 1$ terms to get the correlation functions, we may restrict to $m \geq 1$ in the equation. As proven in Theorem 2.26, $G_{mr}^{\Gamma,\Lambda,\Lambda_0}$

converges as $L \to \infty$, so Q_{mr} converges as well, and both are C^∞ functions of Λ and Λ_0. Moreover, the limit $\varepsilon \to 0$ exists because $C_{\Lambda,\Lambda_0}^{(\Gamma)}(p)$ converges uniformly to the bounded function

$$C_{\Lambda,\Lambda_0}(p) = (K(\frac{p^2}{\Lambda_0^2}) - K(\frac{p^2}{\Lambda^2}))\frac{1}{p^2 + m_\phi{}^2} \tag{3.41}$$

as $\varepsilon \to 0$, and for fixed Λ_0, all integrals are over a finite region only. Thus we have

Theorem 3.4. *The limits $L \to \infty$ and $\varepsilon \to 0$ of $G_{mr}^{\Gamma,\Lambda,\Lambda_0}$ and $Q_{mr}^{\Gamma,\Lambda,\Lambda_0}$ exist, and the limiting functions satisfy*

$$\partial_\Lambda G_{mr}^{\Lambda,\Lambda_0}(k_2,\ldots,k_m) = \frac{1}{2}\mathbb{S}_m\, Q_{mr}^{\Lambda,\Lambda_0}(k_2,\ldots,k_m). \tag{3.42}$$

An explicit solution to this system of equations is given by the expansion in Feynman graphs. If the initial interaction V^{Λ_0} has rotation invariant kernels, the $G_{mr}^{\Lambda,\Lambda_0}$ are rotation invariant, i.e.

$$\forall R \in SO(d): \quad G_{mr}^{\Lambda,\Lambda_0}(k_2,\ldots,k_m) = G_{mr}^{\Lambda,\Lambda_0}(Rk_2,\ldots,Rk_m). \tag{3.43}$$

Remark 3.5. If V^{Λ_0} is even in ϕ, i.e.

$$V^{\Lambda_0}(\phi) = \sum_{m=2}^{\bar m} \int \prod_{i=1}^{m} \mathrm{d}p_i\, \delta_{\Gamma^*}(\sum_{i=1}^{m} p_i, 0)\, \hat v_m^{(\Lambda_0)}(p_2,\ldots,p_m)\Omega_{C_{0,\Lambda_0}}(\prod_{i=1}^{m}\hat\phi(p_i)) \tag{3.44}$$

with $\hat v_m = 0$ if m is odd, then for all Λ,and all $r \geq 1$, $G_{mr}^{\Lambda,\Lambda_0}$ is even in $\hat\phi$, i.e. $G_{mr}^{\Lambda,\Lambda_0} = 0$ if m is odd. If $\hat v_m^{(\Lambda_0)}$ is C^k in all its arguments, the same holds for G_{mr}. IF $\hat v_m$ is even in p, that is, if $\hat v_m^{(\Lambda_0)}$ is invariant under $p_j \to -p_j$, the same holds for G_{mr}.

Remark 3.6. The properties stated in Theorem 3.4 could also have been proven by induction using the flow equation only, without considering all details of the Feynman graph expansion.

3.5 Differential Inequalities

We know now that for a fixed UV cutoff Λ_0, the RG equation

$$\partial_\Lambda G_{mr}^{\Lambda,\Lambda_0}(k_2,\ldots,k_m) = \frac{1}{2}\mathbb{S}_m Q_{mr}^{\Lambda,\Lambda_0}(k_2,\ldots,k_m) \tag{3.45}$$

has a smooth and Euclidean invariant solution if the starting interaction is smooth and Euclidean invariant (since we have taken the limits $\varepsilon \to 0$ and $L \to \infty$, we now drop the superscript Γ). This is the case for $V^{\Lambda_0} = \Omega_{C_{0,\Lambda_0}}(V_0)$ with

$$V_0 = \int \phi(x)^4 \, dx \qquad (3.46)$$

because the kernel in momentum space is simply

$$\hat{v}_4(p_2, p_3, p_4) = 1. \qquad (3.47)$$

We also know that we only have to show that the functions $G_{mr}^{\Lambda_1, \Lambda_0}$ are bounded in compact regions because the Feynman graph expansion then gives us $G_{m,r}^{0,\Lambda_0}$ in terms of C_{0,Λ_1} and the $G_{mr}^{\Lambda_1,\Lambda_0}$.

Definition 3.7. *For a bounded function $F_m : \mathbb{R}^{(m-1)d} \to \mathbb{C}$, let*

$$A_{\Lambda,\eta}(F_m) = \sup\{|F_m(p_2, \ldots, p_m)| : |p_k| \leq \max\{\Lambda, \eta\} \quad \forall k \in \{2, \ldots, m\}\}. \qquad (3.48)$$

$A_{\Lambda,\eta}(\cdot)$ is the sup norm of F_m on a ball whose radius depends on Λ and η (so it is not a norm for functions on \mathbb{R}^d). Obviously, $A_{\Lambda,\eta}(F_m) \leq A_{\Lambda',\eta'}(F_m)$ whenever $\Lambda \leq \Lambda'$ and $\eta \leq \eta'$. To get a differential inequality, we calculate $A_{2\Lambda,\eta}(Q_{mr}^{\Lambda\Lambda_0})$. Thus, in (3.31), we take a supremum over all external momenta k_i over $0 \leq |k_i| \leq \max\{\Lambda, \eta\}$. The integration variables p_1, \ldots, p_l must be in the support of $C_{0,\Lambda}$, so $|p_i| \leq 2\Lambda$ must hold (this is the reason why we took $A_{2\Lambda,\eta}$ and not $A_{\Lambda,\eta}$), and we can thus estimate

$$\sup_{0 \leq |k_i| \leq \max\{2\Lambda,\eta\}} |G_{mr}^{\Lambda\Lambda_0}(p_1, \ldots, p_l, k_1, \ldots, k_{m'-1-l})| \leq A_{2\Lambda,\eta}(G_{mr}^{\Lambda\Lambda_0}). \qquad (3.49)$$

We also bound

$$|\partial_\Lambda C_{\Lambda,\Lambda_0}(p_0)| \leq \|\partial_\Lambda C_{\Lambda,\Lambda_0}\|_\infty \qquad (3.50)$$

then the integral factors into l independent integrals over p_1, \ldots, p_l, each giving

$$\int dp \, |C_{0,\Lambda}(p)| = \|C_{0,\Lambda}\|_1. \qquad (3.51)$$

Thus

$$\begin{aligned}
A_{2\Lambda,\eta}(\partial_\Lambda G_{mr}^{\Lambda\Lambda_0}) &= \frac{1}{2} A_{2\Lambda,\eta}(\mathbb{S}_m Q_{mr}^{\Lambda\Lambda_0}) \leq \frac{1}{2} A_{2\Lambda,\eta}(Q_{mr}^{\Lambda\Lambda_0}) \\
&\leq \frac{1}{2} \int d\kappa_{mr}(m', m'', r', r'', l) \, \|\partial_\Lambda C_{\Lambda,\Lambda_0}\|_\infty \quad (3.52) \\
&\quad (\|C_{0,\Lambda}\|_1)^l \, A_{2\Lambda,\eta}(G_{m'r'}^{\Lambda\Lambda_0}) \, A_{2\Lambda,\eta}(G_{m''r''}^{\Lambda\Lambda_0}).
\end{aligned}$$

Lemma 3.8. *There are constants K_1 and K_2 such that for all $3 \leq \Lambda \leq \Lambda_0$*

$$\|\partial_\Lambda C_{\Lambda,\Lambda_0}\|_\infty \leq \frac{K_1}{\Lambda^3} \qquad (3.53)$$

and

$$\|C_{0,\Lambda}\|_1 \leq K_2 \begin{cases} \Lambda^{d-2} & \text{for } d \geq 3 \\ \log \Lambda & \text{for } d = 2. \end{cases} \qquad (3.54)$$

For $d \geq 3$, K_2 depends only on the dimension d. For $d = 2$, K_2 depends also on m_ϕ, and $K_2 \to \infty$ if $m_\phi \to 0$.

Proof. By (3.10)

$$\partial_\Lambda C_{\Lambda,\Lambda_0}(p) = \frac{2p^2}{\Lambda^3} K'(\frac{p^2}{\Lambda^2}) \frac{1}{p^2 + m_\phi^2} , \tag{3.55}$$

so

$$\|\partial_\Lambda C_{\Lambda,\Lambda_0}\|_\infty \leq \frac{2}{\Lambda^3} \|K'\|_\infty \leq \frac{K_1}{\Lambda^3} \tag{3.56}$$

with $K_1 = 2\|K'\|_\infty$. By (3.3), $K_1 = 2$.

$$\|C_{0,\Lambda}\|_1 = \int d^d p K(\frac{p^2}{\Lambda^2}) \frac{1}{p^2 + m_\phi^2} \leq \int\limits_{|p| \leq 2\Lambda} d^d p \frac{1}{p^2 + m_\phi^2}. \tag{3.57}$$

For $d \geq 3$, $|p|^{d-3}$ is integrable, so we may estimate

$$\frac{|p|^{d-1}}{p^2 + m_\phi^2} \leq |p|^{d-3} \tag{3.58}$$

in the integrand, to get (3.57), with a constant that depends only on the dimension. For $d = 2$,

$$\begin{aligned}
\|C_{0,\Lambda}\|_1 &\leq 2\pi \int\limits_0^{2\Lambda} \frac{p \, dp}{p^2 + m_\phi^2} = \pi \log \sqrt{p^2 + m_\phi^2} \Big|_0^{2\Lambda} \\
&\leq \pi \log(1 + \frac{4\Lambda^2}{m_\phi^2})^{\frac{1}{2}} \tag{3.59} \\
&\leq K_2 |\log \Lambda|
\end{aligned}$$

where K_2 now depends on m_ϕ. As $m_\phi \to 0$, $K_2 \sim |\log m_\phi|$. ∎

By (3.52), we thus have

Lemma 3.9. *For all $d \geq 2$,*

$$A_{2\Lambda,\eta}(\partial_\Lambda G_{mr}^{\Lambda\Lambda_0}) \tag{3.60}$$
$$\leq \frac{1}{2} \frac{K_1}{\Lambda^3} \int d\kappa_{mr} \, (\|C_{0,\Lambda}\|_1)^l \, A_{2\Lambda,\eta}(G_{m'r'}^{\Lambda\Lambda_0}) \, A_{2\Lambda,\eta}(G_{m''r''}^{\Lambda\Lambda_0}).$$

We are going to use these bounds in a Taylor expansion:

$$\begin{aligned}
G_{mr}^{\Lambda\Lambda_0} &= G_{mr}^{\Lambda_0\Lambda_0} - \int\limits_\Lambda^{\Lambda_0} ds \frac{\partial}{\partial s} G_{mr}^{s\Lambda_0} \\
&= G_{mr}^{\Lambda_0\Lambda_0} - \int\limits_\Lambda^{\Lambda_0} ds \frac{1}{2} \mathbb{S}_m Q_{mr}^{s\Lambda_0}. \tag{3.61}
\end{aligned}$$

Inserting (3.60), we get a recursive inequality for the $A_{2\Lambda,\eta}(\partial_\Lambda G_{mr}^{\Lambda\Lambda_0})$. The effective action at scale Λ_0 is simply the initial interaction:

$$G_{mr}^{\Lambda_0\Lambda_0} = V_{mr}^{(\Lambda_0)}, \qquad (3.62)$$

where $V_{mr}^{(\Lambda_0)}$ appears in the polynomial

$$
V^{(\Lambda_0)}(\phi) = \sum_{r=1}^{\infty} \lambda^r \sum_{m=1}^{M(r)} \int dp_1 \ldots dp_m (2\pi)^d \delta(p_1 + \ldots + p_m)
$$
$$
V_{mr}^{(\Lambda_0)}(p_2, \ldots, p_m) \, \Omega_{C_{0,\Lambda_0}}(\hat{\phi}(p_1) \ldots \hat{\phi}(p_m)). \qquad (3.63)
$$

For the unrenormalized ϕ^4-theory, $V^{(\Lambda_0)}(\phi) = -\lambda \int \phi^4 d\underline{x}$, so

$$V_{mr}^{(\Lambda_0)}(p_2, \ldots, p_m) = -\delta_{r1}\delta_{m4}. \qquad (3.64)$$

We thus see that the initial interaction provides the boundary condition for integrating the RG equation, which, in terms of the $G_{mr}^{\Lambda\Lambda_0}$, is a system of first order differential equations in Λ. In every fixed order r in perturbation theory, the system consists of finitely many equations because $G_{mr} = 0$ if $m > \bar{m}(r)$.

3.6 Two Dimensions

We start the investigation of the renormalizability with the two-dimensional case. We shall first show that the $G_{mr}^{\Lambda,\Lambda_0}$ remain bounded in the limit $\Lambda_0 \to \infty$ (boundedness), and then that they converge.

3.6.1 Boundedness

Applying $A_{2\Lambda,\eta}$ to (3.61), and using (3.5), (3.52), and the specific behaviour of $\|C_{0,\Lambda}\|_1$ in two dimensions, we get

$$
A_{2\Lambda,\eta}(G_{mr}^{\Lambda\Lambda_0}) \leq A_{2\Lambda,\eta}(V_{mr}^{(\Lambda_0)}) + \frac{1}{2} \int d\kappa_{mr}(m', m'', r', r'', l) \, K_1 K_2^l
$$
$$
\cdot \int_{\Lambda}^{\Lambda_0} ds \, s^{-3} (\log s)^l \, A_{2s,\eta}(G_{m'r'}^{s\Lambda_0}) A_{2s,\eta}(G_{m''r''}^{s\Lambda_0}) \qquad (3.65)
$$

for $r \geq 2$ and

$$G_{m1}^{\Lambda\Lambda_0} = V_{m1}^{(\Lambda_0)} \quad \text{for all} \quad \Lambda \qquad (3.66)$$

for $r = 1$. If the kernels $V_{mr}^{(\Lambda_0)}$ are all bounded, the simplest ansatz one can make is

$$A_{2\Lambda,\eta}(G_{mr}^{\Lambda\Lambda_0}) \leq \text{const} \qquad (3.67)$$

for all (m,r) and all $\Lambda \in [\Lambda_1, \Lambda_0]$. This ansatz works in $d = 2$ because

$$\int_\Lambda^{\Lambda_0} ds\, s^{-3}(\log s)^l \leq \text{ const } < \infty \tag{3.68}$$

(recall that we assumed $\Lambda_1 \geq 3$). The interaction does not even have to be of ϕ^4 type. Any semibounded polynomial in the fields will do.

Theorem 3.10. *Assume that $V_{mr}^{(\Lambda_0)}$ satisfies $V_{mr}^{(\Lambda_0)} = 0$ for odd m and that $A_{\Lambda_0,\eta}(V_{m,r}^{(\Lambda_0)}) = v_{mr}$ for all m and r and all Λ_0. Then for all $r \geq 1$ and all $m \geq 1$, all $\Lambda_0 \geq \Lambda_1$ and all $\Lambda \in [\Lambda_1, \Lambda_0]$,*

$$A_{2\Lambda,\eta}(G_{mr}^{\Lambda\Lambda_0}) \leq g_{mr} \tag{3.69}$$

(and $G_{mr}^{\Lambda\Lambda_0} = 0$ if m is odd). In other words, all Green functions are uniformly bounded in momentum space uniformly in Λ_0. There can be no infinities as $\Lambda_0 \to \infty$.

Proof. Induction in r. For $r = 1$, the statement holds by (3.66). So assume $r \geq 2$, and (3.69) to hold for all $r' < r$ and all m. Only terms with $r' < r$ and $r'' < r$ appear on the right hand side of (3.65). Therefore, we have

$$A_{2\Lambda,\eta}(G_{mr}^{\Lambda\Lambda_0}) \leq v_{mr} + \frac{1}{2}\int d\kappa_{mr}(m',m'',r',r'',l)\, K_1 K_2^l$$

$$\cdot\; g_{m'r'}\, g_{m''r''} \int_\Lambda^{\Lambda_0} ds\, \frac{(\log s)^l}{s^3}. \tag{3.70}$$

Since $l^l e^{-l} \leq l!$,

$$\int_\Lambda^{\Lambda_0} ds\, \frac{(\log s)^l}{s^3} \leq \sup_{s \geq 1}\frac{(\log s)^l}{s}\int_\Lambda^{\Lambda_0} ds\, \frac{1}{s^2} \leq l! \tag{3.71}$$

and the theorem follows if we put

$$g_{mr} = v_{mr} + \frac{1}{2}\int d\kappa_{mr}(m',m'',r',r'',l)\, K_1 K_2^l\, g_{m'r'} g_{m''r''}\, l!\,. \tag{3.72}$$

This is a recursive definition of g_{mr}. ∎

Already in this simple example, we see some typical features – logarithms may produce factorial growth, and they, as well as $\kappa_{m'm''r}$, determine the precise growth behaviour of g_{mr} in m and r. The dependence of g_{mr} on r is important for the question of convergence of the series in λ (or its resummations).

For $d = 2$, and for ϕ^4-theory, the $l!$ from the logarithm is, however, not really there, and it is instructive for an understanding of the boundary conditions to discuss this point further.

3.6.2 ϕ_2^4

Let us consider ϕ^4 theory, where

$$V_{mr}^{(\Lambda_0)} = -\delta_{m4}\delta_{r1} \qquad (3.73)$$

and look at the $r = 2$ flow. To this order, we have for $m = 6$

$$\partial_\Lambda G_{62}^{\Lambda\Lambda_0}(p_2,\ldots,p_6) = -\partial_\Lambda C_{\Lambda\Lambda_0}(P) \qquad (3.74)$$

with $P = p_1 + p_2 + p_3 - p_4 - p_5$. Integrating from Λ to Λ_0 gives

$$G_{62}^{\Lambda\Lambda_0}(p_2,\ldots,p_6) = C_{\Lambda\Lambda_0}(P) + V_{62}^{(\Lambda_0)}. \qquad (3.75)$$

By (3.73), $V_{62}^{(\Lambda_0)} = 0$, so

$$\|G_{62}^{\Lambda\Lambda_0}\|_\infty \leq \frac{\text{const}}{\Lambda^2} \qquad (3.76)$$

because the infrared cutoff Λ restricts $|P|$ to $|P| \geq \Lambda$ so that $\frac{1}{p^2+m_\phi^2} \leq \frac{1}{\Lambda^2}$. Note that (3.73) was crucial for this argument. The contribution to the $m = 4$ term is

$$\int C_{0\Lambda}(p)\partial_\Lambda C_{\Lambda\Lambda_0}(p+q)dp. \qquad (3.77)$$

At $q = 0$, this is, by supp $K' \subset [1,4]$,

$$
\begin{aligned}
\left|\int C_{0\Lambda}(p)\partial_\Lambda C_{\Lambda\Lambda_0}(p)dp\right| &\leq \int \frac{1}{p^2 + m_\phi^2}\frac{2p^2}{\Lambda^3}|K'(\tfrac{p^2}{\Lambda^2})|\frac{1}{p^2 + m_\phi^2}dp \\
&\leq \frac{2}{\Lambda^3}\|K'\|_\infty \int_\Lambda^{2\Lambda} \rho\,d\rho\frac{1}{(\rho^2 + m_\phi^2)^2}\rho^2 \qquad (3.78) \\
&\leq \frac{2}{\Lambda^3}\|K'\|_\infty \int_\Lambda^{2\Lambda} \frac{d\rho}{\rho} = \frac{2}{\Lambda^3}\|K'\|_\infty \log 2.
\end{aligned}
$$

There is no $\log \Lambda$ term because the lower limit of the ρ-integral is Λ. In any case, a single logarithm does not create factorials.

Similarly, as for $m = 6$, the $r = 3, m = 8$ tree graph requires two propagators, $C_{\Lambda\Lambda_0}^2$ and is therefore $O(\Lambda^{-4})$. This leads us to the ansatz $\Lambda^{-(m-4)}$ for the Green functions.

Theorem 3.11. *For* $V_{mr}^{(\Lambda_0)} = -\delta_{m4}\delta_{r1}$ *(ϕ^4-theory) in two dimensions, there are* $\gamma_{mr} \geq 0$ *such that, for all* $\Lambda \geq 3$ *and for all* η

$$A_{2\Lambda,\eta}(G_{mr}^{\Lambda\Lambda_0}) \leq \begin{cases} \gamma_{mr}\Lambda^{-(m-4)} & m \geq 4 \\ \gamma_{mr} & m \leq 2. \end{cases} \qquad (3.79)$$

Proof. We have already verified the cases $r = 1$ and $r = 2$. To do the inductive step, we may insert (3.79) on the right side of the flow equation, since $r' < r$ and $r'' < r$. We have to bound

$$X = \int_\Lambda^{\Lambda_0} ds \frac{(\log s)^l}{s^3} A_{2s,\eta}(G_{m'r'}^{s\Lambda_0}) A_{2s,\eta}(G_{m''r''}^{s\Lambda_0}). \tag{3.80}$$

For $m' \geq 4$ and $m'' \geq 4$,

$$X \leq \int_\Lambda^{\Lambda_0} ds (\log s)^l \gamma_{m'r'} \gamma_{m''r''} s^{-3-(m'-4)-(m''-4)}. \tag{3.81}$$

The triple (m', m'', l) contributes to $\partial_\Lambda G_{mr}^{\Lambda\Lambda_0}$ only if $m + 2l + 2 = m' + m''$. Inserting this, we have

$$\begin{aligned} X &\leq \gamma_{m'r'} \gamma_{m''r''} \int_\Lambda^{\Lambda_0} ds\, s^{5-(m+2l+2)} (\log s)^l \\ &= \gamma_{m'r'} \gamma_{m''r''} \int_\Lambda^{\Lambda_0} ds\, s^{3-m} (\frac{\log s}{s^2})^l. \end{aligned} \tag{3.82}$$

Since $\Lambda \geq 3$ was assumed, we have for $m \geq 6$

$$X \leq \gamma_{m'r'} \gamma_{m''r''} \int_\Lambda^{\Lambda_0} ds\, s^{3-m} \leq \frac{1}{m-4} \Lambda^{4-m} \gamma_{m'r'} \gamma_{m''r''}. \tag{3.83}$$

For $m = 4$ and $m' \geq 4$, $m'' \geq 4$, $2l = m' + m'' - 6 \geq 2$, so $l \geq 1$, and

$$\begin{aligned} X &\leq \gamma_{m'r'} \gamma_{m''r''} \int_\Lambda^{\Lambda_0} ds\, s^{-1} (\frac{\log s}{s^2})^l \leq \gamma_{m'r'} \gamma_{m''r''} \int_\Lambda^{\Lambda_0} ds\, s^{-2} (\frac{\log s}{s})^l \\ &\leq \gamma_{m'r'} \gamma_{m''r''} \frac{1}{\Lambda}. \end{aligned} \tag{3.84}$$

The restriction $l \geq 1$ if $2 \notin \{m', m''\}$ means graphically that if a four-legged graph is 1-reducible, i.e. if it gets disconnected when one line is removed, this line must join a two-legged piece to one with four legs (if $m = 4$ and $l = 0$, $m' = 2$ and $m'' = 4$ or vice versa).

There remains the case $m' = 2$ or $m'' = 2$. Without loss of generality, let $m' = 2$. Then $m'' = m + 2l$. If $m \geq 4$,

$$X \leq \int_\Lambda^{\Lambda_0} ds \frac{(\log s)^l}{s^3} \gamma_{m'r'} \gamma_{m''r''} s^{-(m''-4)} \tag{3.85}$$

so the power of s that appears is $-3 + 4 - (m + 2l) = 1 - m - 2l$, and

$$X \leq \gamma_{m'r'}\gamma_{m''r''} \int_{\Lambda}^{\Lambda_0} ds\, s^{1-m}\left(\frac{\log s}{s^2}\right)^l \leq \gamma_{m'r'}\gamma_{m''r''} \frac{1}{m-2}\Lambda^{-(m-2)}$$

$$\leq \gamma_{m'r'}\gamma_{m''r''}\frac{1}{m-2}\Lambda^{-(m-4)}. \tag{3.86}$$

There remains $m = 2$, with the cases

(i) $m' = m'' = 2, l = 0$
(ii) $m' \geq 4$ and $m'' \geq 4$, $2l = m' + m'' - 4 \geq 4$, so $l \geq 4$

In case (i), we insert the inductive hypothesis into (3.80), to get

$$X \leq \int_{\Lambda}^{\Lambda_0} ds\, s^{-3}\gamma_{2r'}\gamma_{2r''} \leq \frac{1}{2}\gamma_{2r'}\gamma_{2r''}\Lambda^{-2}. \tag{3.87}$$

In case (ii), we may use (3.81) and replace $m' + m'' = 2l + 4$, to get

$$X \leq \int_{\Lambda}^{\Lambda_0} ds(\log s)^l s^{5-(2l+4)} = \int_{\Lambda}^{\Lambda_0} ds(\log s)^2 s^{-3}\left(\frac{\log s}{s^2}\right)^{l-2}$$

$$\leq \int_{\Lambda}^{\Lambda_0} ds(\log s)^2 s^{-3} \leq \text{const } \frac{1}{\Lambda}. \tag{3.88}$$

Since

$$A_{2\Lambda,\eta}(G_{mr}^{\Lambda\Lambda_0}) \leq A_{2\Lambda,\eta}(V_{mr}^{(\Lambda_0)}) + \frac{1}{2}\sum_{r'+r''=r}\sum_{(m',m'',l)} \kappa_{m'm''l}X, \tag{3.89}$$

the Λ-behaviour of the statement follows from $V_{mr}^{(\Lambda_0)} = 0$ for $m \geq 6$ and the bounds for X. As before, the inequality for the constants is then satisfied by an appropriate recursive definition of the γ_{mr}. ∎

Remark 3.12. The γ_{mr} are independent of η.

3.6.3 Convergence

The above theorems do not imply that the $G_{mr}^{\Lambda\Lambda_0}$ converge as $\Lambda_0 \to \infty$. One might fear that, while remaining bounded, they oscillate strongly for Λ_0 large. It is, however, a simple consequence of the representation in terms of Feynman graphs that the limit exists for the ϕ^4 theory. The argument is as follows. $G_{mr}^{\Lambda\Lambda_0}$ is a finite sum over values of Feynman graphs, so it suffices to consider

every graph separately. The vertex functions are Λ_0-independent (in fact, they are 1). The only Λ_0-dependence appears in the propagators $C_{\Lambda\Lambda_0}(p)$, which are nonnegative. So the integrand is nonnegative and increasing in Λ_0, and the same holds for the integral. Since it is bounded, it must converge.

This argument will not apply in higher dimensions, and it depends on monotonicity properties that do not hold in more general situations. Also, it does not give a rate of convergence. Convergence of the $G_{mr}^{\Lambda\Lambda_0}$ as $\Lambda_0 \to \infty$ can be shown using the RG equation.

The main idea is the same in all dimensions, and we sketch it briefly here. By (3.45) and the boundary condition $G_{mr}^{\Lambda_0\Lambda_0} = V_{mr}^{(\Lambda_0)}$,

$$G_{mr}^{\Lambda\Lambda_0} = V_{mr}^{(\Lambda_0)} - \frac{1}{2}\mathbb{S}_m \int_\Lambda^{\Lambda_0} ds\, Q_{mr}^{s\Lambda_0}, \tag{3.90}$$

so (abbreviating $\frac{\partial}{\partial\Lambda_0}$ by ∂_{Λ_0})

$$\partial_{\Lambda_0} G_{mr}^{\Lambda\Lambda_0} = \partial_{\Lambda_0} V_{mr}^{(\Lambda_0)} - \frac{1}{2}\mathbb{S}_m (Q_{mr}^{\Lambda_0\Lambda_0} + \int_\Lambda^{\Lambda_0} ds\, \partial_{\Lambda_0} Q_{mr}^{s\Lambda_0}). \tag{3.91}$$

Since

$$\partial_{\Lambda_0} C_{\Lambda\Lambda_0}(p) = \frac{2}{\Lambda^3} K'(\frac{p^2}{\Lambda^2}) \frac{p^2}{m_\phi^2 + p^2}, \tag{3.92}$$

we have at $\Lambda = \Lambda_0$

$$\partial_{\Lambda_0} C_{\Lambda\Lambda_0}(p)|_{\Lambda=\Lambda_0} = \frac{2}{\Lambda_0^3} K'(\frac{p^2}{\Lambda_0^2}) \frac{p^2}{m_\phi^2 + p^2} \tag{3.93}$$

so the $Q_{mr}^{\Lambda_0\Lambda_0}$-term is nonzero. In $\partial_{\Lambda_0} Q_{mr}^{s\Lambda_0}$, the derivative ∂_{Λ_0} can act on $G_{m'r'}^{s\Lambda_0}$ or $G_{m''r''}^{s\Lambda_0}$, as well as on $\partial_{\Lambda_0} C_{\Lambda\Lambda_0}(p)$. But by definition of $C_{\Lambda\Lambda_0}$,

$$\partial_{\Lambda_0} \partial_\Lambda C_{\Lambda\Lambda_0}(p) = 0 \quad \text{for all } p. \tag{3.94}$$

Taking $A_{2\Lambda,\eta}$ of (3.91), we get

$$A_{2\Lambda,\eta}(\partial_{\Lambda_0} G_{mr}^{\Lambda\Lambda_0}) \leq A_{2\Lambda,\eta}(\partial_{\Lambda_0} V_{mr}^{(\Lambda_0)}) + \frac{1}{2} A_{2\Lambda,\eta}(Q_{mr}^{\Lambda_0\Lambda_0})$$
$$+ \int d\kappa_{mr}(m', m'', r', r'', l)\, R_{m'm''l}^{\Lambda\Lambda_0} \tag{3.95}$$

with

$$R_{m'm''l}^{\Lambda\Lambda_0} = \int_\Lambda^{\Lambda_0} ds\, \|\partial_s C_{s,\Lambda_0}\|_\infty \|C_{0,s}\|_1^l A_{2s,\eta}(G_{m'r'}^{s\Lambda_0}) A_{2s,\eta}(\partial_{\Lambda_0} G_{m''r''}^{s\Lambda_0}). \tag{3.96}$$

R takes into account all terms because the sum is invariant under exchange of m' and m''.

The first term on the right hand side of (3.95) is the contribution from the explicit Λ_0-dependence of the initial interaction. Since no counterterms are required in two dimensions, $V^{(\Lambda_0)}$ will typically be independent of Λ_0, or decay with Λ_0, so this term will be small. The second term is the influence of the scale Λ_0 on $\partial_{\Lambda_0} G$. Under the stated hypothesis, the factor determining its behaviour is $\partial_\Lambda C_{\Lambda\Lambda_0}|_{\Lambda=\Lambda_0} \sim \frac{1}{\Lambda_0^3}$. This factor also determines the decay of the full G because the last term is covered by the inductive hypothesis. Since $\partial_{\Lambda_0} G_{m''r''}^{\Lambda\Lambda_0}$ appears linearly in there, it will reproduce the inserted behaviour for $\partial_{\Lambda_0} G_{m'r'}^{\Lambda\Lambda_0}$ also for $\partial_{\Lambda_0} G_{mr}^{\Lambda\Lambda_0}$.

Theorem 3.13. *Assume that $V_{mr}^{(\Lambda_0)}$ converges to a finite limit as $\Lambda_0 \to \infty$, such that $A_{2\Lambda_0,\eta}(V_{mr}^{(\Lambda_0)}) \le v_{mr}$ holds for all Λ_0, and that $V_{mr}^{(\Lambda_0)} = 0$ for m odd and $A_{2\Lambda_0,\eta}(\frac{\partial}{\partial \Lambda_0} V_{mr}^{(\Lambda_0)}) \le \frac{w_{mr}}{\Lambda_0^2}$ for all m, r and Λ_0.*
For $d = 2$, the $G_{mr}^{\Lambda\Lambda_0}$ converge as $\Lambda_0 \to \infty$. More precisely, there are constants $\tilde{\gamma}_{mr}$ such that

$$A_{2\Lambda,\eta}(\frac{\partial}{\partial \Lambda_0} G_{mr}^{\Lambda\Lambda_0}) \le \frac{\tilde{\gamma}_{mr}}{\Lambda_0^2}. \tag{3.97}$$

For the ϕ^4 theory $(V_{mr}^{(\Lambda_0)} = -\delta_{m4}\delta_{r1})$, there are $\tilde{\gamma}_{mr}$ such that

$$A_{2\Lambda,\eta}(\frac{\partial}{\partial \Lambda_0} G_{mr}^{\Lambda\Lambda_0}) \le \frac{\log \Lambda_0}{\Lambda_0^3} \tilde{\gamma}_{mr} \tag{3.98}$$

for all $m \ge 1$ and $r \ge 1$.

Proof. For $d = 2$, we use (3.59) and (3.56) to get

$$R_{m'm''l}^{\Lambda\Lambda_0} \le K_1 K_2^l \int_\Lambda^{\Lambda_0} ds \, s^{-3}(\log s)^l$$

$$A_{2s,\eta}(G_{m'r'}^{s\Lambda_0}) A_{2s,\eta}(\partial_{\Lambda_0} G_{m''r''}^{s\Lambda_0}) \tag{3.99}$$

and

$$A_{2\Lambda,\eta}(Q_{mr}^{\Lambda_0\Lambda_0}) \le K_1 K_2^l \Lambda_0^{-3}(\log \Lambda_0)^l A_{2\Lambda_0,\eta}(V_{m'r'}^{(\Lambda_0)}) A_{2\Lambda_0,\eta}(V_{m''r''}^{(\Lambda_0)}) \tag{3.100}$$

If $r = 1$, the statement follows from the hypotheses on $V_{mr}^{(\Lambda_0)}$. Let $r \ge 2$, and the statement be proven for all $r' < r$ and $r'' < r$. Thus by (3.95) and by the known behaviour of $A_{2s,\eta}(G_{m'r'}^{s\Lambda_0})$ (Theorem 3.10)

$$A_{2s,\eta}(\partial_{\Lambda_0} G_{mr}^{s\Lambda_0}) \le \frac{w_{mr}}{\Lambda_0^2} + \sum_{r'+r''=r} \sum_{(m',m'',l)} \kappa_{m'm''l} \cdot$$

$$\{\frac{1}{2} v_{m'r'} v_{m''r''} \frac{(\log \Lambda_0)^l}{\Lambda_0^3} + \int_\Lambda^{\Lambda_0} ds \frac{K_1}{s^3} (\log s)^l g_{m'r'} \frac{\tilde{\gamma}_{m''r''}}{\Lambda_0^2} \}. \tag{3.101}$$

The integral converges, and the coefficient of $\frac{1}{\Lambda_0^2}$ is bounded, so with the usual recursive definition of g_{mr}, the theorem follows.

Part (2) is left as an exercise to the reader. ∎

3.7 Three Dimensions

Wick ordering does not remove all ultraviolet divergences in three dimensions. The value

$$W_2(k) = 4^2\, 3!\, (2\pi)^{-6} \int d^3p\, d^3q\, C_{0,\Lambda_0}(p)\, C_{0,\Lambda_0}(q)\, C_{0,\Lambda_0}(p+q-k) \quad (3.102)$$

of the second order "sunset" graph, drawn in Fig. 3.5, grows like $\log \Lambda_0$ and thus diverges as $\Lambda_0 \to \infty$. We show that there is a logarithmic divergence by proving a similar statement for the integral

$$I_{\Lambda\Lambda_0}(k) = \int d^3p\, d^3q\, C_{0\Lambda}(p)\, C_{0\Lambda}(q)\, \partial_\Lambda C_{\Lambda\Lambda_0}(p+q-k) \quad (3.103)$$

which contributes to $Q_{22}^{\Lambda\Lambda_0}(k)$ (the integral of which gives the full second order two-point function).

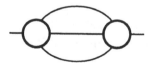

Fig. 3.5. The second order "sunset" graph

Remark 3.14. There are constants $Q_2 \geq Q_1 > 0$ such that for all $\Lambda \geq m$,

$$\frac{Q_1}{\Lambda} \leq I_{\Lambda\Lambda_0}(0) \leq \sup_{k\in\mathbb{R}^d} |I_{\Lambda\Lambda_0}(k)| \leq \frac{Q_2}{\Lambda}, \quad (3.104)$$

hence for all $\Lambda' \geq m$,

$$\int_{\Lambda'}^{\Lambda_0} d\Lambda\, I_{\Lambda\Lambda_0}(0) \geq Q_1 \log \frac{\Lambda_0}{\Lambda'} \quad (3.105)$$

and

$$\int_{\Lambda'}^{\Lambda_0} d\Lambda\, \|I_{\Lambda\Lambda_0}\|_\infty \leq Q_2 \log \frac{\Lambda_0}{\Lambda'}. \quad (3.106)$$

Proof. Estimating $\partial_\Lambda C_{\Lambda\Lambda_0}$ by its sup norm (as usual), we get

$$I_{\Lambda\Lambda_0}(k) \leq \|C_{0\Lambda}\|_1^2 \|\partial_\Lambda C_{\Lambda\Lambda_0}\|_\infty \leq (K_2\Lambda)^2 \frac{K_1}{\Lambda^3} = K_1 K_2^2 \frac{1}{\Lambda} \qquad (3.107)$$

for all k, so the upper bound in (3.104), and (3.106), hold with $Q_2 = K_1 K_2^2$. The lower bound in (3.104) is proven in Appendix A.5. ∎

Of course, this graph will appear as a subgraph in many others, and so its divergence will cause all Green functions to diverge – and we have not even looked at larger graphs yet. In a graphical analysis, one can postpone this problem by considering only graphs without two–legged insertions. We can do the same in the flow equation by truncating the sums on the right hand side so that $m' = 2$ and $m'' = 2$ do not occur. It follows from the graphical representation of the flow equation that this removes two-legged subgraphs from the sum over graphs that solves this truncated flow.

We shall derive bounds for the so obtained functions $\tilde{G}_{mr}^{\Lambda\Lambda_0}$ and turn back to the full problem later. One can apply other truncations, which provide a lot of insight, to the flow equation. One can, e.g., also remove the four-legged insertions from the right hand side, or remove all terms except for $l = 0$ from the sum over l – this truncation of the flow produces tree graphs only. Keeping $l = 1$ only does not lead to one-loop graphs, but to sums over all graphs where every line is part of one loop only, the non-overlapping graphs. In these truncated systems, the flow of the coupling constants can often be calculated (see Sect. 3.9).

3.7.1 Power Counting for the Truncated Equation

We now denote the modification where the sums over m' and m'' both start at 4 by $\int d\tilde{\kappa}_{mr}$. By (3.54), $\|C_{0\Lambda}\|_1 \leq K_2\Lambda$ for $d = 3$, so by (3.60),

$$A_{2\Lambda,\eta}(\partial_\Lambda G_{mr}^{\Lambda\Lambda_0}) \leq \frac{1}{2} \int d\tilde{\kappa}_{mr} \frac{2K_1}{\Lambda^3} \Lambda^l A_{2\Lambda,\eta}(G_{m'r'}^{\Lambda\Lambda_0}) A_{2\Lambda,\eta}(G_{m''r''}^{\Lambda\Lambda_0}). \qquad (3.108)$$

Motivated by the results in $d = 2$, we try the ansatz

$$A_{2\Lambda,\eta}(G_{mr}^{\Lambda\Lambda_0}) \leq \text{const } \Lambda^{a-bm} \qquad (3.109)$$

with $b > 0$. To match the Λ-behaviour we must then have

$$\Lambda^{l-3+a-bm'+a-bm''} \leq \Lambda^{a-bm-1} \qquad (3.110)$$

where the -1 in the exponent on the right hand side is there because we want to get Λ^{a-bm} after integration over Λ, to satisfy the flow inequality. Equating the exponents and using that $m' + m'' = m + 2l + 2$, we see that m drops out of this equation. There remains

$$l(1 - 2b) + a - 2 - 2b \leq 0. \qquad (3.111)$$

By appropriate choice of m', m'', r' and r'', l can be made arbitrarily large, so for this inequality to hold, we must have $b \geq \frac{1}{2}$.

In fact, this inequality would suggest that the larger b, the better. However, we also have to satisfy an induction beginning for the low orders, and this restricts b in a dimension-independent way: (3.111) has to hold for $l = 0$, i.e. for tree graphs, and the scaling behaviour of values of the graphs is $\|C_{\Lambda \Lambda_0}\|_{\infty}^{-n} \sim \Lambda^{-2n}$ where n is the number of internal lines in the graph. This bound is independent of dimensions since there are no loop integrations involved. In low orders, the tree graphs are

$$
\begin{aligned}
r = 2: \quad m = 6, \quad \Lambda^{a-6b} &\sim \Lambda^{-2} \quad (n = 1) \\
r = 3: \quad m = 8, \quad \Lambda^{a-8b} &\sim \Lambda^{-4} \quad (n = 2)
\end{aligned}
\tag{3.112}
$$

From this, we get $b = 1, a = 4$. This simple tree–level–estimate holds for all graphs produced by the truncated flow.

Theorem 3.15. In ϕ^4 theory for $d = 3$, i.e. $V_{mr}^{(\Lambda_0)} = -\delta_{m4}\delta_{r1}$, the unrenormalized truncated flow obtained by removing two-legged insertions satisfies

$$
A_{2\Lambda,\eta}(G_{mr}^{\Lambda \Lambda_0}) \leq \begin{cases} g_{mr}\, \Lambda^{4-m} & m \geq 4 \\ g_{2r}\, \log \Lambda_0 & m = 2. \end{cases}
\tag{3.113}
$$

Proof. $r = 1$ is trivial. Again, we look at

$$
X = \int_{\Lambda}^{\Lambda_0} ds\, A_{2s,\eta}(G_{m'r'}^{s\Lambda_0}) A_{2s,\eta}(G_{m''r''}^{s\Lambda_0})\, s^{-3+l},
\tag{3.114}
$$

with $r' + r'' = r$, but with $\min\{m', m''\} \geq 4$ because of the truncation. By the inductive hypothesis

$$
X \leq g_{m'r'}\, g_{m''r''} \int_{\Lambda}^{\Lambda_0} ds\, s^{-3+l+4-m'+4-m''} = g_{m'r'}\, g_{m''r''} \int_{\Lambda}^{\Lambda_0} ds\, s^{3-m-l}.
\tag{3.115}
$$

Let $m \geq 6$, then $3 - m - l \leq 3 - m \leq -3$, so integration proves the claim.

For $m = 4$, $2l = m' + m'' - m - 2 \geq 4 + 4 - 4 - 2 \geq 2$, so $l \geq 1$, and thus $3 - m - l \leq 2 - m \leq -2$, and $\int_{\Lambda}^{\Lambda_0} s^{-2} ds \leq \Lambda^{-1} \leq m_{\phi}^{-1}$.

For $m = 2$, $2l = m' + m'' - 4 \geq 4$, so $l \geq 2$, and

$$
\int_{\Lambda}^{\Lambda_0} ds\, s^{1-l} \leq \int_{\Lambda}^{\Lambda_0} ds\, s^{-1} = \log \frac{\Lambda_0}{\Lambda} \leq \log \frac{\Lambda_0}{\Lambda_1}.
\tag{3.116}
$$

The constants g_{mr} are again fixed recursively. \blacksquare

Remark 3.16. Closer inspection of this proof shows that

(i) for $r \geq 2$, what we have proven for the four-point function is

$$A_{2\Lambda,\eta}(G_{4r}^{\Lambda\Lambda_0}) \leq g_{4r} \Lambda^{-1} \qquad (3.117)$$

so the $O(\Lambda^0)$ term is only there in first order.

(ii) for $m = 2$, any contribution from $l \geq 3$ is convergent, so the only graphs that can be divergent are of type shown in Fig. 3.5, with the vertices denoting four-point functions. But as we just saw in (i), whenever $r' > 1$, there is an extra decay factor Λ^{-1}, so the only graph whose value can diverge in the limit $\Lambda_0 \to \infty$ is $G_{22}^{\Lambda\Lambda_0}$. We shall show in the following that, indeed, only $G_{22}^{\Lambda\Lambda_0}$ needs to be renormalized in $d = 3$, and we state the better bounds below for the untruncated flow.

3.7.2 Renormalization: A Change of Boundary Conditions

Although we have just argued that in $d = 3$, the value of only one graph is divergent, we keep the following discussion a little bit more general, to prepare for the four-dimensional case. The full argument leading to the logarithmic behaviour of $G_{2r}^{\Lambda\Lambda_0}$ is: integrate the RG equation $\partial_\Lambda G_{2r}^{\Lambda\Lambda_0} = \frac{1}{2} Q_{2r}^{\Lambda\Lambda_0}$ from Λ to Λ_0, and take $A_{2\Lambda,\eta}$. This gives

$$G_{2r}^{\Lambda_0\Lambda_0}(p) - G_{2r}^{\Lambda\Lambda_0}(p) = \frac{1}{2} \int_\Lambda^{\Lambda_0} ds \, Q_{2r}^{s\Lambda_0}(p), \qquad (3.118)$$

$$A_{2\Lambda,\eta}(G_{2r}^{\Lambda\Lambda_0}) \leq A_{2\Lambda,\eta}\left(G_{2r}^{\Lambda_0\Lambda_0} - \frac{1}{2} \int_\Lambda^{\Lambda_0} ds \, Q_{2r}^{s\Lambda_0}\right). \qquad (3.119)$$

Up to now, we had the boundary condition $G_{2r}^{\Lambda_0\Lambda_0} = V_{2r}^{(\Lambda_0)} = 0$ (i.e. only a ϕ^4 term), and the integral over Q grows like $\log \Lambda_0$. We will now try to change the boundary conditions, i.e. choose $G_{2r}^{\Lambda_0\Lambda_0}(p)$ such that it cancels the $\log \Lambda_0$ term in Q. The simplest condition one can put is

$$V_{2r}^{(\Lambda_0)}(p) = \delta\mu_r^{(\Lambda_0)} \qquad (3.120)$$

where $\delta\mu_r^{(\Lambda_0)}$ is chosen as a function of Λ_0 to cancel the terms in $G_{2r}^{\Lambda\Lambda_0}(0)$ that grow with Λ_0, e.g. $\delta\mu_2^{(\Lambda_0)} \sim \log \Lambda_0$. That is, the action now contains a mass counterterm

$$\sum_{r=1}^\infty \lambda^r \delta\mu_r^{(\Lambda_0)} \int \frac{d^3p}{(2\pi)^3} \, \hat\phi(p) \, \hat\phi(-p). \qquad (3.121)$$

But how, and why, should that work? After all, we are only subtracting a constant from a function, and it is not guaranteed that there might not be

any divergences in derivatives of that function as well. However, if we take a derivative

$$\frac{\partial}{\partial k_\alpha} Q_{22}^{\Lambda\Lambda_0}(k) = \int dp_1 dp_2 C_{0\Lambda}(p_1) C_{0\Lambda}(p_2) \frac{\partial}{\partial k_\alpha} \partial_\Lambda C_{\Lambda\Lambda_0}(-p_1 - p_2 + k)$$

(3.122)

we see because of

$$\frac{\partial}{\partial k_\alpha} \partial_\Lambda C_{\Lambda\Lambda_0}(k+p) = \frac{\partial}{\partial k_\alpha} (-\frac{2}{\Lambda^3} K'(\frac{(k+p)^2}{\Lambda^2}) \frac{(k+p)^2}{(k+p)^2 + m_\phi^2})$$

(3.123)

that

$$\|\frac{\partial}{\partial k_\alpha} \partial_\Lambda C_{\Lambda\Lambda_0}\|_\infty \le \frac{const}{\Lambda^4}$$

(3.124)

and hence

$$A_{2\Lambda,\eta}(\frac{\partial}{\partial k_\alpha} Q_{22}^{\Lambda\Lambda_0}) \le const \; \Lambda^2 \Lambda^{-4}$$

(3.125)

Integrating $\frac{\partial}{\partial k_\alpha} \partial_\Lambda G_{22}^{\Lambda\Lambda_0} = \frac{1}{2} \frac{\partial}{\partial k_\alpha} Q_{22}^{\Lambda\Lambda_0}$, we see that the derivative of $G_{22}^{\Lambda\Lambda_0}$ remains finite. Inserting the Taylor expansion

$$Q_{22}^{\Lambda\Lambda_0}(k) = Q_{22}^{\Lambda\Lambda_0}(0) + \int_0^1 dt \frac{\partial}{\partial t} Q_{22}^{\Lambda\Lambda_0}(tk)$$

(3.126)

into the flow equation and integrating we get

$$G_{22}^{\Lambda\Lambda_0}(k) - G_{22}^{\Lambda\Lambda_0}(0) = G_{22}^{\Lambda_0\Lambda_0}(k) - G_{22}^{\Lambda_0\Lambda_0}(0) - \int_\Lambda^{\Lambda_0} ds \int_0^1 dt \, k \cdot \nabla Q_{22}^{s\Lambda_0}(tk).$$

(3.127)

Since $G_{22}^{\Lambda_0\Lambda_0}(k) = \delta\mu_2^{(\Lambda_0)}$, $G_{22}^{\Lambda_0\Lambda_0}(k) - G_{22}^{\Lambda_0\Lambda_0}(0) = 0$,

$$A_{2\Lambda,\eta}(G_{22}^{\Lambda_0\Lambda_0}(k) - G_{22}^{\Lambda_0\Lambda_0}(0))$$

$$\le \int_\Lambda^{\Lambda_0} ds \, A_{2\Lambda,\eta}(\int_0^1 dt \, k \, \nabla Q_{22}^{s\Lambda_0}(tk))$$

$$\le \eta \int_\Lambda^{\Lambda_0} ds \, A_{2\Lambda,\eta}(\nabla Q_{22}^{s\Lambda_0})$$

(3.128)

$$\le \eta \; const \int_\Lambda^{\Lambda_0} ds \, s^{-2} \le const \frac{\eta}{\Lambda_1}.$$

Thus, once $\lim_{\Lambda_0 \to \infty} G_{22}^{\Lambda\Lambda_0}(0)$ is made finite by a subtraction, the function $G_{22}^{\Lambda\Lambda_0}(p)$ is bounded for all p uniformly in Λ_0. This feature is typical for three dimensions. In $d = 4$, a mass counterterm is insufficient and one also needs

a wave function renormalization to get all derivatives up to order 2 finite. Note also that an η appears in the bound. Because of the Taylor expansion, factors of η will appear in the renormalized bounds.

Graphically, one can denote the mass counterterms $\delta\mu_r^{(\Lambda_0)}$ by —■—. They thus appear as new, two-legged vertices in the perturbation expansion. Now, a typical problem in renormalization arises: we have subtracted a divergence by adding a counterterm. This counterterm appears as a new interaction vertex everywhere in graphs, and, since it grows with Λ_0, it might actually spoil the convergence of more complicated graphs. Moreover, how does one put such counterterms consistently to all orders? Enter Wilson's RG, which arranges things in a way that counterterms appear only in combination with the divergence they are to cure. We shall see now that this is particularly simple in the flow equation because the counterterms do not even appear there, so that the RG equation relates only quantities at the same scale Λ. Instead, the counterterms appear only as an initial condition for the flow. Moreover, they can be determined by a renormalization condition imposed on particular Green functions at the final value Λ_1 of the flow. Thus the only place where they are needed is the statement that they can be chosen such as to make $G_{2r}^{\Lambda\Lambda_0}$ finite.

Suppose we demand that

$$G_{2r}^{\Lambda_1\Lambda_0}(0) = \delta\mu_r^R \qquad \text{for all} \quad r \geq 2 \qquad (3.129)$$

where $(\delta\mu_r^{(R)})_{r\geq 1}$ is a given sequence which is independent of Λ_0. (the superscript R only indicates that this is a renormalization condition; R is not a variable). Then (3.128) implies finiteness of $G_{2r}^{\Lambda_1\Lambda_0}(k)$ for all k. This boundary condition determines the mass counterterms at Λ_0 uniquely in the following sense.

Lemma 3.17. *There exists a unique sequence* $(\delta\mu_r^{(\Lambda_0)})_{r\geq 2}$ *such that the effective action* $G_{mr}^{\Lambda_1\Lambda_0}$ *at scale* Λ_1 *obtained from the initial interaction*

$$V_{mr}^{(\Lambda_0)} = \begin{cases} -1 & \text{for } r = 1 \text{ and } m = 4 \\ \delta\mu_r^{(\Lambda_0)} & \text{for } m = 2 \text{ and } r \geq 2 \\ 0 & \text{for all other } m \text{ and } r \end{cases} \qquad (3.130)$$

satisfies (3.129).

Remark 3.18. In fact, a renormalization condition like (3.129) is necessary to fix the counterterms since the requirement that a term growing in Λ_0 is cancelled leaves the freedom of adding finite constants.

Proof. Suppose we have a theory with vertices given by (3.130). Then $G_{mr}^{\Lambda_1\Lambda_0}$ is a sum over values of Feynman graphs with four legged *and* two-legged vertices, such that the orders of the vertices add up to r. This fixes in particular $G_{2r}^{\Lambda_1\Lambda_0}(0)$. For instance

$$G_{21}^{\Lambda_1 \Lambda_0}(0) \;=\; 0, \tag{3.131}$$

$$G_{22}^{\Lambda_1 \Lambda_0}(0) \;=\; \int C_{\Lambda_1 \Lambda_0}(p_1)\, C_{\Lambda_1 \Lambda_0}(p_2)\, C_{\Lambda_1 \Lambda_0}(-p_1 - p_2)\mathrm{d}^3 p_1 \mathrm{d}^3 p_2 + \delta\mu_2^{(\Lambda_0)}.$$

It is obvious that this equation can be solved for $\delta\mu_2^{(\Lambda_0)}$ if $G_{22}^{\Lambda_1 \Lambda_0}(0)$ is given. We now proceed by induction in r. In the induction step, we may assume that the $\delta\mu_{r'}^{(\Lambda_0)}$ have already been fixed for all $r' < r$, in terms of the $G_{2r'}^{\Lambda_1 \Lambda_0}(0)$ with $r' < r$. In order r, we have $G_{2r}^{\Lambda_1 \Lambda_0}(0) = \delta\mu_r^{(\Lambda_0)} +$ a sum over values of graphs with two legs and at least two vertices. In this sum, only $\delta\mu_{r'}^{(\Lambda_0)}$ with $r' < r$ appear. Again, it is trivial to solve for $\delta\mu_r^{(\Lambda_0)}$. ∎

Remark 3.19. There is no first order counterterm because we expand in Wick ordered monomials.

3.7.3 Renormalized ϕ_3^4

Before stating the theorem about renormalized ϕ^4 in $d = 3$, we recall the standard notation for multiindices. For $m \geq 1$, a multiindex α is an $m \cdot d$-tuple

$$\alpha = (\alpha_1, \ldots, \alpha_m) \in (\mathbb{N}_0^d)^m \quad, \quad \alpha_i = (\alpha_{i,1}, \ldots, \alpha_{i,d}) \tag{3.132}$$

We denote

$$|\alpha| = |\alpha_1| + \ldots + |\alpha_m| = \sum_{i=1}^{m} \sum_{j=1}^{d} \alpha_{ij} \tag{3.133}$$

and

$$\alpha! = \alpha_1! \ldots \alpha_m! = \prod_{i=1}^{m} \prod_{j=1}^{d} \alpha_{ij}! \,. \tag{3.134}$$

We write $\alpha' + \alpha'' = \alpha$ if $\alpha'_{ij} + \alpha''_{ij} = \alpha_{ij}$ for all i, j. Note that

$$\alpha' + \alpha'' = \alpha \quad \Longrightarrow \quad |\alpha'| + |\alpha''| = |\alpha|. \tag{3.135}$$

Leibniz's rule,

$$D^\alpha(f_1 \ldots f_p) = \sum_{\alpha^{(1)} + \ldots + \alpha^{(p)} = \alpha} \frac{\alpha!}{\alpha^{(1)}! \ldots \alpha^{(p)}!} D^{\alpha^{(1)}} f_1 \ldots D^{\alpha^{(p)}} f_p, \tag{3.136}$$

is proven by induction.

We shall show inductively that all $G_{mr}^{\Lambda \Lambda_0}$ are C^∞ in the external momenta if the same holds for the initial interaction $V_{mr}^{(\Lambda_0)}$. For this we note that $C_{\Lambda \Lambda_0} \in C^\infty(\mathbb{R}^d, \mathbb{R})$ and

$$D^\alpha \partial_\Lambda C_{\Lambda\Lambda_0}(p) = D^\alpha \left(\frac{2}{\Lambda^3} K'(\frac{p^2}{\Lambda^2}) \frac{p^2}{p^2 + m_\phi^2} \right) \tag{3.137}$$

$$= \frac{2}{\Lambda^3} \sum_{\alpha' + \alpha'' = \alpha} \frac{\alpha!}{\alpha'! \alpha''!} D^{\alpha'} K'(\frac{p^2}{\Lambda^2}) D^{\alpha''} (1 - \frac{m_\phi^2}{p^2 + m_\phi^2}).$$

By the support properties of K, $D^{\alpha'} K'(\frac{p^2}{\Lambda^2}) = 0$ for all α' unless $|p| \in [\Lambda, 2\Lambda]$. Moreover

$$|D^{\alpha''}(1 - \frac{m_\phi^2}{p^2 + m_\phi^2})| \leq \begin{cases} 1 & \text{if } \alpha'' = 0 \\ \frac{K(\alpha'')}{\Lambda^{2+|\alpha''|}} & \text{if } |\alpha''| \geq 1 \end{cases} \tag{3.138}$$

so

$$|D^\alpha \partial_\Lambda C_{\Lambda\Lambda_0}(p)| \leq \frac{K_1(\alpha)}{\Lambda^{3+|\alpha|}} I(\Lambda \leq |p| \leq 2\Lambda) \tag{3.139}$$

with a combinatorial constant K_1 (which depends on m_ϕ).

We also use the following lemma to reduce the number of cases to go through as much as possible.

Lemma 3.20. *Let* $m, m', m'' \in \mathbb{N}, l \in \mathbb{N}_0$ *such that* $m'' + m'' = m + 2l + 2$. *Define*

$$\begin{aligned} D(m) &= \min\{4 - m, 0\} \qquad \text{and} \\ E(m) &= \max\{4 - m, 0\}. \end{aligned} \tag{3.140}$$

Let $\delta \in \mathbb{R}$. *Then*

$$D(m') + D(m'') + l\delta - 3 = D(m) - 1 - (2-\delta)l + E(m) - E(m') - E(m''). \tag{3.141}$$

In particular, if $m \geq 4$,

$$D(m') + D(m'') + l\delta - 3 \leq D(m) - 1 - (2-\delta)l. \tag{3.142}$$

Proof. $D(m) + E(m) = 4 - m$, so

$$\begin{aligned} D(m') + E(m') &+ D(m'') + E(m'') - l\delta - 3 = 4 - m' + 4 - m'' - l\delta - 3 \\ &= 5 - l\delta - (m' + m'') = 5 - l\delta - (m + 2l + 2) \tag{3.143} \\ &= -1 + 4 - m - (2-\delta)l = D(m) - 1 - (2-\delta)l + E(m) \end{aligned}$$

∎

Theorem 3.21. *Let* $d = 3$, $\delta\mu_{R,2} \in \mathbb{R}$ *and* ϕ^4 *with a second-order mass counterterm be given by*

$$V_{mr}^{(\Lambda_0)} = -\delta_{r1}\delta_{m4} + \delta_{r2}\delta\mu_2^{(\Lambda_0)}, \tag{3.144}$$

where $\delta\mu_2^{(\Lambda_0)}$ *is fixed by the condition*

$$G_{22}^{\Lambda_1\Lambda_0}(0) = \delta\mu_2^R. \tag{3.145}$$

Then all $G_{mr}^{\Lambda\Lambda_0}$ with m odd are identically zero,

$$G_{m1}^{\Lambda\Lambda_0}(p_1, p_2, \ldots, p_{m-1}) = -\delta_{m4} \quad \text{for all } p_i, \Lambda, m, \tag{3.146}$$

the $G_{mr}^{\Lambda\Lambda_0}$ for m even, $r \geq 2$ are C^∞ functions of the external momenta (p_1, \ldots, p_{m-1}) and for any multiindex α there are constants $g_{mr|\alpha|}$, depending on $\delta\mu_2^R$, such that for all $\Lambda \in [\Lambda_1, \Lambda_0]$, all $\eta \geq \Lambda_1$, all $m \geq 2$, and all $r \geq 1$,

$$A_{2\Lambda,\eta}(D^\alpha G_{mr}^{\Lambda\Lambda_0}) \leq \begin{cases} g_{220} \log \Lambda & m = r = 2, \alpha = 0 \\ g_{mr|\alpha|} \Lambda^{D(m)-|\alpha|} & \text{otherwise} \end{cases} \tag{3.147}$$

Remark 3.22. 1. More precisely, we prove that

$$A_{2\Lambda,\eta}(G_{mr}^{\Lambda\Lambda_0}) \leq g_{mr|\alpha|} \begin{cases} \log \Lambda & m = r = 2, \alpha = 0 \\ \Lambda^{-|\alpha|} & m = r = 2, |\alpha| \geq 1 \\ \Lambda^{-1} \log \Lambda & m = 2, r > 2, \alpha = 0 \\ \Lambda^{-1-|\alpha|} & m = 2, r > 2, |\alpha| \geq 1 \\ & \text{or } m = 4, r \geq 2 \\ \Lambda^{D(m)-|\alpha|} & m \geq 6. \end{cases} \tag{3.148}$$

2. In Fig. 3.6, the scaling bounds for $G_{mr}^{\Lambda\Lambda_0}$ as a function of Λ are sketched (a) for the truncated unrenormalized Green functions, (b) for the renormalized Green functions. The change of boundary conditions has only shifted the two-point function, by fixing a prescribed value at Λ_1 instead of at Λ_0. The scaling behaviour itself does not change because renormalization does not change the decay properties of the propagator.

3. We shall see in the proof that the boundary condition $V_{mr}^{(\Lambda_0)} = 0$ for $m \geq 6$ is important for these bounds, to ensure that $A_{2\Lambda,\eta}(G_{mr}^{\Lambda\Lambda_0}) \sim \Lambda^{D(m)-|\alpha|}$ for $m \geq 6$. It can be relaxed slightly to a condition $V_{mr}^{(\Lambda_0)} \sim \Lambda_0^{D(m)}$. This is useful for the so-called "improvement", i.e. speeding up the convergence of the $G_{mr}^{\Lambda\Lambda_0}$ to their limits G_{mr}^{Λ} as $\Lambda_0 \to \infty$ by adjusting the boundary condition [30].

Proof. By the flow equation, $D^\alpha \partial_\Lambda G_{mr}^{\Lambda\Lambda_0} = D^\alpha Q_{mr}^{\Lambda\Lambda_0}$. By Leibniz's rule

$$D^\alpha Q_{mr}^{\Lambda\Lambda_0}(\underline{k}) = \int d\kappa_{mr} \int d\omega_\Lambda(\underline{p}) \sum_{\alpha'+\alpha''+\alpha'''=\alpha} \frac{\alpha!}{\alpha'!\alpha''!\alpha'''!} D^{\alpha'} G_{m'r'}^{\Lambda\Lambda_0}(\underline{p}, \underline{k}')$$
$$D^{\alpha''} G_{m''r''}^{\Lambda\Lambda_0}(-\underline{p}, \underline{k}'') D^{\alpha'''} \partial_\Lambda C_{\Lambda\Lambda_0}(p_0) \tag{3.149}$$

so, by (3.139),

$$A_{2\Lambda,\eta}(D^\alpha Q_{mr}^{\Lambda\Lambda_0}) \leq \int d\kappa_{mr} \|C_{0\Lambda}\|_1^l \sum_{\alpha'+\alpha''+\alpha'''=\alpha} \frac{\alpha!}{\alpha'!\alpha''!\alpha'''!} \frac{K_1(\alpha''')}{\Lambda^{3+|\alpha'''|}}$$
$$A_{2\Lambda,\eta}(D^{\alpha'} G_{m'r'}^{\Lambda\Lambda_0}) A_{2\Lambda,\eta}(D^{\alpha''} G_{m''r''}^{\Lambda\Lambda_0}). \tag{3.150}$$

This bound holds in any dimension. Specializing to $d = 3$, we have $\|C_{0\Lambda}\|_1 \leq K_2\Lambda$, so

$$A_{2\Lambda,\eta}(D^\alpha Q_{mr}^{\Lambda\Lambda_0}) \leq \sum_{\alpha'+\alpha''+\alpha'''=\alpha} \frac{\alpha!}{\alpha'!\alpha''!\alpha'''!} \int d\kappa_{mr} K_1(\alpha''') K_2^l \Lambda^{-3-|\alpha'''|+l}$$
$$A_{2\Lambda,\eta}(D^{\alpha'} G_{m'r'}^{\Lambda\Lambda_0}) A_{2\Lambda,\eta}(D^{\alpha''} G_{m''r''}^{\Lambda\Lambda_0}). \tag{3.151}$$

We do the usual induction in r, using (3.148) and $G_{mr}^{\Lambda\Lambda_0} \in C^\infty$ as inductive hypothesis. The case $r = 1$ is trivial by (3.146). We do $r = 2$ first. $G_{m2}^{\Lambda\Lambda_0} = 0$ unless $m \leq 6$; the boundary conditions are

$$V_{62}^{(\Lambda_0)} = V_{42}^{(\Lambda_0)} = 0 \tag{3.152}$$
$$G_{22}^{\Lambda_1\Lambda_0}(0) = \delta\mu_2^R.$$

The lower order Green functions on the right side of (3.151) must have $r' = r'' = 1$, so $m' = m'' = 4$, and by (3.146), $\alpha' = \alpha'' = 0$, and

$$A_{2\Lambda,\eta}(D^\alpha Q_{m2}^{\Lambda\Lambda_0}) \leq \int d\kappa_{m2} K_1(\alpha) K_2^l \Lambda^{-3-|\alpha|+l}. \tag{3.153}$$

The summands m', m'', l in $\int d\kappa_{m2}$ satisfy $m' + m'' = m + 2l + 2$, so $l = \frac{1}{2}(6 - m)$. For $m = 4$ and $m = 6$ and for $m = 2, |\alpha| \geq 1$, we integrate the flow equation from Λ to Λ_0 and use (3.152), to get

$$A_{2\Lambda,\eta}(D^\alpha G_{m2}^{\Lambda\Lambda_0}) \leq K_1(\alpha) K_2^{3-\frac{m}{2}} \text{const} \int_\Lambda^{\Lambda_0} ds \, s^{-|\alpha|-\frac{m}{2}}$$
$$\leq K_1(\alpha) K_2^{3-\frac{m}{2}} \text{const} \, \Lambda^{-|\alpha|-\frac{m}{2}+1} \tag{3.154}$$

which satisfies (3.148). For $r = 2$, $\alpha = 0$, we integrate from Λ_1 to Λ_0, to get

$$|G_{22}^{\Lambda\Lambda_0}(p)| \leq |\delta\mu_2^R| + \int_{\Lambda_1}^\Lambda ds \, |Q_{22}^{s\Lambda_0}(p)|$$
$$\leq |\delta\mu_2^R| + K_1 K_2^2 \text{const} \int_{\Lambda_1}^\Lambda \frac{ds}{s} \tag{3.155}$$
$$\leq |\delta\mu_2^R| + K_1 K_2^2 \log \frac{\Lambda}{\Lambda_1}.$$

Since $\Lambda \geq 3$, $\log \Lambda \geq 1$, so we can bound this sum by $g_{220} \log \Lambda$, with the same constant g_{220}.

Let $r \geq 3$, and assume (3.148) for all $r' < r$. We integrate the flow equation from Λ to Λ_0 and use $V_{mr}^{(\Lambda_0)} = 0$ for all m (since $r \geq 3$), and the inductive hypothesis (3.148), to get

$$A_{2\Lambda,\eta}(D^\alpha G_{mr}^{\Lambda\Lambda_0}) \leq \int d\kappa_{mr} \sum_{\alpha'+\alpha''+\alpha'''=\alpha} \frac{\alpha!}{\alpha'!\alpha''!\alpha'''!}$$
$$K_1(\alpha''') K_2^l g_{m'r'|\alpha'|} g_{m''r''|\alpha''|} X_{\Lambda\Lambda_0} \tag{3.156}$$

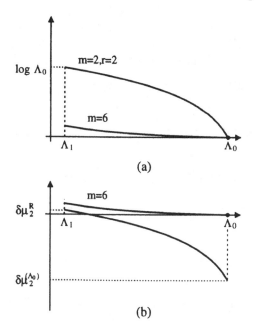

Fig. 3.6. Scaling bounds: (a) unrenormalized; (b) renormalized. The scaling behaviour is unchanged because it depends on the propagator. The boundary condition is changed under renormalization.

with

$$
\begin{aligned}
X_{\Lambda\Lambda_0} &= X_{\Lambda\Lambda_0}(m', m'', l, r', r'', |\alpha'|, |\alpha''|, |\alpha'''|) \\
&= \int_{\Lambda}^{\Lambda_0} ds \, s^{-3+l-|\alpha'''|} \psi_{m'r'|\alpha'|}(s)
\end{aligned}
\tag{3.157}
$$

where

$$
\psi_{mra}(s) = \begin{cases}
\log s & m = r = 2, \alpha = 0 \\
s^{-a} & m = r = 2, |\alpha| \geq 1 \\
s^{-1-a}(\log s)^{\delta_{a0}} & m = 2, r > 2 \\
s^{-1-a} & m = 4, r > 2 \\
s^{4-m-a} & m \geq 6.
\end{cases}
\tag{3.158}
$$

We can write this more compactly as

$$
\psi_{mra}(s) = s^{D(m)-a-\Gamma_{mr}}(\log s)^{\delta_{m2}\delta_{a0}}
\tag{3.159}
$$

where

$$
\Gamma_{mr} = \delta_{m2}(1 - \delta_{r2}) - \delta_{m4}(1 - \delta_{r1}).
\tag{3.160}
$$

It is clear from (3.156) that if we can show that $X_{\Lambda\Lambda_0}$ has the Λ-behaviour of (3.148), then the Theorem is proven because $g_{mr|\alpha|}$ can be determined

recursively. We verify this now by going through all cases of (m', m'', r', r'', l) that can occur in $\int d\kappa_{mr}$. Recall in particular that $m' + m'' = m + 2l + 2$, and that $r' + r'' = r \geq 3$, so one of r' and r'' has to be larger than 1.

1. Let $2 \notin \{m', m''\}$. By (3.159), the integrand for $X_{\Lambda\Lambda_0}$ is s^γ with

$$
\begin{aligned}
\gamma &= -3 + l - |\alpha'''| + D(m') - |\alpha'| + D(m'') - |\alpha| - \Gamma \\
&= -3 + l + D(m') + D(m'') - |\alpha| - \Gamma
\end{aligned} \tag{3.161}
$$

and where $\Gamma = \Gamma_{m'r'} + \Gamma_{m''r''} \geq 0$.
1.1 $m \geq 4$: By (3.142) (with $\delta = 1$)),

$$
\gamma \leq D(m) - 1 - l - |\alpha| - \Gamma = 3 - m - |\alpha| - l - \Gamma \tag{3.162}
$$

so

$$
X_{\Lambda\Lambda_0} \leq \int_\Lambda^{\Lambda_0} ds \, s^{3-m-|\alpha|-l-\Gamma}. \tag{3.163}
$$

For $m \geq 6$, we bound this using $l \geq 0$ and $\Gamma \geq 0$ by

$$
X_{\Lambda\Lambda_0} \leq \int_\Lambda^{\Lambda_0} ds \, s^{3-m-|\alpha|} \leq \frac{1}{m + |\alpha| - 4} \Lambda^{4-m-|\alpha|} \tag{3.164}
$$

so (3.148) holds. Let $m = 4$. Then $2l = m' + m'' - m - 2 \geq 4 + 4 - 4 - 2 = 2$, so $l \geq 1$, and

$$
X_{\Lambda\Lambda_0} \leq \int_\Lambda^{\Lambda_0} ds \, s^{3-4-1-|\alpha|} \leq \frac{1}{1 + |\alpha|} \Lambda^{-1-|\alpha|} \tag{3.165}
$$

also implies (3.148).
1.2 $m = 2$. By (3.141) (with $E(m) = 2, D(m) = 0$, and $\delta = 1$), the integrand is s^γ with

$$
\gamma = 1 - l - |\alpha| - E(m') - E(m'') - \Gamma. \tag{3.166}
$$

In the present case, we have assumed $m' \geq 4$ and $m'' \geq 4$, so $E(m') = E(m'') = 0$, and

$$
\gamma \leq 1 - |\alpha| - (\Gamma + l). \tag{3.167}
$$

Counting the numbers of external legs,

$$
2l = m' + m'' - 2 - 2 = m' + m'' - 4 \tag{3.168}
$$

so $l = 2$ if $m' = m'' = 4$, and $l \geq 3$ otherwise.
Now $r' + r'' = r \geq 3$, so $r' = 2$ or $r'' = 2$, which implies that for $m' = m'' = 4$, $\Gamma \leq 1$ must hold.

Thus, in all cases, $\Gamma + l \geq 3$, and therefore

$$
\begin{aligned}
X_{\Lambda\Lambda_0} &\leq \int_\Lambda^{\Lambda_0} ds\, s^{1-|\alpha|-(\Gamma+l)} \leq \int_\Lambda^{\Lambda_0} ds\, s^{-(2+|\alpha|)} \\
&\leq \frac{1}{1+|\alpha|} \Lambda^{-1-|\alpha|} \quad\quad\quad\quad (3.169)
\end{aligned}
$$

which proves (3.148) for all $|\alpha| \geq 0$, even without the logarithm.

2. $m' = 2$ or $m'' = 2$. Since everything is symmetric under exchange $(m', r', \alpha') \leftrightarrow (m'', r'', \alpha'')$, we may assume $m' = 2, m'' \geq 2$.
The sum over l in $\int d\kappa_{mr}$ runs from $l = 0$ to $l = \min\{m', m''\} - 1$, so $m' = 2$ implies $l \in \{0, 1\}$. Thus there are the cases

2.1 $m' = 2, m'' = m, l = 0$ (tree graph)
2.2 $m' = 2, m'' = m + 2, l = 1$ (one loop)

2.1 Using (3.159) and $\Gamma_{m'r'} \geq 0$ and $\Gamma_{m''r''} \geq 0$, we get

$$
X_{\Lambda\Lambda_0} \leq \int_\Lambda^{\Lambda_0} ds\, s^{-3-|\alpha|+D(m'')} (\log s)^\beta \quad\quad (3.170)
$$

where $\beta = 2$ if $m'' = 2$, and $\beta = 1$ if $m'' \geq 4$. If $m'' \geq 4$,

$$
\begin{aligned}
X_{\Lambda\Lambda_0} &\leq \int_\Lambda^{\Lambda_0} ds\, s^{1-m''-|\alpha|} \log s \\
&\leq \sup_{t\geq 1} \frac{\log t}{t} \int_\Lambda^{\Lambda_0} ds\, s^{2-m''-|\alpha|} \quad\quad (3.171) \\
&\leq \frac{1}{m''+|\alpha|-3} \Lambda^{3-m''-|\alpha|} \leq \Lambda^{D(m)-|\alpha|-1}
\end{aligned}
$$

If $m'' = 2$,

$$
\begin{aligned}
X_{\Lambda\Lambda_0} &\leq \int_\Lambda^{\Lambda_0} ds\, s^{-3-|\alpha|} (\log s)^2 \\
&\leq \sup_{t\geq 1} \frac{(\log t)^2}{t} \int_\Lambda^{\Lambda_0} ds\, s^{-2-|\alpha|} \leq 2\Lambda^{-1-|\alpha|}. \quad (3.172)
\end{aligned}
$$

Both these bounds are better than the one stated in (3.148). The next case is one from which one gets the logarithm.

2.2 Since $m'' = m + 2$, $m'' \geq 4$. The exponent of s is

$$-3 + 1 + D(m'') - |\alpha| - \Gamma = -2 + 4 - (m+2) - |\alpha| - \Gamma = -m - |\alpha| - \Gamma$$

so, since $\Gamma \geq 0$,

$$X_{\Lambda\Lambda_0} = \int_\Lambda^{\Lambda_0} ds\, s^{-m-|\alpha|} \log s. \tag{3.173}$$

For $m \geq 4$,

$$X_{\Lambda\Lambda_0} \leq \sup_{t \geq 1} \frac{\log t}{t} \int_\Lambda^{\Lambda_0} ds\, s^{1-m-|\alpha|} \tag{3.174}$$

$$\leq \frac{1}{m + |\alpha| - 2} \Lambda^{2-m-|\alpha|} \leq \frac{1}{m + |\alpha| - 2} \Lambda^{D(m)-|\alpha|-2}.$$

For $m = 2$, then $m'' = 4$. We distinguish between further cases. If $r'' \geq 2$, then $\Gamma \geq \Gamma_{42} = 1$, so

$$X_{\Lambda\Lambda_0} \leq \int_\Lambda^{\Lambda_0} ds\, s^{-3-|\alpha|} \log s \tag{3.175}$$

$$\leq \sup_{t \geq 1} \frac{\log t}{t} \int_\Lambda^{\Lambda_0} ds\, s^{-2-|\alpha|} \leq \Lambda^{-1-|\alpha|}.$$

If $r'' = 1$, the function $G_{m''r''}^{\Lambda\Lambda_0}$ is the bare vertex function, which does not depend on momentum at all. Moreover, no other factor of the integrand in this contribution to $Q_{2r}^{\Lambda\Lambda_0}$ depends on the external momentum to either - this is most easily seen in the graphical representation shown in Fig. 3.7. Thus any momentum derivative D^α of this function vanishes, and hence fulfills (3.148). So there remains only the case $r' = 1, \alpha = 0$. There we have

$$X_{\Lambda\Lambda_0} \leq \int_\Lambda^{\Lambda_0} ds\, s^{-2} \log s \leq \int_\Lambda^\infty ds\, s^{-2} \log s$$

$$= \frac{\log \Lambda}{\Lambda} + \int_\Lambda^\infty \frac{ds}{s^2} \leq 2 \frac{\log \Lambda}{\Lambda} \tag{3.176}$$

which is the bound stated in (3.148).

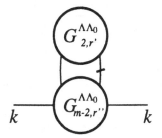

Fig. 3.7. A generalized tadpole graph

Remark 3.23. It is instructive to look how the logarithm from the second order graph proliferates to all orders: In third order, only $r' = 2$ makes such a logarithm in the graph drawn in Fig. 3.8(a). The value of this graph now reappears in the fourth order graph 3.8(b) etc. So the logarithms come from the so-called generalized tadpole graphs.

(a) (b)

Fig. 3.8. Some of the graphs producing the logarithms in $d = 3$. (a) Third order (b) Fourth order

3.8 Four Dimensions

In four dimensions, there is an infinite number of divergent values of graphs, not just one as in three dimensions, because the theory is just renormalizable. However, the scaling ansatzes to bound the G_{mr} will actually be a little simpler as in $d = 3$, and so the proofs will be shorter.

3.8.1 Counterterms in Second Order

For $d = 4$,

$$\|C_{0\Lambda}\|_1 \leq K_2 \Lambda^2 \qquad (3.177)$$

and the right hand side gives the correct behaviour of $\|C_{0\Lambda}\|_1$ at large Λ. In second order ($r = 2$) of unrenormalized perturbation theory, we thus have the following results, corresponding to the graphs in Fig. 3.9. For $m = 6$,

Fig. 3.9. Second order graphs

$$A_{2\Lambda,\eta}(Q_{62}^{\Lambda\Lambda_0}) \leq \frac{\text{const}}{\Lambda^3} \implies A_{2\Lambda,\eta}(G_{62}^{\Lambda\Lambda_0}) \leq \frac{\text{const}}{\Lambda^2}. \tag{3.178}$$

For $m = 4$,

$$Q_{42}^{\Lambda\Lambda_0} = \text{const} \int d^4p \, C_{0\Lambda}(p) \, \partial_\Lambda C_{\Lambda\Lambda_0}(p+q) \tag{3.179}$$

so

$$A_{2\Lambda,\eta}(Q_{42}^{\Lambda\Lambda_0}) \leq \text{const} \, K_2 \Lambda^2 \frac{K_1}{\Lambda^3} = \frac{\text{const}}{\Lambda} \tag{3.180}$$

and

$$A_{2\Lambda,\eta}(G_{42}^{\Lambda\Lambda_0}) \leq \text{const} \int_\Lambda^{\Lambda_0} \frac{ds}{s} = \text{const} \log \frac{\Lambda_0}{\Lambda}. \tag{3.181}$$

Similarly for $m = 2$,

$$A_{2\Lambda,\eta}(G_{22}^{\Lambda\Lambda_0}) \leq \int_\Lambda^{\Lambda_0} ds \, s^{-3} (s^2)^2 = \frac{1}{2}(\Lambda_0^2 - \Lambda^2). \tag{3.182}$$

Thus, in four dimensions, the four point function diverges like $\log \Lambda_0$ and the two-point function even like Λ_0^2 as $\Lambda_0 \to \infty$. With the same bounds applied to derivatives of these functions, we would get

$$A_{2\Lambda,\eta}(D^\alpha G_{22}^{\Lambda\Lambda_0}) \leq \text{const} \begin{cases} \Lambda_0 & \text{for } |\alpha| = 1 \\ \log \Lambda_0 & \text{for } |\alpha| = 2 \\ \frac{1}{\Lambda} & \text{for } |\alpha| \geq 3. \end{cases} \tag{3.183}$$

By analogy to the three-dimensional case, we may expect to need four types of counterterms – one for the four–point function, and three for the two–point function, to make $G_{22}^{\Lambda\Lambda_0}(0)$, as well as the first two derivatives of $G_{22}^{\Lambda\Lambda_0}$, converge as $\Lambda_0 \to \infty$.

However, if the initial interaction $V_{mr}^{(\Lambda_0)}$ is even in p, i.e. invariant under $p \to -p$, one needs only three counterterms because this symmetry poses automatically a renormalization condition on $G_{22}^{\Lambda\Lambda_0}$: since $C_{\Lambda\Lambda_0}(-p) = C_{\Lambda\Lambda_0}(p)$, $V_{mr}^{(\Lambda_0)}$ being differentiable and even in p implies

$$G_{2r}^{\Lambda\Lambda_0}(-p) = G_{2r}^{\Lambda\Lambda_0}(p) \tag{3.184}$$

and therefore

$$(\frac{\partial}{\partial p_\mu} G_{2r}^{\Lambda\Lambda_0})(0) = 0. \tag{3.185}$$

The usual Taylor expansion

$$\frac{\partial}{\partial p_\mu} G_{2r}^{\Lambda_1 \Lambda_0}(p) = (\frac{\partial}{\partial p_\mu} G_{2r}^{\Lambda_1 \Lambda_0})(0) + \sum_{\nu=1}^d p_\nu \int_0^1 dt \frac{\partial^2}{\partial p_\mu \partial p_\nu} G_{2r}^{\Lambda \Lambda_0}(tp) \quad (3.186)$$

then implies that once the logarithmic divergence in the second derivative of $G_{2r}^{\Lambda \Lambda_0}$ has been removed by renormalization, the first derivative is automatically finite, by (3.185) with $\Lambda = \Lambda_1$.

We shall start with an initial interaction that is rotation-invariant, i.e. for all $R \in SO(d)$,

$$V_{mr}^{(\Lambda_0)}(p_1, \ldots, p_{m-1}) = V_{mr}^{(\Lambda_0)}(Rp_1, \ldots, Rp_{m-1}). \quad (3.187)$$

This invariance, together with the invariance $C_{\Lambda \Lambda_0}(Rp) = C_{\Lambda \Lambda_0}(p)$ for all R, implies that there is a function $g : \mathbb{R}_0^+ \to \mathbb{R}$ such that

$$G_{2r}^{\Lambda \Lambda_0}(p) = g(p^2) \quad (3.188)$$

Differentiating this equation, we get

$$\frac{\partial}{\partial p_\mu} G_{2r}^{\Lambda \Lambda_0}(p) = 2p_\mu g'(p^2)$$

$$\frac{\partial^2}{\partial p_\mu \partial p_\nu} G_{2r}^{\Lambda \Lambda_0}(p) = 2\delta_{\mu\nu} g'(p^2) + 4p_\mu p_\nu g''(p^2) \quad (3.189)$$

which restricts the matrix structure of the second derivative of $G_{2r}^{\Lambda \Lambda_0}$ to

$$\frac{\partial^2}{\partial p_\mu \partial p_\nu} G_{2r}^{\Lambda \Lambda_0}(p) = \delta_{\mu\nu} g_1(p^2) + p_\mu p_\nu g_2(p^2). \quad (3.190)$$

In particular, we see that to get a finite second derivative at $p = 0$, it suffices to put a counterterm proportional to $\delta_{\mu\nu}$.

In summary, the second order analysis and symmetry considerations suggest that we need three counterterms, to make $G_{42}^{\Lambda_1 \Lambda_0}(0,0,0)$, $G_{22}^{\Lambda_1 \Lambda_0}(0)$ and $g_1(0)$ finite. In other words, we need terms of the form a constant times

$$\phi^4, \quad \phi^2, \quad \text{and} \quad p^2\phi^2 \quad (3.191)$$

to renormalize. These are exactly the terms that were in the original action.

3.8.2 Power Counting (Skeleton Flow)

Inserting (3.177) into (3.150), we get

$$A_{2\Lambda,\eta}(D^\alpha Q_{mr}^{\Lambda\Lambda_0}) \leq \int d\kappa_{mr} \sum_{\substack{\alpha'+\alpha'' \\ +\alpha'''=\alpha}} \frac{\alpha!}{\alpha'!\alpha''!\alpha'''!} K_2^l K_1(\alpha''') \Lambda^{-3+2l-|\alpha'''|}$$

$$A_{2\Lambda,\eta}(D^{\alpha'} G_{m'r'}^{\Lambda\Lambda_0}) A_{2\Lambda,\eta}(D^{\alpha''} G_{m''r''}^{\Lambda\Lambda_0}). \qquad (3.192)$$

Again, we consider a truncated flow obtained by a restriction on m' and m'' in the sum occuring in $\int d\kappa_{mr}$, and as before, we remove all those m' and m'' corresponding to the above divergences. That is, we leave out all terms with $m' = 2$ or $m'' = 2$, and those where $m' = 4$ and $r' > 1$ or where $m'' = 4$ and $r'' > 1$. This removes two- or four-legged proper (and nontrivial, i.e. consisting of more than one four-legged vertex) subgraphs from the graphical expansion. Denoting the solution to the truncated equation by $\overline{G}_{mr}^{\Lambda\Lambda_0}$, we make the ansatz that for all $m \geq 6$,

$$A_{2\Lambda,\eta}(D^\alpha \overline{G}_{mr}^{\Lambda\Lambda_0}) \leq g_{mr} \Lambda^{4-m-|\alpha|}. \qquad (3.193)$$

This bound is obvious for $r = 1$ and $m \geq 6$, since $\overline{G}_{m1}^{\Lambda\Lambda_0} = 0$ for those m. For $m = 4$ and $r = 1$, it holds because of the initial condition $V_{41}^{(\Lambda_0)} = -1$. Because of the truncation, other terms do not occur on the right hand side of the RG equation.

We now proceed inductively in r. Since $m' + m'' = m + 2l + 2$,

$$\int_\Lambda^{\Lambda_0} ds\, s^{-3+2l-|\alpha'''|} A_{2\Lambda,\eta}(D^{\alpha'} \overline{G}_{m'r'}^{s\Lambda_0}) A_{2\Lambda,\eta}(D^{\alpha''} \overline{G}_{m''r''}^{s\Lambda_0})$$

$$\leq g_{m'r'} g_{m''r''} \int_\Lambda^{\Lambda_0} ds\, s^{-3+2l+4-m'+4-m''-|\alpha|}$$

$$= g_{m'r'} g_{m''r''} \int_\Lambda^{\Lambda_0} ds\, s^{5-m-2-|\alpha|} \qquad (3.194)$$

$$\leq g_{m'r'} g_{m''r''} \begin{cases} \frac{\Lambda^{4-m-|\alpha|}}{m+|\alpha|-4} & m+|\alpha| \geq 5 \\ \log \Lambda_0 & m = 4, \alpha = 0 \quad \text{or} \quad m = 2, |\alpha| = 2 \\ \Lambda_0^{2-|\alpha|} & m = 2, |\alpha| \leq 1. \end{cases}$$

We have thus proven the following lemma.

Lemma 3.24. *In unrenormalized, truncated (skeleton) ϕ^4, i.e. with*

$$V_{mr}^{(\Lambda_0)}(p_1, p_2, p_3) = -\delta_{r1} \delta_{m4}$$

and with the truncation $m' \geq 6$ and $m'' \geq 6$ in $\int d\kappa_{mr}$, there are constants $g_{mr|\alpha|}$ such that

$$A_{2\Lambda,\eta}(D^\alpha \overline{G}_{mr}^{\Lambda\Lambda_0}) \leq g_{mr|\alpha|} \begin{cases} \Lambda^{4-m-|\alpha|} & m+|\alpha| \geq 5 \\ \log \Lambda_0 & m+|\alpha| = 4 \\ \Lambda_0^{2-|\alpha|} & m = 2, |\alpha| \leq 1. \end{cases} \qquad (3.195)$$

Remark 3.25. It is no contradiction to calculate the behaviour of \overline{G}_{4r} and omit all these terms on the right side of the RG equation. This is clear operationally, since this simply corresponds to a particular modification of the RG equation. In terms of the expansion in Feynman graphs, \overline{G}_{4r} is the order r contribution to the four-point function from all those graphs that do not contain two– or four–legged subgraphs. Thus, leaving out these terms on the right hand side only removes the four-legged subgraphs that contain more than one vertex, and thus its feedback for the other Green functions; this truncation does not make the four-point function zero.

Remark 3.26. In contrast to $d = 3$, there is no extra decay s^{-l} in the integral in (3.194), because $\|C_{0\Lambda}\|_1 \sim \Lambda^2$ produces a $2l$ in the exponent instead of an l. In other words, the power counting becomes independent of l. For $d \geq 5$, $\|C_{0\Lambda}\|_1 \sim \Lambda^{d-2}$, so there remains a factor $s^{(d-4)l}$ in the integral. The theory is then perturbatively non-renormalizable.

3.8.3 The Boundary Conditions for Renormalization

Recall that the original action of the model was

$$S = \int (\frac{1}{2}(\nabla\phi)^2 + \frac{m_\phi^2}{2}\phi^2 + \lambda\phi^4)d^4x \qquad (3.196)$$

and that we had put the quadratic terms into the Gaussian measure and the ϕ^4 term as the initial interaction $V_{mr}^{(\Lambda_0)} = -\delta_{r1}\delta_{m4}$ (corresponding to $e^{-\lambda \int \phi^4 dx}$). To incorporate the counterterms, we write an action

$$S_b = \int (\frac{Z}{2}(\nabla\phi)^2 + \frac{m_\phi^2 + \delta m_\phi^2}{2}\phi^2 - (\lambda + \delta\lambda)\phi^4)d^4x \qquad (3.197)$$

where $Z - 1$, δm_ϕ^2 and $\delta\lambda$ are the formal power series in λ, starting at order λ^2, i.e. , $r \geq 2$. In other words, we put

$$G_{mr}^{\Lambda_0\Lambda_0} = V_{mr}^{(\Lambda_0)} = 0 \qquad \text{if } m \geq 6 \text{ or } m \text{ is odd}$$

$$G_{4r}^{\Lambda_0\Lambda_0}(p_1,p_2,p_3) = V_{4r}^{(\Lambda_0)}(p_1,p_2,p_3) = \begin{cases} -1 & \text{if } r = 1 \\ c_r^{\Lambda_0} & \text{if } r \geq 2 \end{cases} \qquad (3.198)$$

$$G_{2r}^{\Lambda_0\Lambda_0}(p) = V_{2r}^{(\Lambda_0)}(p) = b_r^{\Lambda_0}p^2 + a_r^{\Lambda_0} \qquad r \geq 2.$$

This means that as formal power series,

$$\frac{\delta m_\phi^2}{2} = \sum_{r=2}^{\infty} b_r^{\Lambda_0} \lambda^r$$

$$\frac{Z-1}{2} = \sum_{r=2}^{\infty} a_r^{\Lambda_0} \lambda^r \qquad (3.199)$$

$$\delta\lambda = \sum_{r=2}^{\infty} c_r^{\Lambda_0} \lambda^r.$$

Although $Z - 1$ and δm_ϕ^2 multiply quadratic terms in ϕ, these terms are not put into the measure, but left as interaction terms. The formal series start at $r = 2$ because of Wick ordering. More explicitly, we are now expanding

$$\int d\mu_C(\phi) \, e^{-\lambda V^{(\Lambda_0)}(\phi+\psi)} \qquad (3.200)$$

in λ, where $V^{(\Lambda_0)}$ is the formal power series

$$V^{(\Lambda_0)} = \Omega_{C_{0,\Lambda_0}} \left(\int dx \left(\frac{1}{2}((Z-1)(\nabla\phi)^2 + \delta m_\phi^2 \phi^2) + (\lambda + \delta\lambda)\phi^4 \right) \right). \qquad (3.201)$$

To fix the counterterm, we impose renormalization conditions at $\Lambda = \Lambda_1$. They are

$$\begin{aligned}
G_{41}^{\Lambda_1 \Lambda_0}(0,0,0) &= -1 \\
G_{4r}^{\Lambda_1 \Lambda_0}(0,0,0) &= c_r^R & \forall r \geq 2 \\
G_{2r}^{\Lambda_1 \Lambda_0}(0) &= a_r^R & \forall r \geq 2 \qquad (3.202) \\
(\frac{\partial^2}{\partial p_\mu \partial p_\nu} G_{2r}^{\Lambda_1 \Lambda_0})(0) &= 2\delta_{\mu\nu} b_r^R & \forall r \geq 2.
\end{aligned}$$

The following lemma states that the boundary conditions (3.202) fix the counterterms uniquely to all orders in λ, provided the form of the action is fixed by (3.198).

Lemma 3.27. *Let $(a_r^R, b_r^R, c_r^R)_{r\geq 2}$ be a sequence in \mathbb{R}^3. There is exactly one sequence $(a_r^{\Lambda_0}, b_r^{\Lambda_0}, c_r^{\Lambda_0})_{r\geq 2}$ such that the effective action $G^{\Lambda_1 \Lambda_0}$ at scale Λ_1 obtained from the initial condition (3.198) has expansion coefficients $G_{mr}^{\Lambda_1 \Lambda_0}$ obeying (3.202).*

Proof. The proof is a trivial generalization of that of Lemma 3.17. ∎

Remark 3.28. By Wick ordering,

$$G_{41}^{\Lambda_1 \Lambda_0}(p_1, p_2, p_3) = -1 \iff G_{41}^{\Lambda_0 \Lambda_0}(p_1, p_2, p_3) = -1. \qquad (3.203)$$

The renormalization conditions can be thought of as being fixed by experiment. For instance, setting $c_r^R = 0$ $\forall r \geq 2$, i.e. , imposing

$$G_4^{\Lambda_1 \Lambda_0}(0,0,0) = \sum_{r=1}^{\infty} \lambda^r G_{4r}^{\Lambda_1 \Lambda_0}(0,0,0) = -\lambda \qquad (3.204)$$

means fixing the *renormalized coupling constant* λ by the four-point function at $(0,0,0)$ (which can be related to scattering amplitudes). The *bare coupling constant*

$$\lambda_b = \lambda + \delta\lambda = \lambda + \sum_{r=2}^{\infty} c_r^{\Lambda_0} \lambda^r \qquad (3.205)$$

will be seen to diverge for $\Lambda_0 \to \infty$. As we shall discuss later, one nice point about the RG equation is that it allows us to write all quantities in terms of renormalized parameters, so that the bare parameters never appear explicitly.

3.8.4 Renormalized ϕ^4 Theory

We define renormalized ϕ^4 theory as the sequence of effective actions $G_{mr}^{\Lambda \Lambda_0}$ in $d = 4$, satisfying

$$G_{mr}^{\Lambda_0 \Lambda_0} = V_{mr}^{(\Lambda_0)} = 0 \qquad \text{for} \quad m \geq 6 \quad \text{or} \quad m \quad \text{odd},$$

and satisfying the renormalization condition (3.202). By Lemma 3.27, the initial interaction $V_{mr}^{(\Lambda_0)}$ is fixed completely by this to be given by (3.198), and in particular, $V_{mr}^{(\Lambda_0)}$ is C^{∞} in all momentum variables, invariant under Euclidean rotations, and even in p. Consequently, the $G_{mr}^{\Lambda \Lambda_0}$ have the same properties for all $\Lambda \in [0, \Lambda_0]$.

Theorem 3.29. *In renormalized ϕ^4 theory, for any multiindex α, all $\eta \geq 1$, all $m \geq 1$ and $r \geq 1$, there are constants $g_{mr|\alpha|} > 0$ and $E_{mr} \in \mathbb{N}_0$ such that for all $\Lambda \geq \Lambda_1 \geq 3$,*

$$A_{2\Lambda,\eta}(D^\alpha G_{mr}^{\Lambda \Lambda_0}) \leq g_{mr|\alpha|} \Lambda^{4-m-|\alpha|} (\log \Lambda)^{E_{mr}}. \qquad (3.206)$$

Remark 3.30. This bound is far from optimal, as far as the coefficients and the powers of the logarithm are concerned. A better bound which tracks down the polynomial behaviour in $\log \Lambda$ in more detail is necessary for good bounds of the $g_{mr|\alpha|}$ as functions of r.

Proof. Recall $\partial_\Lambda D^\alpha G_{mr}^{\Lambda \Lambda_0} = \frac{1}{2} S_m D^\alpha Q_{mr}^{\Lambda \Lambda_0}$, (3.150), and (3.192). Again, we use induction in r; the case $r = 1$ is trivial.

$r \geq 2$: We insert the inductive hypothesis to get

$$A_{2\Lambda,\eta}(D^\alpha Q_{mr}^{\Lambda \Lambda_0}) \leq \sum \frac{\alpha!}{\alpha'!\alpha''!\alpha'''!} K(\alpha''') \sum_{(m',m'',l)} K_2^l g_{m'r'|\alpha'|} g_{m''r''|\alpha''|}$$
$$\Lambda^{2l-3-|\alpha'''|+4-m'+4-m''} (\log \Lambda)^{E_{m'r'}+E_{m''r''}}. \qquad (3.207)$$

Using $m' + m'' = m + 2l + 2$, $|\alpha'| + |\alpha''| + |\alpha'''| = |\alpha|$, we have

$$A_{2\Lambda,\eta}(D^\alpha Q_{mr}^{\Lambda\Lambda_0}) \le q_{mr|\alpha|} \Lambda^{3-m-|\alpha|} (\log \Lambda)^{\tilde{E}_{mr}} \qquad (3.208)$$

$$\tilde{E}_{mr} = \max_{\substack{m',r',m'',r'': \\ r'+r''=r}} (E_{m'r'} + E_{m''r''}).$$

"Irrelevant terms": $m + |\alpha| \ge 5$: Then $D^\alpha V_{mr}^{(\Lambda_0)} = 0$ (for $m = 2$, this holds because $D^\alpha p^2 = 0$ for $|\alpha| \ge 3$)

$$D^\alpha G_{mr}^{\Lambda\Lambda_0}(\underline{k}) = D^\alpha V_{mr}^{(\Lambda_0)}(\underline{k}) - \int_\Lambda^{\Lambda_0} ds \, D^\alpha Q_{mr}^{s\Lambda_0}(\underline{k}). \qquad (3.209)$$

The first term is zero. Applying $A_{2\Lambda,\eta}$, its monotonicity in Λ, and (3.208), we get

$$A_{2\Lambda,\eta}(D^\alpha G_{mr}^{\Lambda\Lambda_0}) \le q_{mr|\alpha|} \int_\Lambda^{\Lambda_0} ds \, s^{3-(m+|\alpha|)} (\log s)^{\tilde{E}_{mr}}. \qquad (3.210)$$

Since this converges for $\Lambda_0 \to \infty$, it is

$$\le q_{mr|\alpha|} \int_\Lambda^\infty ds \, s^{3-(m+|\alpha|)} (\log s)^{\tilde{E}_{mr}}. \qquad (3.211)$$

Using that for $w \ge 2$, $E \ge 0$

$$\int \Lambda^{-w} (\log \Lambda)^E \, d\Lambda = -E! \, \Lambda^{-(w-1)} \sum_{p=0}^E \frac{(\log \Lambda)^p}{p!} \frac{1}{(w-1)^{E+1-p}}, \qquad (3.212)$$

we have

$$A_{2\Lambda,\eta}(D^\alpha G_{mr}^{\Lambda\Lambda_0}) \qquad (3.213)$$

$$\le q_{mr|\alpha|} \tilde{E}_{mr}! \, \Lambda^{4-m-|\alpha|} \sum_{p=0}^{\tilde{E}_{mr}} \frac{(\log \Lambda)^p}{p!} \frac{1}{(m+|\alpha|-4)^{\tilde{E}_{mr}-p+1}}$$

$$\le \frac{q_{mr|\alpha|}e}{m+|\alpha|-4} (\log \Lambda)^{\tilde{E}_{mr}} \Lambda^{4-m-|\alpha|} \tilde{E}_{mr}! \,.$$

(the last inequality is in fact only a very crude bound, concerning the combinatorial behaviour. One has to retain the polynomial in the logs to get the Borel summability). It is now obvious that for $m + |\alpha| \ge 5$, the constants $g_{mr|\alpha|}$ and E_{mr} can be defined recursively. For $m = 4$, we define $E_{mr} = \tilde{E}_{mr} + 1$. For $m = 2$, we define $E_{2r} = \tilde{E}_{2r} + 1$.

Marginal and relevant terms: $m + |\alpha| \le 4$

1. $m = 4, \alpha = 0$. Because of

$$G_{4r}^{\Lambda\Lambda_0}(0,0,0) - G_{4r}^{\Lambda_1\Lambda_0}(0,0,0) = \int_{\Lambda_1}^{\Lambda} d\Lambda' \, \partial_{\Lambda'} G_{4r}^{\Lambda'\Lambda_0}(0,0,0) \qquad (3.214)$$

and the bound for $A_{2\Lambda,\eta}(Q_{mr}^{\Lambda\Lambda_0})$, we have

$$|G_{4r}^{\Lambda\Lambda_0}(0,0,0)| \leq |c_r^R| + \int_{\Lambda_1}^{\Lambda} ds \, q_{mr0} \, s^{3-4} (\log s)^{\tilde{E}_{4r}}$$

$$\leq |c_r^R| + (\log \Lambda)^{\tilde{E}_{4r}} q_{mr} \int_{\Lambda_1}^{\Lambda} \frac{ds}{s} \qquad (3.215)$$

$$\leq |c_r^R| + (\log \Lambda)^{\tilde{E}_{4r}+1} q_{mr0},$$

because $\int_{\Lambda_1}^{\Lambda} \frac{ds}{s} = \log \frac{\Lambda}{\Lambda_1} \leq \log \frac{4}{3} \leq \log \Lambda$. By Taylor expansion,

$$G_{4r}^{\Lambda\Lambda_0}(p_1, p_2, p_3) - G_{4r}^{\Lambda\Lambda_0}(0,0,0) = \qquad (3.216)$$

$$= \int_0^1 dt \, (p_1 \nabla_{p_1} + p_2 \nabla_{p_2} + p_3 \nabla_{p_3}) G_{4r}^{\Lambda\Lambda_0}(tp_1, tp_2, tp_3).$$

By (3.213), and since $A_{2\Lambda,\eta}(D^1 G_{4r}^{\Lambda\Lambda_0}) \leq g_{4r1}\Lambda^{4-4-1}(\log \Lambda)^{E_{4r}}$,

$$A_{2\Lambda,\eta}(G_{4r}^{\Lambda\Lambda_0}) \leq |c_r^R| + (\log \Lambda)^{\tilde{E}_{4r}+1}$$

$$+ \ 3\max\{2, \frac{\eta}{\Lambda}\} g_{4r1} (\log \Lambda)^{E_{4r}} \qquad (3.217)$$

$$\leq \ g_{mr} (\log \Lambda)^{E_{4r}}.$$

Note that E_{4r} was already fixed previously to be $\geq \tilde{E}_{4r} + 1$.

2. $m = 2, |\alpha| = 2$:

$$(\frac{\partial^2}{\partial p_\mu \partial p_\nu} G_{2r}^{\Lambda\Lambda_0})(0) = (\frac{\partial^2}{\partial p_\mu \partial p_\nu} G_{2r}^{\Lambda_1\Lambda_0})(0)$$

$$+ \int_{\Lambda_1}^{\Lambda} ds \, \frac{1}{2} \frac{\partial^2}{\partial p_\mu \partial p_\nu} Q_{2r}^{s\Lambda_0}(0) \qquad (3.218)$$

so

$$\left| \frac{\partial^2 G_{2r}^{\Lambda\Lambda_0}}{\partial p_\mu \partial p_\nu}(0) \right| \leq |2\delta_{\mu\nu} b_r^R| + \int_{\Lambda_1}^{\Lambda} ds \, \frac{1}{2} A_{2s,\eta}\left(\frac{\partial^2 Q_{2r}^{s\Lambda_0}}{\partial p_\mu \partial p_\nu} \right)$$

$$\leq |2b_r^R| + q_{2r2} \int_{\Lambda_1}^{\Lambda} ds \, s^{-3-2-2} (\log s)^{\tilde{E}_{2r}}$$

$$\leq |2b_r^R| + q_{2r2} (\log \Lambda)^{\tilde{E}_{2r}+1} \qquad (3.219)$$

(the rotational symmetry of G_2 is only essential if one poses the renormalization condition at some $\bar{p} \neq 0$; this can be done [31]). Thus

$$\frac{\partial^2}{\partial p_\mu \partial p_\nu} G_{2r}^{A A_0}(p) = \frac{\partial^2}{\partial p_\mu \partial p_\nu} G_{2r}^{A A_0}(0) + \int\limits_0^1 dt \, p_\xi \, \frac{\partial^3}{\partial_\mu \partial_\nu \partial_\xi} G_{2r}^{A A_0}(tp), \quad (3.220)$$

and

$$A_{2A,\eta}(D^2 G_{2r}^{A A_0}) \leq 2\delta_{\mu r}|b_r^R| + q_{2r2}(\log A)^{\bar{E}_{2r}+1}$$
$$+ \max\{2A, \eta\} \, A_{2A,\eta}(D^3 G_{2r}^{A A_0}) \quad (3.221)$$

Since $A_{2A,\eta}(D^3 G_{2r}^{A A_0}) \leq g_{2r3} A^{4-2-3} (\log A)^{E_{2r}}$, this is

$$\leq 2|b_r^R| + q_{2r2}(\log A)^{\bar{E}_{2r}+1} + \max\{2, \frac{\eta}{A}\} \, g_{2r3}(\log A)^{E_{2r}}. \quad (3.222)$$

3. $m = 2, |\alpha| = 1$: By the invariance of $V_{mr}^{(A_0)}$ under $p \to -p$, we have $(\frac{\partial}{\partial p_\mu} G_{2r}^{A A_0})(0) = 0$. So

$$\frac{\partial}{\partial p_\mu} G_{2r}^{A A_0}(p) = \int\limits_0^1 dt \, p_\nu \, \frac{\partial^2}{\partial p_\mu \partial p_\nu} G_{2r}^{A A_0}(tp) \quad (3.223)$$

and

$$\begin{aligned} A_{2A,\eta}(D^1 G_{2r}^{A A_0}) &\leq \max\{2A, \eta\} \, A_{2A,\eta}(D^2 G_{2r}^{A A_0}) \\ &\leq \max\{2A, \eta\} \, g_{2r2} \, A^{4-2-2} (\log A)^{E_{2r}} \quad (3.224) \\ &\leq g_{2r1} \, A \, (\log A)^{E_{2r}} \end{aligned}$$

where we have defined $g_{2r1} = \max\{2, \frac{\eta}{A_1}\} \, g_{2r2}$.

4. $m = 2, |\alpha| = 0$: The by-now standard procedure gives

$$G_{2r}^{A A_0}(0) = G_{2r}^{A_1 A_0}(0) + \int\limits_{A_1}^A ds \, \partial_A G_{2r}^{s A_0}(0) \quad (3.225)$$

$$\begin{aligned} |G_{2r}^{A A_0}(0)| &\leq |a_r^R| + \int\limits_{A_1}^A ds \, q_{2r0} \, s^{3-2}(\log s)^{E_{2r}} \\ &\leq |a_r^R| + q_{2r0} \, \frac{A^2}{2}(\log s)^{E_{2r}} \quad (3.226) \end{aligned}$$

$$G_{2r}^{A A_0}(p) = G_{2r}^{A A_0}(0) + \int\limits_0^1 dt \, p_\mu \, \frac{\partial}{\partial p_\mu} G_{2r}^{A A_0}(tp) \quad (3.227)$$

so

$$A_{2A,\eta}(G_{2r}^{AA_0}) \leq |G_{2r}^{AA_0}(0)| + \max\{2A, \eta\}\, A_{2A,\eta}(D^1 G_{2r}^{AA_0}). \tag{3.228}$$

Since $A_{2A,\eta}(D^1 G_{2r}^{AA_0}) \leq g_{2r1}\, A^{4-2-1}\, (\log A)^{E_{2r}}$,

$$A_{2A,\eta}(G_{2r}^{AA_0}) \leq |G_{2r}^{AA_0}(0)| + \max\{2, \frac{\eta}{A}\}\, g_{2r1}\, A^2\, (\log A)^{E_{2r}}. \tag{3.229}$$

Defining g_{2r0} suitably, we have proven the theorem.

■

Remark 3.31. 1. The proof is a little lengthy, but the arguments are all similar – a combination of power counting and Taylor expansion using the renormalization conditions.
2. Applying $\frac{\partial}{\partial A_0}$ to the RG equation, one can prove in a similar way that

$$A_{2A,\eta}\Big(\frac{\partial}{\partial A_0} G_{mr}^{AA_0}\Big) \leq h_{mr}\, \frac{(\log A_0)^{E_0}}{A_0^2} \tag{3.230}$$

so that the limit $A_0 \to \infty$ exists. This proof is a relatively straightforward generalization of the one given for $d = 2$ with application of renormalization conditions, so we omit it here.

3.9 The RG Flow in the Ladder Approximation

We have shown that ϕ^4 is renormalizable in perturbation theory, that is, that the renormalized Green functions exist as formal power series in the bare coupling λ. We have not discussed convergence of the series in λ. With the flow equation methods, one can show $r!$-bounds, so that the perturbation theory is locally Borel summable [33]. One should, actually, not expect convergence since even the one–dimensional integral $\int_{\mathbb{R}} e^{-\lambda x^4 - x^2}\, dx$ is not analytic in λ in a disk around zero (in fact, it diverges for $\mathrm{Re}\lambda < 0$). One may still ask if the perturbation series is Borel summable, i.e. if the Borel transform is maybe analytic in a strip around the real axis, not just in a disk around zero. This has been proven for $d \leq 3$, but in $d = 4$ it is believed not to be true. In the following, we shall give a heuristic argument why there is a problem. In the course of this argument, we shall see another aspect of renormalization, namely that of running coupling constants.

We have proven that the limit $A_0 \to \infty$ is well-defined as a formal power series. Now we study what a simple truncation can tell us about convergence. Instead of the exact flow, we consider an approximate flow. The first approximation we shall make is to discard the two-legged Green functions. Then the lowest terms contributing to $\partial_A G_{4r}^{AA_0}$ are those with $m' = m'' = 4$. The second approximation is to discard all $m' \geq 6$ (or $m'' \geq 6$) terms. This gives

$$\partial_\Lambda G_{4r}^{\Lambda\Lambda_0}(\underline{k}) = \tag{3.231}$$

$$\sum_{r'+r''=r} M \quad \mathbb{S}_4 \int d^d p \, C_{0\Lambda}(p) \partial_\Lambda C_{\Lambda\Lambda_0}(p+k_1+k_2)$$

$$G_{4r'}^{\Lambda\Lambda_0}(k_1,k_2,p) \, G_{4r''}^{\Lambda\Lambda_0}(-p,k_1+k_2+p,k_3)$$

(with a combinatorial constant M) so when summing up the series again and defining

$$G_4^{\Lambda\Lambda_0}(\lambda;\underline{k}) = \sum_{r=1}^{\infty} G_{4r}^{\Lambda\Lambda_0}(\underline{k}) \, \lambda^r \tag{3.232}$$

we get the integral equation

$$\partial_\Lambda G_4^{\Lambda\Lambda_0}(\lambda;\underline{k}) = M \int d^d p \, C_{0\Lambda}(p) \, \partial_\Lambda C_{\Lambda\Lambda_0}(p+k_1+k_2)$$

$$G_4^{\Lambda\Lambda_0}(\lambda;k_1,k_2,p) \, G_4^{\Lambda\Lambda_0}(\lambda;-p,k_1+k_2+p,k_3) \tag{3.233}$$

(see Fig. 3.10).

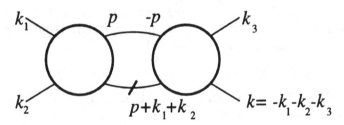

Fig. 3.10. The bubble graph

Recalling that

$$\partial_\Lambda C_{\Lambda\Lambda_0}(p) = -\partial_\Lambda C_{0\Lambda}(p) \tag{3.234}$$

and taking the limit $\Lambda_0 \to \infty$, we get

$$\partial_\Lambda G_4^\Lambda(\lambda;\underline{k}) = M \int d^d p \, C_{0\Lambda}(p)(-\partial_\Lambda C_{0\Lambda})(p+k_1+k_2)$$

$$G_4^\Lambda(\lambda;k_1,k_2,p) \, G_4^\Lambda(\lambda;-p,k_1+k_2+p,k_3). \tag{3.235}$$

For the boundary condition that we imposed,

$$G_4^{\Lambda_1}(\lambda;0,0,0) = -\lambda, \tag{3.236}$$

the coupling we expand in is now the renormalized coupling at scale Λ_1. (Note that by our definition, $v(\Lambda) > 0$, since we want to have convergent integrals, i.e. the ϕ^4 term has the "right" sign). If we had imposed the condition (3.236) at a different scale $\Lambda_2 > \Lambda_1$, we would have got the same Green functions, but expressed in terms of a different renormalized coupling. Note that in (3.235),

Λ_0 is removed: the RG equation relates only the Green functions at scales Λ, without any reference to the cutoff scale.

The semigroup property of the flow implies that

$$\lambda_2 = \lambda_1 + \sum_{r=2}^{\infty} \beta_r \lambda_1^r \tag{3.237}$$

where the β_r are given as sums over Feynman graphs with propagators $C_{\Lambda_1 \Lambda_2}$, as discussed. In other words, if we define a scale-dependent coupling $v(\Lambda)$ by defining $v(\Lambda_1)$ by (3.236) and setting

$$v(\Lambda) = -G_4^{\Lambda}(\lambda; 0, 0, 0), \tag{3.238}$$

and approximate the G_4^{Λ} in (3.235) by $v(\Lambda)$ (as suggested by the power counting behaviour of derivatives) then the relation between all these different choices of Λ is given by

$$\partial_{\Lambda} v(\Lambda) = \beta_2(\Lambda) \, v(\Lambda)^2. \tag{3.239}$$

where the coefficient

$$\beta_2(\Lambda) = M \int d^d p \, C_{0\Lambda}(p) \, \partial_{\Lambda} C_{0\Lambda}(p) \tag{3.240}$$

is the lowest order term in the *beta function* of the RG. Since

$$\beta_2(\Lambda) = M \int d^d p \, \frac{1}{(p^2 + m_\phi^2)^2} \, (-\frac{2p^2}{\Lambda^3}) \, K'(\frac{p^2}{\Lambda^2}) K(\frac{p^2}{\Lambda^2}), \tag{3.241}$$

we have (again for $\Lambda > m$)

$$\underline{\beta} \Lambda^{d-5} \le \beta_2(\Lambda) \le \overline{\beta} \Lambda^{d-5}. \tag{3.242}$$

So, to get the behaviour of $v(\Lambda)$, we need to solve

$$\partial_{\Lambda} v(\Lambda) = \beta \Lambda^{d-5} v(\Lambda)^2 \tag{3.243}$$

where β is a some constant. For $d < 4$, this integrates to

$$\frac{1}{v(\Lambda_1)} - \frac{1}{v(\Lambda_2)} = \tilde{\beta}(\Lambda_1^{d-4} - \Lambda_2^{d-4}) \tag{3.244}$$

with $\tilde{\beta} = \frac{\beta}{4-d}$, so

$$v(\Lambda_2) = \frac{v(\Lambda_1)}{1 - v(\Lambda_1)\tilde{\beta}(\Lambda_1^{d-4} - \Lambda_2^{d-4})}. \tag{3.245}$$

If at fixed Λ_1, $v(\Lambda_1)\tilde{\beta}\Lambda_1^{d-4} < 1$, the limit $\Lambda_2 \to \infty$ exists, and we have

$$\lim_{\Lambda \to \infty} v(\Lambda) = \frac{v(\Lambda_1)}{1 - v(\Lambda_1)\tilde{\beta}\Lambda_1^{d-4}} \tag{3.246}$$

(all this because $d - 4 < 0$).

However, for $d = 4$, the same calculation gives

$$\frac{1}{v(\Lambda_1)} - \frac{1}{v(\Lambda_2)} = \beta \log \frac{\Lambda_2}{\Lambda_1} \tag{3.247}$$

so

$$v(\Lambda_2) = \frac{v(\Lambda_1)}{1 - v(\Lambda_1)\beta \log \frac{\Lambda_2}{\Lambda_1}}. \tag{3.248}$$

The limit $\Lambda_2 \to \infty$ cannot be taken because $v(\Lambda_2)$ already has a pole at

$$\Lambda_2 = \Lambda_1 \, e^{\frac{1}{v(\Lambda_1)\beta}}, \tag{3.249}$$

the so-called Landau pole. This is an indication that the UV limit of the theory does not exist, since the scale-dependent coupling diverges at a finite scale already.

The approximations involved cannot easily be justified (one would like to argue that the terms we left out are irrelevant by power counting, but they aren't really if one looks at the details because the theory is just renormalizable). Moreover, when increasing Λ_2 in (3.248), one leaves the weak-coupling region where a perturbative analysis may be meaningful long before one gets to the singularity. The arguments for triviality known today are independent of, and go far beyond, the arguments given here [10, 11, 12, 13].

In the next chapter, we shall investigate a system where it can be proven by a more detailed "improved" power counting theorem that the ladder approximation indeed gives the leading contribution. The improvements to power counting arise there because the singularity is on a curved surface instead of at a point.

Had we started with the "wrong" sign of the coupling ($v(\Lambda) < 0$) we would have got asymptotic freedom. Similarly, the flow going into the infrared is asymptotically free: at fixed $v(\Lambda_2) > 0$,

$$v(\Lambda_1) = \frac{v(\Lambda_2)}{1 + v(\Lambda_2) \log \frac{\Lambda_2}{\Lambda_1}} \xrightarrow[\Lambda_1 \to 0]{} 0. \tag{3.250}$$

This behaviour is the basis for a nonperturbative construction [35, 36, 37] of infrared ϕ_4^4. There are several excellent accounts of nonperturbative renormalization with bosonic fields. Readers who want to continue in that direction should consult [7], [38, 39], or [40].

4. The Fermi Surface Problem

In this chapter, we consider an infrared problem with fermions, which is physically more realistic than the ultraviolet problem in ϕ^4 theory. In infrared problems, the perturbation expansion is singular because of the slow decay of the free propagator at large distances, not because of a short-distance singularity.

We shall again restrict rigorous proofs to perturbation theory to all orders. Nonperturbative bounds for these systems, given in $d = 1$ in [59, 60] and in $d = 2$ dimensions in [55, 56, 61], are still the subject of ongoing research and thus at the moment beyond the scope of an introductory text. Still, the perturbative analysis done here, which is based on [45, 46, 47] and [53], justifies certain resummations of perturbation theory, which we then perform in order to discuss the onset of superconductivity. It also gives a precise relation between the Fermi surface geometry and the power counting behaviour of the Green functions, and thus shows why these systems behave differently in one and more than one dimension. In particular, basic regularity properties of the selfenergy are proven, and their role in distinguishing between Fermi liquid and Luttinger liquid behaviour is discussed briefly.

4.1 Physical Motivation

The models that we shall analyze in the following describe electrons in a crystal, and we are particularly interested in the case where these systems show metallic behaviour.

We shall not investigate the reasons why crystals exist at all. Given a system of nuclei and electrons, it is a basic problem to show that, at sufficiently low temperatures, they arrange to ions situated at the sites of a periodic crystal and remaining electrons that move in the field of these ions. To this day, there is no rigorous quantum mechanical proof that the continuous translation symmetry gets broken to a discrete one at low temperatures by the formation of a crystal. Kennedy and Lieb have proved a discrete version of such a crystallization theorem for the Falicov-Kimball model, which is a lattice model. They showed that the symmetry under the lattice translation group remains unbroken at high temperatures but gets broken down to a smaller group when the temperature is low enough [41]. This is up to now the

only rigorous result concerning crystallization derived from first principles in quantum mechanics.

Once one assumes the existence of such a crystal as given, there is a standard way of obtaining a model Hamiltonian. In a first step, one usually neglects the (residual) interaction between those electrons that are not bound in the ions. The dynamics of such independent electrons in a static background of ions is given by a Schrödinger operator in a periodic potential. The effect of the periodic potential is that the relation between the (quasi)momentum p of an electron and its energy E is given by a number of band functions. For a mathematical periodic Schrödinger operator, the number of bands is infinite, but in a real crystal, only finitely many of them occur (if the energy of a particle gets very large, it will leave the crystal; if the temperature gets too large, the material will melt).

To this "free part" of the Hamiltonian, one adds the interaction between the electrons. This interaction may be a screened Coulomb interaction. If the crystal is not taken to be static, the electrons also interact with the lattice vibrations . In the harmonic approximation, this is described by phonon interactions. These phonon interactions are responsible for superconductivity in ordinary superconductors.

A popular, and in many cases realistic, approximation to the free Hamiltonian is the tight-binding approximation where, as a first approximation, one assumes the electrons to belong to a single atom as before, and then builds Bloch states from a superposition of these localized states in the natural way. One then gets a model with electrons that can occupy sites of a lattice, and hop from one site to another. Thus the kinetic energy takes the form of a hopping term for the electrons, with the hopping parameters given by overlap integrals [43]. The simplest such model is the Hubbard model, introduced about 40 years ago [42], in which the hopping is short-range and the potential is on-site. Many of the features of this model are still not completely understood, in particular in two dimensions.

4.2 Many-Fermion Systems on a Lattice

Again, we take a finite lattice as our starting point. In fact, the crystals of solid state physics are always finite, so this is the right starting point. However, the applications we are interested in involve macroscopic pieces of crystal, that is, sidelengths L which are huge when measured in terms of the lattice spacing. For such systems, it is the *correct idealization* to look at the thermodynamic limit $L \to \infty$. In the presence of symmetry breaking, some care is required in distinguishing between a large but finite and an infinite system. For the discussion of Fermi liquid properties, however, we may take the thermodynamic limit.

Let $\varepsilon > 0$ and let $L \in \mathbb{R}$ such that $\frac{L}{2\varepsilon} \in \mathbb{N}$ is a very large number, and let $\Lambda = \varepsilon \mathbb{Z}^d / L \mathbb{Z}^d$ be the cubic lattice of spacing ε and sidelength L. One need

not take a cubic lattice - any other lattice of maximal rank will do. Again, Λ is a torus, that is, we use periodic boundary conditions for convenience. For the questions considered below, the boundary conditions do not matter.

Let $\Gamma = \Lambda \times \{-1, 1\}$ and $\mathcal{H} = L^2(\Gamma, \mathbb{C})$, with the scalar product

$$\langle f | g \rangle = \int_\Gamma d\xi \, \overline{f(\xi)} g(\xi) = \varepsilon^d \sum_{\mathbf{x} \in \Lambda} \sum_{\sigma \in \{-1,1\}} \overline{f(\mathbf{x}, \sigma)} g(\mathbf{x}, \sigma). \qquad (4.1)$$

The variable σ labels the spin of the electron in units of $\hbar/2$. Since Γ is a finite set, \mathcal{H} is simply a set of sequences of finite length $|\Gamma|$, hence has dimension $|\Gamma|$.

The Hilbert space for a many-particle system is the fermionic Fock space $\mathcal{F}_\Gamma = \bigoplus_{n=0}^\infty \bigwedge^n \mathcal{H}$ (see Appendix B.1). Because \mathcal{H} is finite-dimensional, $\bigwedge^n \mathcal{H} = 0$ if $n > \dim \mathcal{H}$, so \mathcal{F}_Γ is finite-dimensional, too. For every vector f in the one-particle Hilbert space \mathcal{H}, there is a creation (annihilation) operator $a^*(f)$ $(a(f))$ acting on states of \mathcal{F}_Γ, which has the effect of adding (removing) a particle in state f. The map $f \mapsto a(f)$ is conjugate linear, and $f \mapsto a^*(f)$ is linear in f. The operators $a^{(*)}(f)$ obey the canonical anticommutation relations (CAR)

$$\{a(f), a^*(g)\} = a(f)a^*(g) + a^*(g)a(f) = \langle f \mid g \rangle. \qquad (4.2)$$

Let $(\chi_{\mathbf{x},\sigma})_{(\mathbf{x},\sigma) \in \Gamma}$ be the ONB of \mathcal{H} given by

$$\chi_{\mathbf{x},\sigma}(\mathbf{y}, \tau) = \varepsilon^{-d/2} \delta_{\mathbf{x},\mathbf{y}} \delta_{\sigma,\tau}. \qquad (4.3)$$

We denote the corresponding creation and annihilation operators by

$$a_{\mathbf{x},\sigma}^{(*)} = a^{(*)}(\chi_{\mathbf{x},\sigma}); \qquad (4.4)$$

they obey the CAR with the normalization $\{a_{\mathbf{x},\sigma}, a_{\mathbf{x}',\sigma'}^*\} = \delta_{\mathbf{x},\mathbf{x}'} \delta_{\sigma,\sigma'}$.

A basis better suited for the continuum limit $\varepsilon \to 0$ is $(\tilde{\chi}_{\mathbf{x},\sigma})_{(\mathbf{x},\sigma) \in \Gamma}$, given by

$$\tilde{\chi}_{\mathbf{x},\sigma}(\mathbf{y}, \tau) = \varepsilon^{-d} \delta_{\mathbf{x},\mathbf{y}} \delta_{\sigma,\tau}. \qquad (4.5)$$

The corresponding operators $\tilde{a}_{\mathbf{x},\sigma}^{(*)} = a^{(*)}(\tilde{\chi}_{\mathbf{x},\sigma})$ satisfy

$$\{a_{\mathbf{x},\sigma}, a_{\mathbf{x}',\sigma'}^*\} = \varepsilon^{-d} \delta_{\mathbf{x},\mathbf{x}'} \delta_{\sigma,\sigma'} = \delta_\Gamma(\mathbf{x}, \mathbf{x}') \delta_{\sigma,\sigma'}. \qquad (4.6)$$

4.2.1 The Hamiltonian

The Hamiltonian of the system is $H = H_0 + \lambda V$, where

$$H_0 = \int_\Gamma d\xi \int_\Gamma d\eta \, T(\xi, \eta) \tilde{a}_\xi^* \tilde{a}_\eta = \sum_{\mathbf{x},\mathbf{y},\sigma,\tau} \varepsilon^d T(\mathbf{x}, \sigma, \mathbf{y}, \tau) a_{\mathbf{x},\sigma}^* a_{\mathbf{y},\tau} \qquad (4.7)$$

is quadratic in the creation and annihilation operators, $\lambda \in \mathbb{R}$ is the coupling constant, and

$$V = \int_\Gamma d\xi \int_\Gamma d\eta \, V(\xi, \eta) \tilde{a}_\xi^* \tilde{a}_\xi \tilde{a}_\eta^* \tilde{a}_\eta = \sum_{\mathbf{x}, \mathbf{y}, \sigma, \tau} V(\mathbf{x}, \sigma, \mathbf{y}, \tau) a_{\mathbf{x}, \sigma}^* a_{\mathbf{x}, \sigma} a_{\mathbf{y}, \tau}^* a_{\mathbf{y}, \tau}$$

$$(4.8)$$

is the interaction.

The quadratic part of the Hamiltonian, H_0, describes hopping of electrons from position \mathbf{y}, spin τ, to position \mathbf{x}, spin σ, with an amplitude $\varepsilon^d T(\mathbf{x}, \sigma, \mathbf{y}, \tau)$. The condition

$$T(\mathbf{y}, \tau, \mathbf{x}, \sigma) = \overline{T(\mathbf{x}, \sigma, \mathbf{y}, \tau)} \tag{4.9}$$

makes H_0 hermitian. The interaction V couples the local densities $n_{\mathbf{x}, \sigma} = a_{\mathbf{x}, \sigma}^* a_{\mathbf{x}, \sigma}$ and $n_{\mathbf{y}, \tau} = a_{\mathbf{y}, \tau}^* a_{\mathbf{y}, \tau}$. Since, by the CAR, $[n_{\mathbf{x}, \sigma}, n_{\mathbf{y}, \tau}] = 0$ for all $\mathbf{x}, \mathbf{y}, \sigma$, and τ, we assume that $v(\mathbf{y}, \tau, \mathbf{x}, \sigma) = v(\mathbf{x}, \sigma, \mathbf{y}, \tau)$. To have V hermitian, we require that $v(\mathbf{x}, \sigma, \mathbf{y}, \tau)$ is real.

In the detailed analysis of the model, we shall only treat the case where both T and v are translation invariant, and where T is diagonal in, and v is independent of, the spin index. That is,

$$T(\mathbf{x}, \sigma, \mathbf{y}, \tau) = \delta_{\sigma, \tau} \Theta(\mathbf{x} - \mathbf{y}), \quad \Theta(-\mathbf{x}) = \overline{\Theta(\mathbf{x})} \tag{4.10}$$

and

$$V(\mathbf{x}, \sigma, \mathbf{y}, \tau) = v(\mathbf{x} - \mathbf{y}), \quad v(-\mathbf{x}) = v(\mathbf{x}) \in \mathbb{R}. \tag{4.11}$$

In this case, V can be written as

$$V = \sum_{\mathbf{x}, \mathbf{y}} n(\mathbf{x}) v(\mathbf{x} - \mathbf{y}) n(\mathbf{y}) \tag{4.12}$$

where $n(\mathbf{x}) = n_{\mathbf{x}, +} + n_{\mathbf{x}, -}$ is the charge density operator at \mathbf{x}. These assumptions imply that the model is translation invariant and that it has a *global spin rotation invariance*, i.e. H is invariant under the replacement $a(\mathbf{x}) \mapsto U a(\mathbf{x})$, where

$$a(\mathbf{x}) = \begin{pmatrix} a_{\mathbf{x}, -} \\ a_{\mathbf{x}, +} \end{pmatrix} \tag{4.13}$$

and U is an arbitrary SU(2) matrix. The symmetry is global because U is the same for all \mathbf{x}.

There are more complicated interactions that have this spin rotation invariance, for instance a spin-spin interaction $\sum_{\mathbf{x}, \mathbf{y}} w(\mathbf{x}, \mathbf{y}) \mathbf{S}(\mathbf{x}) \cdot \mathbf{S}(\mathbf{y})$, with $\mathbf{S}(\mathbf{x}) = a_{\mathbf{x}}^* \frac{\sigma}{2} a_{\mathbf{x}}$ the local spin operator (σ is the vector of Pauli matrices). These cases can be dealt with in a similar way as we do here; we restrict to density-density interactions for simplicity.

The discretized Laplacian corresponds to nearest-neighbour hopping,

$$T(\mathbf{x}, \sigma, \mathbf{y}, \tau) = \delta_{\sigma, \tau} \frac{1}{\varepsilon^2} \sum_{\nu=1}^d (2\delta_\Gamma(\mathbf{x}, \mathbf{y}) - \delta_\Gamma(\mathbf{x}, \mathbf{y} + \varepsilon \mathbf{e}_\nu) - \delta_\Gamma(\mathbf{x}, \mathbf{y} - \varepsilon \mathbf{e}_\nu))$$

$$(4.14)$$

with e_ν the unit vector in ν direction.

The model on the unit lattice ($\varepsilon = 1$) with $H = tH_0 + \frac{U}{2}V - (\frac{U}{2} + 2td)N$, H_0 given by the hopping amplitude (4.14), and $v(\mathbf{x}, \mathbf{y})$ vanishing for $\mathbf{x} \neq \mathbf{y}$ is the Hubbard model with nearest-neighbour hopping t and on-site interaction U. The subtraction of the term proportional to the number operator $N = \sum_{\mathbf{x}} n(\mathbf{x})$ implies that

$$H_{\text{hubb}} = -t \sum_{\substack{\mathbf{x}, \mathbf{y}, \sigma \\ |\mathbf{x}-\mathbf{y}|=1}} a^*_{\mathbf{x},\sigma} a_{\mathbf{y},\sigma} + U \sum_{\mathbf{x}} n_{\mathbf{x},+} n_{\mathbf{x},-}. \tag{4.15}$$

Terms proportional to N can be subtracted from H without essential changes because, as discussed below, N commutes with H.

With our assumptions on T, the quadratic part H_0 of the Hamiltonian is diagonal in Fourier space. To make this explicit, we introduce the ONB $(\varphi_{\mathbf{k},\sigma})_{(\mathbf{k},\sigma) \in \Gamma^*}$, where $\Gamma^* = \Lambda^* \times \{-1, 1\}$ with

$$\Lambda^* = \tfrac{2\pi}{L} \mathbb{Z}^d / \tfrac{2\pi}{\varepsilon} \mathbb{Z}^d, \tag{4.16}$$

given by

$$\varphi_{\mathbf{k},\sigma}(\mathbf{y}, \tau) = \delta_{\sigma,\tau} e^{i\mathbf{k}\mathbf{y}} (\varepsilon L)^{-d/2}, \tag{4.17}$$

and define

$$c^{(*)}_{\mathbf{k},\sigma} = a^{(*)}(\varphi_{\mathbf{k},\sigma}). \tag{4.18}$$

Then $\{c_{\mathbf{k},\sigma}, c^*_{\mathbf{k}',\sigma'}\} = \delta_{\mathbf{k},\mathbf{k}'} \delta_{\sigma,\sigma'}$.

For taking the thermodynamic limit, the basis consisting of $\tilde{\varphi}_{\mathbf{k},\sigma} = L^{d/2} \varphi_{\mathbf{k},\sigma}$ is more natural; the corresponding creation and annihilation operators obey

$$\{\tilde{c}_{\mathbf{k},\sigma}, \tilde{c}^*_{\mathbf{k}',\sigma'}\} = L^d \delta_{\mathbf{k},\mathbf{k}'} \delta_{\sigma,\sigma'} = \delta_{\Lambda^*}(\mathbf{k}, \mathbf{k}') \delta_{\sigma,\sigma'}. \tag{4.19}$$

By Fourier transformation,

$$\chi_{x,\sigma} = L^{-d/2} \sum_{\mathbf{k} \in \Lambda^*} e^{i\mathbf{k}\mathbf{x}} \varphi_{\mathbf{k},\sigma} \tag{4.20}$$

and, similarly, $\tilde{\chi}_{x,\sigma} = L^{-d} \sum_{\mathbf{k} \in \Lambda^*} e^{i\mathbf{k}\mathbf{x}} \tilde{\varphi}_{\mathbf{k},\sigma}$. By the (anti)linearity of the creation and annihilation operators, they obey the Fourier transformation formula (4.20), with $\chi_{x,\sigma}$ replaced by $a^*_{\mathbf{x},\sigma}$ and $\varphi_{\mathbf{k},\sigma}$ replaced by $c^*_{\mathbf{k},\sigma}$.

Under the assumption (4.10),

$$H_0 = \sum_{\mathbf{p},\sigma} \hat{\Theta}(\mathbf{p}) c^*_{\mathbf{p},\sigma} c_{\mathbf{p},\sigma} = L^{-d} \sum_{\mathbf{p},\sigma} \hat{\Theta}(\mathbf{p}) \tilde{c}^*_{\mathbf{p},\sigma} \tilde{c}_{\mathbf{p},\sigma} \tag{4.21}$$

where $\hat{\Theta}(\mathbf{p}) = \varepsilon^d \sum_{\mathbf{x}} \Theta(\mathbf{x}) e^{i\mathbf{p}\mathbf{x}}$ is real. The number operators $\hat{n}_{\mathbf{p},\sigma} = c^*_{\mathbf{p},\sigma} c_{\mathbf{p},\sigma}$ all commute, so they can be diagonalized simultaneously. The interpretation of H_0 in Fourier space is simply that the total energy is the sum of the one-particle energies, where a particle in state (\mathbf{p}, σ) gets the energy $\hat{\Theta}(\mathbf{p})$.

The discretized Laplacian has $\hat{\Theta}(\mathbf{p}) = \frac{2}{\varepsilon^2} \sum_\mu (1 - \cos(\varepsilon p_\mu))$, which has the limit \mathbf{p}^2 as $\varepsilon \to 0$.

4.2.2 The Grand Canonical Ensemble

The grand canonical ensemble corresponding to the Hamiltonian H, the inverse temperature β and the chemical potential μ is given by the linear functional $\langle \cdot \rangle$, where $\mathcal{L}(\mathcal{F}, \mathcal{F}) \to \mathbb{C}$, $\mathcal{O} \mapsto \langle \mathcal{O} \rangle$,

$$\langle \mathcal{O} \rangle = \frac{1}{Z_\Lambda} \operatorname{Tr}\left(e^{-\beta(H-\mu N)} \mathcal{O} \right), \tag{4.22}$$

with $Z_\Lambda = \operatorname{Tr} e^{-\beta(H-\mu N)}$, and N is the number operator. Because

$$N = \int d\xi\, \bar{a}_\xi^* \bar{a}_\xi = \sum_{\mathbf{x},\sigma} n_{\mathbf{x},\sigma} = \sum_{\mathbf{k},\sigma} \hat{n}_{\mathbf{k},\sigma}, \tag{4.23}$$

and because all the local number operators $n_{\mathbf{x},\sigma}$ commute, N commutes with V. By the Fourier representation of H_0, and because all $\hat{n}_{\mathbf{k},\sigma}$ commute, N also commutes with H_0. Thus $[H, N] = 0$. The expected particle number is given by

$$\langle N \rangle = \frac{1}{\beta}\frac{\partial}{\partial \mu} \log Z_\Lambda. \tag{4.24}$$

By (4.23), N is a sum of positive terms, so $\langle N \rangle \geq 0$ by the following lemma.

Lemma 4.1. *For all $\beta \geq 0$ and all $\mu \in \mathbb{R}$, Z_Λ is nonzero, so $\langle \cdot \rangle$ is well-defined. Moreover,*

$$\langle \mathcal{O} \rangle \leq \|\mathcal{O}\| \tag{4.25}$$

where $\|\mathcal{O}\|$ denotes the operator norm of \mathcal{O}, and for all linear operators B on \mathcal{F}_Γ,

$$\langle B^* B \rangle \geq 0. \tag{4.26}$$

Proof. \mathcal{F}_Γ has an ONB of eigenvectors η_α of $H - \mu N$. The eigenvalues ϵ_α of $H - \mu N$ are real because H and N are selfadjoint and μ is real. Thus

$$Z_\Lambda = \sum_\alpha \langle \eta_\alpha \mid e^{-\beta(H-\mu N)} \eta_\alpha \rangle = \sum_\alpha e^{-\beta\epsilon_\alpha} > 0, \tag{4.27}$$

$$Z_\Lambda |\langle \mathcal{O} \rangle| \leq \sum_\alpha e^{-\beta\epsilon_\alpha} |\langle \eta_\alpha \mid \mathcal{O}\eta_\alpha \rangle| \leq \sum_\alpha e^{-\beta\epsilon_\alpha} \|\mathcal{O}\| = Z_\Lambda \|\mathcal{O}\|, \tag{4.28}$$

and

$$Z_\Lambda \langle B^* B \rangle = \sum_\alpha e^{-\beta\epsilon_\alpha} |\langle \eta_\alpha \mid B^* B \eta_\alpha \rangle| = \sum_\alpha e^{-\beta\epsilon_\alpha} \|B\eta_\alpha\|^2 \geq 0. \tag{4.29}$$

∎

Corollary 4.2. *For all $\mu \in \mathbb{R}$, $\frac{\partial}{\partial \mu}\langle N \rangle \geq 0$.*

Proof.

$$\frac{\partial}{\partial\mu}\langle N\rangle = \frac{1}{\beta}\frac{\partial^2}{\partial\mu^2}\log Z_\Lambda = \frac{\partial}{\partial\mu}\frac{1}{Z_\Lambda}\,\mathrm{Tr}\left(e^{-\beta(H-\mu N)}N\right)$$

$$= -\beta\langle N\rangle^2 + \beta\langle N^2\rangle = \beta\langle(N-\langle N\rangle)^2\rangle \tag{4.30}$$

$$= \beta\langle(N-\langle N\rangle)^*(N-\langle N\rangle)\rangle \geq 0.$$

In the second line we used that $\langle \mathbb{1}\rangle = 1$. ∎

Thus $\langle N\rangle$ is an increasing function of μ. This is not unexpected because of the term $-\mu N$ that μ contributes to the energy.

Typical observables that we want to study are the correlation functions $\langle\prod a^*_{\mathbf{x},\sigma}\prod a_{\mathbf{y},\tau}\rangle$. In particular, we are interested in the behaviour of the two-point function $\langle a^*_{\mathbf{x},\sigma}a_{\mathbf{y},\tau}\rangle$.

By spin symmetry, we have $\langle a^*_{\mathbf{x},\sigma}a_{\mathbf{y},\tau}\rangle = \delta_{\sigma\tau}\langle a^*_{\mathbf{x},\sigma}a_{\mathbf{y},\sigma}\rangle$ and, by (4.18), (4.20), and the linearity of $\langle\,\cdot\,\rangle$,

$$\langle a^*_{\mathbf{x},\sigma}a_{\mathbf{y},\sigma}\rangle = \varepsilon^d\sum_{\mathbf{k}\in\Lambda^*}e^{i\mathbf{k}(\mathbf{x}-\mathbf{y})}\langle c^*_{\mathbf{k},\sigma}c_{\mathbf{k},\sigma}\rangle \tag{4.31}$$

because, by translation invariance, $\langle c^*_{\mathbf{k},\sigma}c_{\mathbf{k}',\sigma}\rangle = 0$ for $\mathbf{k}\neq\mathbf{k}'$. In terms of the other bases, the relation reads

$$\langle\tilde{a}^*_{\mathbf{x},\sigma}\tilde{a}_{\mathbf{y},\sigma}\rangle = L^{-d}\sum_{\mathbf{k}\in\Lambda^*}e^{i\mathbf{k}(\mathbf{x}-\mathbf{y})}\langle\tilde{c}^*_{\mathbf{k},\sigma}\tilde{c}_{\mathbf{k},\sigma}\rangle. \tag{4.32}$$

4.2.3 The Fermi Gas

Ultimately, we want to study the system with a weak interaction (λ small) by means of a perturbation expansion in λ. It is thus natural to look at the case of no interaction ($\lambda = 0$) first. For $\lambda = 0$, the fermions are independent ("free"), and calculating the partition function and $\langle c^*_{\mathbf{k},\sigma}c_{\mathbf{k},\sigma}\rangle$ is easy: we evaluate the trace for Z_Λ as a sum over the ONB $(\Phi_A)_{A\subset\Gamma^*}$ given by $\Phi_A = \prod_{(\mathbf{x},\sigma)\in A}c^*_{\mathbf{k},\sigma}\Omega$ (we fix an ordering on Γ^* and take the product in the ordering induced on A). Recall that $H_0 = \sum_{\mathbf{k},\sigma}\hat{\Theta}(\mathbf{k})\hat{n}_{\mathbf{k},\sigma}$ with $\hat{n}_{\mathbf{k},\sigma} = c^*_{\mathbf{k},\sigma}c_{\mathbf{k},\sigma}$. The $\hat{n}_{\mathbf{k},\sigma}$ fulfil the relations

$$[\hat{n}_{\mathbf{k},\sigma},\hat{n}_{\mathbf{k}',\sigma'}] = 0, \quad \hat{n}^2_{\mathbf{k},\sigma} = \hat{n}_{\mathbf{k},\sigma}, \quad [\hat{n}_{\mathbf{k},\sigma},c^*_{\mathbf{k}',\sigma'}] = \delta_{\mathbf{k}\mathbf{k}'}\delta_{\sigma,\sigma'}c^*_{\mathbf{k},\sigma}. \tag{4.33}$$

Thus

$$e^{-\beta(H_0-\mu N)} = \prod_{(\mathbf{k},\sigma)\in\Gamma^*}e^{-\beta E(\mathbf{k})\hat{n}_{\mathbf{k},\sigma}} \tag{4.34}$$

where

$$E(\mathbf{k}) = \hat{\Theta}(\mathbf{k}) - \mu. \tag{4.35}$$

For $(\mathbf{k},\sigma)\neq(\mathbf{k}',\sigma')$, $\hat{n}_{\mathbf{k},\sigma}$ and $c^*_{\mathbf{k}',\sigma'}$ commute, and $\hat{n}_{\mathbf{k},\sigma}\Omega = 0$, so

$$e^{-\beta(H_0-\mu N)}\Phi_A = \prod_{(\mathbf{k},\sigma)\in A} e^{-\beta E(\mathbf{k})\hat{n}_{\mathbf{k},\sigma}} c_{\mathbf{k},\sigma}^* \prod_{(\mathbf{k},\sigma)\notin A} e^{-\beta E(\mathbf{k})\hat{n}_{\mathbf{k},\sigma}} \Omega$$

$$= \prod_{(\mathbf{k},\sigma)\in A} e^{-\beta E(\mathbf{k})\hat{n}_{\mathbf{k},\sigma}} c_{\mathbf{k},\sigma}^* \Omega, \tag{4.36}$$

and all factors in the last product acting on Ω anticommute. An expansion of the exponential and (4.33) imply that

$$e^{-\beta E(\mathbf{k})\hat{n}_{\mathbf{k},\sigma}} = 1 + (e^{-\beta E(\mathbf{k})} - 1)\hat{n}_{\mathbf{k},\sigma}. \tag{4.37}$$

Therefore

$$e^{-\beta E(\mathbf{k})\hat{n}_{\mathbf{k},\sigma}} c_{\mathbf{k},\sigma}^* \Omega = c_{\mathbf{k},\sigma}^* \Omega + (e^{-\beta E(\mathbf{k})} - 1)\hat{n}_{\mathbf{k},\sigma} c_{\mathbf{k},\sigma}^* \Omega. \tag{4.38}$$

By (4.33), $n_{\mathbf{k},\sigma} c_{\mathbf{k},\sigma}^* \Omega = c_{\mathbf{k},\sigma}^* \Omega$, so

$$e^{-\beta(H_0-\mu N)}\Phi_A = \prod_{(\mathbf{k},\sigma)\in A} e^{-\beta E(\mathbf{k})}\Phi_A. \tag{4.39}$$

This equation has the simple interpretation that for independent electrons, the state Φ_A, which contains particles in the states (\mathbf{k},σ), for $(\mathbf{k},\sigma) \in A$, has total energy $\sum_{(\mathbf{k},\sigma)\in A} E(\mathbf{k})$. By the binomial theorem,

$$Z_\Lambda = \sum_{A\subset\Gamma^*} \langle\Phi_A \mid e^{-\beta(H_0-\mu N)}\Phi_A\rangle$$

$$= \sum_{A\subset\Gamma^*} \prod_{(\mathbf{k},\sigma)\in A} e^{-\beta E(\mathbf{k})} \tag{4.40}$$

$$= \prod_{(\mathbf{k},\sigma)\in\Gamma^*} \left(1 + e^{-\beta E(\mathbf{k})}\right).$$

The number density is calculated in the same way:

$$\langle\hat{n}_{\mathbf{p},\tau}\rangle = \frac{1}{Z_\Lambda} \operatorname{Tr}\left(e^{-\beta(H_0-\mu N)}\hat{n}_{\mathbf{p},\tau}\right) \tag{4.41}$$

so

$$Z_\Lambda \langle\hat{n}_{\mathbf{p},\tau}\rangle = \sum_{A\subset\Gamma^*} \langle\Phi_A \mid e^{-\beta(H_0-\mu N)}\hat{n}_{\mathbf{p},\tau}\Phi_A\rangle. \tag{4.42}$$

Since $\hat{n}_{\mathbf{p},\tau}\Omega = 0$, $\hat{n}_{\mathbf{p},\tau}\Phi_A = 0$ unless $(\mathbf{p},\tau) \in A$. If $(\mathbf{p},\tau) \in A$, then $\hat{n}_{\mathbf{p},\tau}\Phi_A = \Phi_A$. Thus

$$Z_\Lambda\langle\hat{n}_{\mathbf{p},\tau}\rangle = \sum_{\substack{A\subset\Gamma^* \\ (\mathbf{p},\tau)\in A}} \prod_{(\mathbf{k},\sigma)\in A} e^{-\beta E(\mathbf{k})}. \tag{4.43}$$

The constraint in the sum that (\mathbf{p},τ) must be in A can be enforced by a factor $\prod_{(\mathbf{k},\sigma)\notin A} \mathcal{N}_{\mathbf{p},\tau}(\mathbf{k},\sigma)$ in the summand, where

$$\mathcal{N}_{\mathbf{p},\tau}(\mathbf{k},\sigma) = \begin{cases} 0 & \text{if } \mathbf{k} = \mathbf{p} \text{ and } \sigma = \tau \\ 1 & \text{otherwise.} \end{cases} \tag{4.44}$$

Again, by the binomial theorem,

$$
\begin{aligned}
Z_\Lambda \langle \hat{n}_{\mathbf{p},\tau} \rangle &= \sum_{A \subset \Gamma^*} \prod_{(\mathbf{k},\sigma) \in A} e^{-\beta E(\mathbf{k})} \prod_{(\mathbf{k},\sigma) \notin A} \mathcal{N}_{\mathbf{p},\tau}(\mathbf{k},\sigma) \\
&= \prod_{(\mathbf{k},\sigma) \in \Gamma} \left(\mathcal{N}_{\mathbf{p},\tau}(\mathbf{k},\sigma) + e^{-\beta E(\mathbf{k})} \right) \qquad (4.45) \\
&= e^{-\beta E(\mathbf{p})} \prod_{\substack{(\mathbf{k},\sigma) \in \Gamma^* \\ (\mathbf{k},\sigma) \neq (\mathbf{p},\tau)}} \left(1 + e^{-\beta E(\mathbf{k})} \right).
\end{aligned}
$$

Dividing out Z_Λ, we get

$$
\langle \hat{n}_{\mathbf{p},\tau} \rangle = \frac{e^{-\beta E(\mathbf{p})}}{1 + e^{-\beta E(\mathbf{p})}} = f_\beta(E(\mathbf{p})) \qquad (4.46)
$$

where

$$
f_\beta(E) = \frac{1}{1 + e^{\beta E}} \qquad (4.47)
$$

is the *Fermi-Dirac distribution*. It has the zero-temperature limit

$$
\lim_{\beta \to \infty} f_\beta(E) = \begin{cases} 1 & \text{if } E < 0 \\ \frac{1}{2} & \text{if } E = 0 \\ 0 & \text{if } E > 0. \end{cases} \qquad (4.48)
$$

The interpretation of this is that in the Fermi gas at $T = 0$, all states with energy $E < 0$ are filled and all states with $E > 0$ are empty. The set of \mathbf{p} for which $E(\mathbf{p}) = 0$ is the surface of the occupied region, the *Fermi surface*

$$
\mathcal{S} = \{ \mathbf{p} \in \mathcal{B} : E(\mathbf{p}) = 0 \}. \qquad (4.49)
$$

Here $\mathcal{B} = \mathbb{R}^d / \frac{2\pi}{\varepsilon} \mathbb{Z}^d$ is the first Brillouin zone for the infinite lattice. On a finite lattice, also Λ^* is a finite set, so $\{ E(\mathbf{p}) : \mathbf{p} \in \Lambda^* \}$ is finite, and $\mathcal{S} \cap \Lambda^* \neq \emptyset$ can happen only for finitely many μ.

The discontinuity of the zero-temperature Fermi-Dirac distribution at the Fermi surface is the characteristic feature of the Fermi gas; it is a direct consequence of the Pauli principle: the energy of different particles simply adds up, no state can be occupied twice, and thus in the ground state, only the states with lowest possible energies are filled.

On the other hand, (4.31) implies that the position space two-point function is simply the inverse Fourier transform of the number density, so, in the infinite volume limit $L \to \infty$,

$$
\langle a^*_{\mathbf{x},\sigma} a_{\mathbf{y},\sigma} \rangle = \int_\mathcal{B} \frac{d^d \mathbf{k}}{(2\pi)^d} e^{i\mathbf{k}(\mathbf{x}-\mathbf{y})} f_\beta(E(\mathbf{k})). \qquad (4.50)
$$

The discontinuity of the Fermi distribution at $T = 0$ implies a slow decay of the two-point function in infinite volume – if the two-point function were

in L^1, its Fourier transform would be continuous by the Riemann-Lebesgue lemma.

As an explicit example, we calculate the two-point function for $d = 3$ in the continuum limit $\varepsilon \to 0$, where $\Theta(\mathbf{k}) = \mathbf{k}^2/2m$: in infinite volume and at zero temperature $(T \downarrow 0)$, with $k_F = \sqrt{2m\mu}$,

$$\langle a^*_{\mathbf{x},\sigma} a_{\mathbf{y},\sigma}\rangle = \int\limits_{|\mathbf{k}| \leq k_F} \frac{d^d k}{(2\pi)^d} e^{i\mathbf{k}(\mathbf{x}-\mathbf{y})}. \tag{4.51}$$

Because of rotational invariance, we may assume that $\mathbf{x} - \mathbf{y} = |\mathbf{x} - \mathbf{y}|\mathbf{e}_1$ and introduce polar coordinates. This gives

$$
\begin{aligned}
\langle a^*_{\mathbf{x},\sigma} a_{\mathbf{y},\sigma}\rangle &= 2\pi \int\limits_0^{k_F} r^2 dr \int\limits_0^{\pi} d\theta\, \sin\theta\, e^{ir|\mathbf{x}-\mathbf{y}|\cos\theta} \\
&= \frac{4\pi}{|\mathbf{x} - \mathbf{y}|} \int\limits_0^{k_F} r\, dr\, \sin(r|\mathbf{x} - \mathbf{y}|) \\
&= -\frac{4\pi k_F}{|\mathbf{x} - \mathbf{y}|^2} \cos(k_F|\mathbf{x} - \mathbf{y}|) + \frac{4\pi}{|\mathbf{x} - \mathbf{y}|^3} \sin(k_F|\mathbf{x} - \mathbf{y}|).
\end{aligned}
\tag{4.52}
$$

It is obvious from this form that the function is not in $L^1(\mathbb{R}^3)$. As we shall see, this slow decay leads to divergences in perturbation theory, very roughly speaking in a similar way as the singularity at coinciding points led to the ultraviolet singularities in the ultraviolet ϕ^4 problem. However, the treatment of these singularities differs from the previous one: we will not be able to argue that putting counterterms to cure the divergences only changes the parameters in the action at an unobservable cutoff scale. In infrared renormalization, the inclusion of the counterterms changes the model, and it requires a separate argument to show that things can be arranged such that one gets back a statement about the original model.

Note that at positive temperature, the Fermi distribution f_β is C^∞, so the two-point function decays exponentially in the distance. Although this decay is nonuniform in the temperature, and thus not useful for proving statements about the behaviour as a function of temperature, it means that in the model, there is a *natural physical cutoff*, namely positive temperature. Whenever one encounters an ill-defined quantity at zero temperature, the correct interpretation, if any, is to look at its positive-temperature counterpart and see if the limit exists. In fact, the physical definition of zero-temperature correlation functions is that one has to take the limit $T \to 0$ of the ones at positive temperature.

If one takes $\mathbf{x} = \mathbf{y}$ and sums over σ in (4.31), one gets the local number density $\langle n_{\mathbf{x}}\rangle$, which, by translation invariance, does not depend on \mathbf{x} and hence equals the density $\rho = 2L^{-d}\langle N\rangle$. By (4.50), we thus have that the

filling factor, defined as the density divided by the maximal possible density in the band (hence proportional to ρ), is, in the temperature zero limit, equal to the d-dimensional volume enclosed by the Fermi surface. For instance, in the Hubbard model on the square lattice, the filling factor is $\nu = L^{-d}\langle N \rangle/2$ because the maximal number of electrons at each site is 2 by the Pauli principle.

4.2.4 The Functional Integral Representation

We introduce the functional integral as a Grassmann integral, to set up a flow of effective actions similar to the bosonic case. The idea is to use the formula

$$\operatorname{Tr} e^{-\beta H} = \lim_{n \to \infty} \operatorname{Tr}\left((1 - \tfrac{\beta H}{n})^n\right) \tag{4.53}$$

and to rewrite the n^{th} power and the trace in terms of Grassmann integrals. On a finite lattice, H is just a matrix, so the right hand side of (4.53) converges. The functional integral is discussed in Appendix B.5; here we just state the result. The correlation functions can be obtained as derivatives with respect to source terms η of the limit $n_\tau \to \infty$ of the Grassmann integral

$$Z_{n_\tau, L}(\eta) = \int d\mu_C(\psi) e^{-\lambda \mathcal{V}(\psi) + (\eta, \psi)} \tag{4.54}$$

with the following specifications. The fermionic fields ψ are labelled by $X = (x, \sigma, j) \in \Gamma$, where Γ is the set $\Gamma = \Lambda \times \{-1, 1\} \times \{1, 2\}$, where $\Lambda = \mathbb{T} \times \Lambda$, and

$$\mathbb{T} = \{\tau_k = -\tfrac{\beta}{2} + \tfrac{k}{n_\tau}\beta : k \in \{0, \dots, n_\tau - 1\} \tag{4.55}$$

is a discretization of the interval $[-\beta/2, \beta/2)$. We require for convenience that n_τ is even. The variable τ is a discrete Euclidean time, σ is the spin index and j is a "charge" index that distinguishes between fields $\bar\psi$ and ψ that replace the creation and annihilation operators in the Grassmann integral, that is, for $x = (\tau, \mathbf{x}) \in \Lambda$,

$$\bar\psi_\sigma(x) = \psi(x, \sigma, 1) \quad \text{and} \quad \psi_\sigma(x) = \psi(x, \sigma, 2). \tag{4.56}$$

The fields ψ have antiperiodic boundary conditions on \mathbb{T}, that is, they are actually defined on a discrete torus of sidelength 2β, with the restriction

$$\psi(\tau + \beta, \mathbf{x}, \sigma, j) = -\psi(\tau, \mathbf{x}, \sigma, j). \tag{4.57}$$

One can rewrite the integral (4.54) for $Z_{n_\tau, L}(\eta)$ in terms of the product measure $\prod_{(x,\sigma)} d\bar\psi(x, \sigma) d\psi(x, \sigma)$ times the exponential of the action

$$S(\psi) = \int_{-\beta/2}^{\beta/2} d\tau \left[\langle \bar\psi(\tau), \partial_\tau \psi(\tau) \rangle_\Gamma - H_\Lambda(\bar\psi(\tau), \psi(\tau)) \right]. \tag{4.58}$$

Here $\langle \bar{\psi}(\tau), \partial_\tau \psi(\tau) \rangle_\Gamma = \int_\Gamma d\xi \bar{\psi}(\tau, \xi) \partial_\tau \psi(\tau, \xi)$, the notation $H_\Lambda(\bar{\psi}(\tau), \psi(\tau))$ means the normal ordered polynomial of H, with a^* replaced by $\bar{\psi}$ and a replaced by ψ. The integral over τ is shorthand for a sum $\varepsilon_\tau \sum_{\tau \in \mathbf{T}}$, with $\varepsilon_\tau = \beta/n_\tau$, and δ_τ is a discretized derivative with respect to τ. Details of this are discussed in Appendix B; here we shall only need the properties of the Gaussian measure and its covariance, which we summarize now.

Because the Grassmann variables anticommute, the Grassmann bilinear form

$$(\eta, \psi) = \int_\Gamma dX \eta(X) \psi(X) = \int_{\mathbf{T}} d\tau \int_\Lambda dx \sum_\sigma \sum_j \eta(\tau, \mathbf{x}, \sigma, j) \psi(\tau, \mathbf{x}, \sigma, j)$$

(4.59)

is antisymmetric: $(\eta, \psi) = -(\psi, \eta)$.

The 'Grassmann Gaussian measure' is not really a measure but just a continuous linear functional. It is determined by its characteristic function

$$\int d\mu_C(\psi) e^{(\eta, \psi)} = e^{\frac{1}{2}(\eta, C\eta)}.$$

(4.60)

This looks formally identical to bosonic Gaussian integrals, but the crucial difference is that the Grassmann bilinear form is antisymmetric. Accordingly, the propagator C is a skew symmetric matrix,

$$C(X, X') = -C(X', X)$$

(4.61)

given by

$$C(X, X') = \varepsilon_{jj'} \delta_{\sigma\sigma'} \int_{\Lambda^*} dp \, e^{ip(x-x')} \hat{C}\left((-1)^j p\right)$$

(4.62)

where $\varepsilon_{12} = -\varepsilon_{21} = 1$, $\varepsilon_{11} = \varepsilon_{22} = 0$, $\Lambda^* = \mathbf{M}_{n_\tau} \times \Lambda^*$, with

$$\mathbf{M}_{n_\tau} = \{\tfrac{\pi}{\beta}(2n+1) : -\tfrac{n_\tau}{2} \le n < \tfrac{n_\tau}{2}\}$$

(4.63)

is the dual to Λ, and for $p = (\omega, \mathbf{p}) \in \Lambda^*$,

$$\hat{C}(\omega, \mathbf{p}) = \frac{1}{i\hat{\omega} - E(\mathbf{p})},$$

(4.64)

with

$$\hat{\omega} = \tfrac{1}{i\varepsilon_\tau}(e^{i\varepsilon_\tau\omega} - 1),$$

(4.65)

and E the dispersion relation given by H_0 (see (4.35)). (4.61) can be seen by transforming the summation variable to $k = (-1)^j p$ in the integral. k also runs over Λ^*, and $p(x - x') = k((-1)^j x + (-1)^{j'} x')$ is symmetric under exchange of X and X'. The antisymmetry of C now follows from the antisymmetry of $\varepsilon_{jj'}$.

The interaction \mathcal{V} is the Grassmann polynomial

$$\mathcal{V}(\psi) = \int_\mathbb{T} d\tau \, V_A(\bar{\psi}(\tau), \psi(\tau)), \tag{4.66}$$

that is, $\bar{\psi}(\tau)$ replaces a^* and $\psi(\tau)$ replaces a.

In the limit $n_\tau \to \infty$, the time variable becomes continuous, $\hat{\omega} \to \omega$, and the set of Matsubara frequencies becomes $\mathbb{M} = \{\frac{\pi}{\beta}(2l+1) : l \in \mathbb{Z}\}$. The Grassmann integral then becomes infinite-dimensional. Although infinite-dimensional Grassmann algebras and Grassmann Gaussian integrals are well-defined objects, we shall take the limit $n_\tau \to \infty$ only at a point where we do not need to refer to Grassmann integrals any more. Thus no infinite-dimensional integrals are needed here. Since \mathbb{M} is an infinite set and because $\omega_l \propto l$, the sum $\sum 1/\omega$ diverges, and correct limit $n_\tau \to \infty$ of the propagator is the boundary value

$$\hat{C}(\omega, \mathbf{p}) = \frac{e^{i\omega 0^+}}{i\omega - E(\mathbf{p})}. \tag{4.67}$$

This can also be seen more directly – the Fermi distribution $(1 + e^{\beta E})^{-1}$ is periodic with period $2\pi/\beta$ in the imaginary part of E, and its poles are at $i\omega$, where $\omega \in \mathbb{M}$ are the Matsubara frequencies. Fourier transformation (see Appendix B.5.5) implies that

$$\frac{1}{1 + e^{\beta E}} = \lim_{\varepsilon \downarrow 0} \frac{1}{\beta} \sum_{\omega \in \mathbb{M}} \frac{e^{i\varepsilon \omega}}{i\omega - E}. \tag{4.68}$$

In the following, we study $\log Z_{n_\tau, L}$, and the effective action, by perturbation theory in λ. Perturbation theory for fermionic models is derived similarly to the bosonic case, but there are extra signs as compared to the bosonic case because of the anticommutation relations. It is proven in Appendix B.6 that these extra signs do not disrupt the polymer formulas, so that the theorems about connected functions extend to fermions. As mentioned, the singularity of the propagator in momentum space, related to the slow decay in position space, makes renormalization necessary. We introduce a renormalization flow in a way similar to the bosonic case.

4.2.5 RG Flow: Energy Scales

The infrared divergences are caused by the singularity of \hat{C} for $\omega = 0$ and $\mathbf{p} \in S$. Therefore we introduce the renormalization group flow by a cutoff that forbids small energies. We label energy scales by a parameter $t \geq 0$, where the energy scale associated to t,

$$\epsilon_t = \epsilon_0 e^{-t}, \tag{4.69}$$

is a decreasing function of t. ϵ_0 is a fixed energy scale, which, however, depends on the filling region one wants to study (this is discussed in Sect. 4.2.8). Since the cutoff was not present in the original model, we want to take the limit

$t \to \infty$ after rearranging everything suitably by renormalization. The cutoff is put in by replacing the propagator C by the t-dependent cutoff propagator C_t, which has the Fourier transform

$$\tilde{C}_t(k) = \frac{1}{i\hat{\omega} - E(\mathbf{k})} \, \chi_2\left(\epsilon_t^{-2}|i\hat{\omega} - E(\mathbf{k})|^2\right). \tag{4.70}$$

Here χ_1 and χ_2 are a partition of unity, $\chi_1 + \chi_2 = 1$, $\chi_i \in C^\infty(\mathbb{R}_0^+, [0,1])$, with

$$\chi_1(x) = \begin{cases} 1 & \text{if } x \le \frac{1}{4} \\ 0 & \text{if } x \ge 1, \end{cases} \tag{4.71}$$

$\chi_1'(x) < 0$ for all $x \in (\frac{1}{4}, 1)$, and $\|\chi_1'\|_\infty \le 2$. We shall also use a Wick ordering covariance D_t which is defined similarly, only that χ_2 is replaced by χ_1 in (4.70). We define the effective action at $t \ge 0$ as

$$\mathcal{G}(t, \psi) = \log \int d\mu_{C_t}(\chi) \, e^{\mathcal{G}(0, \chi + \psi)}. \tag{4.72}$$

4.2.6 Model Assumptions

We first state the assumptions and then discuss their significance and give some examples of models where they are fulfilled.

Let \mathcal{B} be the dual of the infinite-volume limit of Λ^*; for the cubic lattice $\Lambda = \epsilon\mathbb{Z}^d$, \mathcal{B} is the torus $\mathcal{B} = \mathbb{R}^d / \frac{2\pi}{\epsilon}\mathbb{Z}^d$.

1. *Assumptions on the dispersion relation and the Fermi surface:* There is $k_0 \ge 2$ such that $E \in C^{k_0}(\mathcal{B}, \mathbb{R})$, and for all $\mathbf{p} \in \mathcal{B}$, $E(-\mathbf{p}) = E(\mathbf{p})$ holds. The Fermi surface $\mathcal{S} = \{\mathbf{p} \in \mathcal{B} : E(\mathbf{p}) = 0\}$ has a nonempty interior $\mathcal{I} = \{\mathbf{p} \in \mathcal{B} : E(\mathbf{p}) < 0\}$, and 0 is at the center of \mathcal{I}. There is $g_0 > 0$ such that for all \mathbf{p} on \mathcal{S}, $|\nabla E(\mathbf{p})| \ge g_0$ holds. The Fermi surface is a subset of an ϵ-independent bounded region of momentum space (hence \mathcal{S} is compact). \mathcal{S} has positive curvature everywhere (thus \mathcal{I} and $\mathcal{I} \cup \mathcal{S}$ are strictly convex sets). There is $\mathcal{V}_1 > 0$ such that for all L and ϵ

$$\int_{\Lambda^*} d\mathbf{k} \, \mathbf{1}\left(|E(\mathbf{k})| \le 2\right) \le \mathcal{V}_1. \tag{4.73}$$

 The constant

$$E_{\max} = \sup_{\mathbf{p} \in \mathcal{B}} |E(\mathbf{p})| \tag{4.74}$$

 is independent of ϵ.

2. *Assumptions on the interaction:* The interaction \hat{v} is a C^{k_0} function from $\mathbb{R} \times \mathcal{B}$ to \mathbb{C}, all its derivatives up to order k_0 are bounded functions on $\mathcal{B} \times \mathbb{R}$, $\hat{v}(-p_0, \mathbf{p}) = \overline{\hat{v}(p_0, \mathbf{p})}$, and the limit $p_0 \to \infty$ of \hat{v} exists and is C^{k_0} in \mathbf{p}.

A priori, the statement that E is symmetric under $\mathbf{p} \to -\mathbf{p}$ does not look sensible on a torus. By the above we mean that there is a point $\mathbf{p}^{(0)}$ such that for all $\mathbf{p} \in B$, $E(R_0\mathbf{p}) = E(\mathbf{p})$, where R_0 denotes the point reflection at $\mathbf{p}^{(0)}$. We then choose coordinates such that $\mathbf{p}^{(0)} = 0$, then the reflection symmetry condition simply reads $E(\mathbf{p}) = E(-\mathbf{p})$.

The above hypotheses allow us to introduce natural coordinates in a neighbourhood of the Fermi surface: there is $r_0 > 0$ and a C^2-diffeomorphism π from $(-2r_0, 2r_0) \times S^{d-1}$ to an open neighbourhood of the Fermi surface S in B, $(\rho, \theta) \mapsto \pi(\rho, \theta)$, such that

$$E(\pi(\rho, \theta)) = \rho \quad \text{and} \quad |\partial_\rho \pi(\rho, \theta)| \leq \frac{2}{g_0}. \tag{4.75}$$

The coordinates ρ and θ are easily constructed by first going to polar coordinates (r, θ) and then making a change of variables from r to $\rho = E(\mathbf{p}(r, \theta))$. The latter is regular by our assumptions on the Fermi surface. Coordinates (ρ, θ) satisying (4.75) exist under much weaker assumptions on S (see [45], Lemma 2.1, and [46], Sect. 2.2). Let $J(\rho, \theta) = |\det \pi'(\rho, \theta)|$ be the Jacobian of this change of variables and denote

$$J_0 = \sup_{\substack{|\rho| \leq \epsilon_0 \\ \theta \in S^{d-1}}} J(\rho, \theta) \quad \text{and} \quad J_1 = \sup_{|\rho| \leq \epsilon_0} \int_{S^{d-1}} d\theta\, J(\rho, \theta). \tag{4.76}$$

The density of states $N(E)$ in infinite volume is defined by

$$N(E) = \int_B \frac{d\mathbf{p}}{(2\pi)^d} \delta(E - E(\mathbf{p})). \tag{4.77}$$

With the above change of variables we see that

$$N(E) = \frac{1}{(2\pi)^d} \int_{S^{d-1}} d\theta\, J(E, \theta). \tag{4.78}$$

Thus the Jacobian plays the role of a local density of states, and its integral over θ is the density of states at energy E. So, under our assumptions, the density of states is a finite C^{k_0-1} function of the energy E in a fixed energy interval $(-r_0, r_0)$ around the Fermi surface $(E = 0)$.

The energy ϵ_0 that sets the initial scale of the RG flow will be chosen smaller than or equal to r_0; we shall discuss the role of ϵ_0 below.

We can now also be more precise about the local integrability properties of the propagator in the limit $T \to 0$. With the coordinates ρ and θ, we see that

$$I_\alpha = \int_{|ip_0 - E(\mathbf{p})| \leq \epsilon} \frac{d^{d+1}p}{(2\pi)^{d+1}} \frac{1}{|ip_0 - E(\mathbf{p})|^\alpha} = \int \frac{dp_0}{2\pi} \int dE \frac{N(E)}{|ip_0 - E|^\alpha}. \tag{4.79}$$

Introducing polar coordinates in the (E, p_0) plane, we see that

$$I_\alpha = \int\limits_0^\varepsilon \frac{r\,\mathrm{d}r}{r^\alpha}\nu(r) \tag{4.80}$$

with $\nu(r) = \int_0^{2\pi} \frac{\mathrm{d}\theta}{2\pi} N(E\cos\theta)$. Since $\nu(0) \neq 0$, I_α diverges for all $\alpha \geq 2$. The problems with unrenormalized perturbation theory come from graphs where, because of many two-legged insertions, high powers of C get integrated [44]. These contributions diverge for $T \to 0$.

4.2.7 The Physical Significance of the Assumptions

The assumptions on the interaction mean that it has to decay fast enough in position space. It need not decay exponentially fast; a power law decay with a large enough power suffices to make all derivatives of \hat{v} to order k_0 bounded. Thus the analysis applies to the Hubbard model and all its generalizations obtained from making the interaction spread out a bit. It also applies to screened Coulomb interactions, but not to the Coulomb interaction itself, which has Fourier transform $1/\mathbf{k}^2$. Screening in Coulomb systems is in general not exponential. Another interesting case is that of interactions with acoustical phonons, which are p_0-dependent and of the form

$$\hat{v}(p_0, \mathbf{p}) = \frac{c^2\mathbf{p}^2}{p_0^2 + c^2\mathbf{p}^2}, \tag{4.81}$$

where c, the velocity of sound, is a small constant. The \hat{v} given by (4.81) is bounded, but its derivatives are not. Treating such interactions requires a slight generalization of the analysis done here, which is, however, not very difficult. The condition $\hat{v}(-p_0, \mathbf{p}) = \overline{\hat{v}(p_0, \mathbf{p})}$ implies that the two-point function $I_2(\omega, \mathbf{p})$ (defined below) is real for $\omega = 0$.

The condition that E is C^{k_0} is fulfilled for the band functions arising from Schrödinger operators with L^2 potentials if the levels do not cross. It is fulfilled in lattice models if the hopping amplitudes decay sufficiently fast with the distance. For lattice models with finite-range hopping, the dispersion relation is even analytic in \mathbf{p}. However, the dispersion relation itself gets changed by selfenergy effects, so one needs to do the analysis with assumptions that are as weak as possible.

The geometric conditions on the Fermi surface S impose a restriction on the filling factor $\nu = \langle N\rangle/(2L^d)$. They hold in all cases where the band function has a nondegenerate minimum and where the Fermi energy μ is just above this minimum of the band. In that case, the interior of the Fermi surface is strictly convex because the level sets of a strictly convex function are strictly convex. If the quadratic form at the minimum is positive definite, the Fermi surface has positive curvature if μ is just above the minimum.

A typical example is the discretized Laplacian

$$E(\mathbf{k}) = t\frac{2}{\varepsilon^2}\sum_{\nu=1}^{d}(1 - \cos(\varepsilon k_\nu)) - \mu. \qquad (4.82)$$

For $\varepsilon \to 0$ and $t = \frac{1}{2m}$, m the electron mass, $E(\mathbf{k}) \to \mathbf{k}^2/2m - \mu$, the dispersion relation of the jellium model, which satisfies the above hypotheses for all $\mu \neq 0$ if large $|\mathbf{k}|$ are cut off. For $\varepsilon = 1$, one gets the tight-binding dispersion relation with hopping parameter t, which satisfies the above hypotheses if $\mu \neq 2td$ (half-filling). The bounds on the curvature κ of the Fermi surface and the lower bound g_0 for the gradient of E depend on μ.

The Fermi surfaces of real metals can be much more complicated than the ones discussed here [43]. Part of the analysis can be extended to rather general Fermi surfaces (not necessarily convex or connected) [45].

4.2.8 The Role of the Initial Energy Scale

We assume that $\epsilon_0 \leq 1$ and (for convenience in stating some bounds) that β is so large that $\beta\epsilon_0 = \frac{\epsilon_0}{k_B T} \geq 6$. With the units chosen in a natural way, i.e., with typical bandwidths of electron volts, this corresponds to temperatures T up to 1000 Kelvin if ϵ_0 is of order one, which seems a sufficient temperature range to study conduction in crystals. However, ϵ_0 is not an arbitrary parameter, and, as discussed in the following, it may need to be chosen rather small.

If C_0 were equal to zero, the renormalization group flow for the effective action $\mathcal{G}(t)$ defined in (4.72) would have the initial condition $\mathcal{G}(0) = -\lambda V$. However, with our definition of C_t, $C_0 \neq 0$, so $\mathcal{G}(0)$ differs from the original interaction. It already contains the integral over fields with momenta away from the Fermi surface. The focus on the fields with momenta near to S was motivated by the observation in (4.80) that singularities only come from the singularity on the Fermi surface. More generally, for the behaviour at long distances, only the behaviour near to S, i.e. at low energies is important. The vagueness of the notions "long distance" and "low energies" is evident. For a complete analysis of the system, it is important to be more precise about these issues – energy scales correspond to temperature scales, and values of the coupling constant for which a certain behaviour is proven also depends on these energy scales via constants.

These issues are, in the present setup, connected to the choice of the initial energy scale ϵ_0. The above assumptions on S and on the behaviour of E on S are stable in the sense that the same conditions hold for the level sets $S_\epsilon = \{\mathbf{p} \in \mathcal{B} : E(\mathbf{p}) = \epsilon\}$, if $\epsilon \leq \epsilon_0$ and ϵ_0 is chosen small enough (depending on r_0 and g_0). The renormalization group flow for $t \geq 0$ takes place at energies $\epsilon_t \leq \epsilon_0$. Although the integration over fields with energies $\geq \epsilon_0$ cannot produce any singularities, it does determine the initial condition $\mathcal{G}(0)$ for the flow. This initial interaction may differ substantially from the original interaction of the model. For instance, even if the original interaction is repulsive, $\mathcal{G}(0)$ may have attractive parts and therefore induce superconductivity

at low temperatures. This is the basis for the Kohn-Luttinger effect; see the discussion in Sect. 4.5.4.

If one gets near to half-filling in the Hubbard model, the scale ϵ_0 becomes smaller and smaller, so that the analysis done here applies only in lower and lower energy ranges, and the initial condition becomes more and more important. An analysis of the effects of approaching $\mu = 0$ is possible, but will not be given here. Some heuristic remarks will follow in Sect. 4.8.

4.3 The Renormalization Group Differential Equation

In this section we derive the RGDE for fermionic models. The special setup of the previous sections is not necessary: the RG equation can be formulated more generally on any finite-dimensional Grassmann algebra.

4.3.1 The Effective Action

Let Γ be a finite set, for a function $X \mapsto F(X)$ from Γ to any linear space let $\int_\Gamma dX\, F(X) = \epsilon_\Gamma \sum_{X \in \Gamma} F(X)$, where $\epsilon_\Gamma > 0$ is a constant, let $\delta(X, X') = \epsilon_\Gamma^{-1} \delta_{XX'}$, and define the bilinear form $(f, g) = \int_\Gamma dX\, f(X)g(X)$. Let \mathcal{A} be the finite-dimensional Grassmann algebra generated by the generators $(\psi(X), \chi(X), \eta(X))_{X \in \Gamma}$ and let $\frac{\delta}{\delta\psi(X)} = \epsilon^{-1} \frac{\partial}{\partial\psi(X)}$; recall that the fermionic derivatives anticommute.

For $t \geq 0$ let C_t be an invertible, antisymmetric linear operator acting on functions defined on Γ, i.e. $(C_t f)(X) = \int_\Gamma dX C_t(X, X')f(X')$ with

$$C_t(X', X) = -C_t(X, X'). \tag{4.83}$$

Let C_t be continuously differentiable in t; denote $\frac{\partial C_t}{\partial t} = \dot{C}_t$. Let $d\mu_{C_t}$ be the linear functional (Grassmann Gaussian measure) with characteristic function

$$\int d\mu_{C_t}(\psi)\, e^{(\eta, \psi)_\Gamma} = e^{\frac{1}{2}(\eta,\, C_t\, \eta)_\Gamma}. \tag{4.84}$$

The integrals of arbitrary monomials are obtained from this formula by taking derivatives with respect to η. The measure is normalized: $\int d\mu_{C_t}(\psi) = 1$.

Let $V(\psi) \in \mathcal{A}$ be an element of the even subalgebra and have no constant part, $\lambda \in \mathbb{C}$, and $\mathcal{G}(0, \psi) = \lambda V(\psi)$. The *effective action* at $t \geq 0$ is

$$\mathcal{G}(t, \psi) = \log \int d\mu_{C_t}(\chi)\, e^{\mathcal{G}(0, \chi + \psi)}. \tag{4.85}$$

Because the measure is normalized, $e^{\mathcal{G}(t,\psi)} = 1$ for $\lambda = 0$. By the nilpotency of the Grassmann variables, $e^{\mathcal{G}(t,\psi)}$ is a polynomial in λ (the degree of which grows with $|\Gamma|$). Thus $\mathcal{G}(t, \psi)$ is analytic in λ for $|\lambda| < \lambda_0(\Gamma)$.

4.3.2 The RG Equation

The RGDE is obtained by taking a derivative with respect to t of the defining equation (4.85) of the effective action. The Boltzmann factor $e^{\mathcal{G}(t,\psi)}$ is given by a convolution of the initial Boltzmann factor $e^{\mathcal{G}(0,\psi)}$ with a Gaussian measure, just as the solution to a heat equation is the convolution of the initial function with the heat kernel. And indeed, the differential equation satisfied by the Boltzmann factor is simply the associated heat equation.

Proposition 4.3. *Let*

$$
\Delta_{C_t} = \frac{1}{2}\left(\frac{\delta}{\delta\psi}, C_t \frac{\delta}{\delta\psi}\right)_\Gamma = \frac{1}{2}\int_\Gamma dX \int_\Gamma dX' \frac{\delta}{\delta\psi(X)} C_t(X,X') \frac{\delta}{\delta\psi(X')}.
$$

$$(4.86)$$

Then

$$
\frac{\partial}{\partial t}\, e^{\mathcal{G}(t,\psi)} = \Delta_{\dot{C}_t}\, e^{\mathcal{G}(t,\psi)}
$$

$$(4.87)$$

and

$$
e^{\mathcal{G}(t,\psi)} = e^{\Delta_{C_t}}\, e^{\mathcal{G}(0,\psi)}.
$$

$$(4.88)$$

If $\mathcal{G}(0,\psi)$ is an element of the even subalgebra, then for all $t > 0$, $\mathcal{G}(t,\psi)$ is an element of the even subalgebra, and it satisfies the renormalization group equation

$$
\frac{\partial}{\partial t}\mathcal{G}(t,\psi) = \dot{\Delta}_{C_t}\mathcal{G}(t,\psi) + \frac{1}{2}\left(\frac{\delta\mathcal{G}(t,\psi)}{\delta\psi}, \dot{C}_t \frac{\delta\mathcal{G}(t,\psi)}{\delta\psi}\right).
$$

$$(4.89)$$

Proof. For any $F(\psi) \in \mathcal{A}$, define $F(\frac{\delta}{\delta\eta})$ by replacing every factor $\psi(X)$ by $\frac{\delta}{\delta\eta(X)}$ in the polynomial expression for F. Then $F(\psi) = [F(\frac{\delta}{\delta\eta})\, e^{(\eta,\psi)_\Gamma}]_{\eta=0}$ (the derivatives $\frac{\delta}{\delta\eta}$ also generate a finite-dimensional Grassmann algebra, so the expansion for F terminates at some power). Since Grassmann integration is a continuous operation, and by (4.84),

$$
\begin{aligned}
e^{\mathcal{G}(t,\psi)} &= \left[e^{\mathcal{G}(0,\frac{\delta}{\delta\eta})} \int d\mu_{C_t}(\chi)\, e^{(\eta,\chi+\psi)_\Gamma}\right]_{\eta=0} \\
&= \left[e^{\mathcal{G}(0,\frac{\delta}{\delta\eta})}\, e^{\frac{1}{2}(\eta,C_t\eta)_\Gamma}\, e^{(\eta,\psi)_\Gamma}\right]_{\eta=0}.
\end{aligned}
$$

$$(4.90)$$

For any formal power series $f(z) = \sum f_k z^k$,

$$
f(\Delta_{C_t})\, e^{(\eta,\psi)_\Gamma} = f\left(\frac{1}{2}(\eta,\, C_t\,\eta)_\Gamma\right) e^{(\eta,\psi)_\Gamma},
$$

$$(4.91)$$

so $e^{\frac{1}{2}(\eta,\, C_t\,\eta)_\Gamma + (\eta,\psi)_\Gamma} = e^{\Delta_{C_t}}\, e^{(\eta,\psi)_\Gamma}$. Since Δ_{C_t} is bilinear in the derivatives, it commutes with all factors that depend only on η and can be taken out in front in (4.90). This implies (4.88). Since Δ_{C_t} also commutes with $\dot{\Delta}_{C_t}$, (4.87) follows.

If $\mathcal{G}(0, \psi)$ is an element of the even subalgebra, the same holds for $\mathcal{G}(t, \psi)$ by (4.88), since every application of Δ_{C_t} removes two fields. Thus performing the derivatives with respect to ψ gives (4.89). ∎

Let the Grassmann polynomials $\mathcal{G}_r(t, \psi)$ be defined by

$$\mathcal{G}(t, \psi) = \sum_{r=1}^{\infty} \lambda^r \mathcal{G}_r(t, \psi). \tag{4.92}$$

Assume that $C = \lim_{t \to \infty} C_t$ exists and let $D_t = C - C_t$ so that $\dot{C}_t = -\dot{D}_t$. D_t is the covariance of the unintegrated fields.

Again, we expand in the Wick ordered monomials

$$\Omega_{D_t}(\psi(X_1) \ldots \psi(X_p)) = e^{-\Delta_{D_t}} \psi(X_1) \ldots \psi(X_p). \tag{4.93}$$

They form a basis of the Grassmann algebra because the relation (4.93) can simply be inverted by multiplying by $e^{\Delta_{D_t}}$, and they inherit the antisymmetry from the ordinary monomials:

$$\Omega_{D_t}(\psi(X_{\pi(1)}) \ldots \psi(X_{\pi(p)})) = \varepsilon(\pi) \Omega_{D_t}(\psi(X_1) \ldots \psi(X_p)). \tag{4.94}$$

The polynomial $\mathcal{G}_r(t, \psi)$ has the representation (here $\underline{X} = (X_1, \ldots, X_m)$)

$$\mathcal{G}_r(t, \psi) = \sum_{m=0}^{\bar{m}(r)} \int_{\Gamma^m} d\underline{X} \, G_{mr}(t \mid \underline{X}) \, \Omega_{D_t} \left(\prod_{k=1}^{m} \psi(X_k) \right). \tag{4.95}$$

where $G_{mr}(t \mid \underline{X})$ is the *connected, amputated, Wick-ordered m–point Green function.* Using Wick ordered monomials provides even greater simplifications for the fermionic models than for the ϕ^4 case.

The $G_{mr}(t \mid X_1, \ldots, X_m)$ are assumed to be totally antisymmetric, that is, for all $\pi \in S_m$, $G_{mr}(t \mid X_{\pi(1)}, \ldots, X_{\pi(m)}) = \varepsilon(\pi) G_{mr}(t \mid X_1, \ldots, X_m)$, because any part of G that is not antisymmetric would cancel in (4.95).

Application of $\frac{\partial}{\partial t}$ to (4.95) gives a sum of two terms since two factors depend on t. By (4.93) and $\dot{C}_t = -\dot{D}_t$, the derivative of the Wick ordered monomial is $\frac{\partial}{\partial t} \Omega_{D_t}(\prod_k \psi(X_k)) = \Delta_{\dot{C}_t} \Omega_{D_t}(\prod_k \psi(X_k))$. This implies that the term linear in \mathcal{G} drops out of (4.89) by Wick ordering with respect to D_t, and (4.89) now reads

$$\sum_{m=0}^{\bar{m}(r)} \int_{\Gamma^m} d\underline{X} \, \Omega_{D_t} \left(\prod_{k=1}^{m} \psi(X_k) \right) \frac{\partial}{\partial t} G_{mr}(t \mid \underline{X}) = \frac{1}{2} \mathcal{Q}_r(t, \psi) \tag{4.96}$$

where $\mathcal{Q}_r(t, \psi)$ is defined by

$$\left(\frac{\delta \mathcal{G}(t, \psi)}{\delta \psi}, C_t \frac{\delta \mathcal{G}(t, \psi)}{\delta \psi} \right) = \sum_{\lambda=1}^{\infty} \lambda^r \mathcal{Q}_r(t, \psi). \tag{4.97}$$

Being an element of the Grassmann algebra, $Q_r(t, \psi)$ has the representation

$$Q_r(t, \psi) = \sum_{m=0}^{\bar{m}(r)} \int_{\Gamma^m} d\underline{X}\, Q_{mr}(t \mid \underline{X})\, \Omega_{D_t}\left(\prod_{k=1}^{m} \psi(X_k)\right). \qquad (4.98)$$

Because G is in the even subalgebra and $\dot{C}_t = -\dot{D}_t$,

$$\dot{Q}_r(t, \psi) = \frac{1}{2}\left[\left(\tfrac{\delta}{\delta\chi}G(t, \chi), \dot{C}_t\tfrac{\delta}{\delta\psi}G(t, \psi)\right)\right]_{\chi=\psi}$$

$$= -\frac{1}{2}\left[\dot{H}_t G(t, \chi)G(t, \psi)\right]_{\chi=\psi} \qquad (4.99)$$

with $H_t = (\tfrac{\delta}{\delta\chi}, D_t\tfrac{\delta}{\delta\psi})$. Upon inserting the expansion of G, we have to reorder expressions of the form

$$P = -\frac{1}{2}\dot{H}_t \Omega_{D_t}\left(\prod_{k=1}^{m} \chi(X_k)\right)\Omega_{D_t}\left(\prod_{l=1}^{n}\psi(Y_l)\right) \qquad (4.100)$$

By definition (4.93) of the Wick ordered monomials,

$$P = -\frac{1}{2}\dot{H}_t e^{-\Delta_{D_t}^{(\chi)}}\prod_{k=1}^{m}\chi(X_k)\, e^{-\Delta_{D_t}^{(\psi)}}\prod_{l=1}^{n}\psi(Y_l). \qquad (4.101)$$

Here the superscripts on the Laplacians indicate with respect to which variables the derivatives are taken. All differential operators are bilinear, hence commute with one another. Moreover, $\Delta_{D_t}^{(\psi)}$ commutes with all polynomials in χ. Thus

$$P = -\frac{1}{2}e^{-\Delta_{D_t}^{(\chi)}-\Delta_{D_t}^{(\psi)}}\dot{H}_t\prod_{k=1}^{m}\chi(X_k)\prod_{l=1}^{n}\psi(Y_l). \qquad (4.102)$$

Since

$$-\Delta_{D_t}^{(\chi)} - \Delta_{D_t}^{(\psi)} = -\Delta_{D_t}^{(\chi,\psi)} + H_t, \qquad (4.103)$$

where

$$\Delta_{D_t}^{(\chi,\psi)} = \frac{1}{2}\left(\tfrac{\delta}{\delta\chi} + \tfrac{\delta}{\delta\psi}, D_t(\tfrac{\delta}{\delta\chi} + \tfrac{\delta}{\delta\psi})\right) \qquad (4.104)$$

we get

$$P = \frac{1}{2}e^{-\Delta_{D_t}^{(\chi,\psi)}}\left(-\tfrac{\partial}{\partial t}e^{H_t}\right)\prod_{k=1}^{m}\chi(X_k)\prod_{l=1}^{n}\psi(Y_l). \qquad (4.105)$$

Because

$$\left[e^{-\Delta_{D_t}^{(\chi,\psi)}}\prod_{k\in K}\chi(X_k)\prod_{l\in L}\psi(Y_l)\right]_{\chi=\psi} = \Omega_{D_t}\left(\prod_{k\in K}\psi(X_k)\prod_{l\in L}\psi(Y_l)\right)$$

$$(4.106)$$

and because $e^{-\Delta_{D_t}^{(\chi,\psi)}}$ is a linear differential operator, it now suffices to calculate the action of e^{H_t} on the monomial of ψ and χ.

Lemma 4.4.

$$e^{H_t} \prod_{k=1}^{m} \chi(X_k) \prod_{l=1}^{n} \psi(Y_l) = \sum_{s \geq 0} (-1)^{ms + \frac{s(s+1)}{2}} \sum_{\substack{A \subseteq \underline{m}, B \subseteq \underline{n} \\ |A| = |B| = s}} v(A)v(B)$$

$$\det {}_{AB}(D_t) \prod_{p \notin A} \chi(X_p) \prod_{q \notin B} \psi(Y_q). \quad (4.107)$$

Here $\underline{m} = \{1, \ldots, m\}$,

$$\det {}_{AB}(D_t) = \sum_{\pi \in S_s} \varepsilon(\pi) \prod_{k=1}^{s} D_t(X_{a_k}, Y_{b_{\pi(k)}}), \quad (4.108)$$

where $\varepsilon(\pi)$ is the sign of the permutation π, and $v(\emptyset) = 1$, and for $A = \{a_1, \ldots, a_s\}$ with $a_1 < \ldots < a_s$,

$$v(A) = (-1)^{\sum_k a_k}. \quad (4.109)$$

Proof. We abbreviate $\chi_k = \chi(X_k)$, $\psi_l = \psi(Y_l)$, and $\tilde{D}_{kl} = D_t(X_k, Y_l)$. Then

$$H_t = \sum_{k=1}^{m} \sum_{l=1}^{n} \frac{\partial}{\partial \chi_k} \tilde{D}_{kl} \frac{\partial}{\partial \psi_l}. \quad (4.110)$$

The expansion of the exponential gives

$$e^{H_t} = \sum_{S \subseteq \underline{m} \times \underline{n}} \prod_{(k,l) \in S} \frac{\partial}{\partial \chi_k} \tilde{D}_{k,l} \frac{\partial}{\partial \psi_l}. \quad (4.111)$$

A set S with $s \geq 1$ elements contributing a nonzero term to this sum is of the form $S = \{(a_k, b_{\pi(k)}) : k \in \{1, \ldots, s\}\}$, where $A = \{a_1, \ldots, a_s\} \subseteq \underline{m}$ with $a_1 < \ldots < a_s$ and $B = \{b_1, \ldots, b_s\} \subseteq \underline{n}$ with $b_1 < \ldots < b_s$, and where π is a permutation of $\{1, \ldots, s\}$. The map $S \mapsto (A, B, \pi)$ is bijective. Thus

$$e^{H_t} = 1 + \sum_{s \geq 1} \sum_{\substack{A \subseteq \underline{m}, B \subseteq \underline{n} \\ |A| = |B| = s}} \sum_{\pi \in S_s} \prod_{k=1}^{s} \frac{\partial}{\partial \chi_{a_k}} \tilde{D}_{a_k, b_{\pi(k)}} \frac{\partial}{\partial \psi_{b_{\pi(k)}}}, \quad (4.112)$$

and we need to calculate its action on $\chi_1 \cdots \chi_m \psi_1 \cdots \psi_n$. All sign factors now come from reordering the Grassmann monomials:

$$\prod_{k=1}^{s} \frac{\partial}{\partial \chi_{a_k}} \frac{\partial}{\partial \psi_{b_{\pi(k)}}} = (-1)^{s(s-1)/2} \prod_{k=1}^{s} \frac{\partial}{\partial \chi_{a_k}} \prod_{l=1}^{s} \frac{\partial}{\partial \psi_{b_{\pi(l)}}}$$

$$= (-1)^{s(s-1)/2} \varepsilon(\pi) \prod_{k=1}^{s} \frac{\partial}{\partial \chi_{a_k}} \prod_{l=1}^{s} \frac{\partial}{\partial \psi_{b_l}} \quad (4.113)$$

$$= (-1)^{s(s-1)/2} \varepsilon(\pi) \prod_{k=0}^{s-1} \frac{\partial}{\partial \chi_{a_{s-k}}} \prod_{l=0}^{s-1} \frac{\partial}{\partial \psi_{b_{s-l}}}$$

and

$$\chi_1 \cdots \chi_m = v(A)(-1)^{s(s+1)/2} \prod_{k=1}^{s} \chi_{a_k} \prod_{k \notin A} \chi_k \qquad (4.114)$$

(the factors in the last product are ordered according to increasing k), and similarly for the ψ. Commuting the product of derivatives with respect to ψ with $\chi_1 \cdots \chi_m$ gives a factor $(-1)^{sm}$. There are no additional sign factors. ∎

Lemma 4.4 and a comparison of the coefficients give the component form of the RGE

$$\frac{\partial}{\partial t} G_{mr}(t \mid X_1, \ldots, X_m) = \frac{1}{2} \mathbb{A}_m Q_{mr}(t \mid X_1, \ldots, X_m) \qquad (4.115)$$

where \mathbb{A}_m is the antisymmetrization operator

$$(\mathbb{A}_m f)(X_1, \ldots, X_m) = \frac{1}{m!} \sum_{\pi \in S_m} \varepsilon(\pi) f(X_{\pi(1)}, \ldots, X_{\pi(m)}), \qquad (4.116)$$

and $Q_{mr}(t \mid \underline{X})$ is given by the following expression.

Proposition 4.5. $Q_{m1}(t \mid \underline{X}) = 0$, and for $r \geq 2$,

$$Q_{mr}(t \mid \underline{X}) = \int d\kappa_{mr} \int_{\Gamma^i} d\underline{V} \int_{\Gamma^i} d\underline{W} \left(-\frac{\partial}{\partial t} \det \mathcal{D}_t^{(i)}(\underline{V}, \underline{W}) \right)$$

$$G_{m_1 r_1}(t \mid \underline{X_1}, \underline{V}) \, G_{m_2 r_2}(t \mid \underline{\tilde{W}}, \underline{X_2}), \qquad (4.117)$$

where $\int d\kappa_{mr}$ stands for the sum

$$\int d\kappa_{mr} F = \sum_{\substack{r_1, r_2 \geq 1 \\ r_1 + r_2 = r}} \sum_{\substack{(m_1, m_2, i) \\ \in \mathcal{M}_{r_1 r_2 m}}} \kappa_{m_1 m_2 i} \, F(r_1, m_1, r_2, m_2, i) \qquad (4.118)$$

with positive weights $\kappa_{m_1 m_2 i} = \binom{m_1}{i} \binom{m_2}{i}$, so that $d\kappa_{mr}$ is a positive measure. $\mathcal{M}_{r_1 r_2 m}$ is the set of (m_1, m_2, i) such that $i \geq 1$, $1 \leq m_1 \leq \bar{m}(r_1)$, $1 \leq m_2 \leq \bar{m}(r_2)$, $m_1 + m_2 = m + 2i$, and m_1 and m_2 are even. $\underline{X_1} = (X_1, \ldots, X_{m_1 - i})$, $\underline{X_2} = (X_{m_1 - i + 1}, \ldots, X_m)$, $\underline{V} = (V_1, \ldots, V_i)$, $\underline{W} = (W_1, \ldots, W_i)$, $\underline{\tilde{W}} = (W_i, \ldots, W_1)$, and $\mathcal{D}_t^{(i)}(\underline{V}, \underline{W})$ is the $i \times i$ matrix $(\mathcal{D}_t^{(i)}(\underline{V}, \underline{W}))_{kl} = D_t(V_k, W_l)$.

One important feature of (4.117) that distinguishes it from bosons is that the determinant of the propagators appears in this equation. The Gram bound for this determinant improves the combinatorics by a factorial (see [53, 54]).

4.3.3 The Component RGE in Fourier Space

Let $\Lambda = \mathbb{T} \times \Lambda$ and $\Lambda^* = \mathbb{M}_{n_r} \times \Lambda^*$, as given in Sect. 4.2.4. Let $\Gamma = \Lambda \times \{-1, 1\} \times \{1, 2\}$ and $\Gamma^* = \Lambda^* \times \{-1, 1\} \times \{1, 2\}$. The Fourier transforms of the $G_{mr}(t \mid \cdot)$ are

$$\hat{G}_{mr}(t \mid P_1, \ldots, P_m) = \int_{\Lambda^m} \prod_{k=1}^m dx_k \, e^{-ip_k x_k} \, G_{mr}(t \mid X_1, \ldots, X_m). \quad (4.119)$$

For $K = (k, \sigma, j) \in \Gamma^*$ let $\sim K = (-k, \sigma, 3 - j)$, and define $\delta_{\Gamma^*}(K, K') = \delta_{\Lambda^*}(k, k')\delta_{\sigma\sigma'}\delta_{jj'}$. The Fourier transform of the propagator D_t is of the form $\tilde{D}_t(K, K') = \delta_{\Gamma^*}(K, \sim K') \, \bar{D}_t(K')$, where

$$\bar{D}_t(K) = (-1)^j D_t((-1)^j k). \quad (4.120)$$

This combination of signs implies that $D_t(X, Y) = -D_t(Y, X)$. The propagator D_t for the many-fermion system was given in Sect. 4.2.5. The Fourier transform of $Q_{mr}(t \mid \cdot)$ is

$$Q_{mr}(t \mid P_1, \ldots, P_m) =$$

$$\int d\hat{\kappa}_{mr} \int dK_1 \ldots dK_i \left(-\frac{\partial \bar{D}_t(K_1)}{\partial t}\right) \prod_{j=2}^i \bar{D}_t(K_j) \quad (4.121)$$

$$\hat{G}_{m_1 r_1}(t \mid \underline{P}^{(1)}, K_1, \ldots, K_i) \, \hat{G}_{m_2 r_2}(t \mid \sim K_i, \ldots, \sim K_1, \underline{P}^{(2)}).$$

Here $\int d\hat{\kappa}_{mr} = \int d\kappa_{mr} i \, i!$, $\underline{P}^{(1)} = (P_1, \ldots, P_{m_1-i})$, and $\underline{P}^{(2)} = (P_{m_1-i+1}, \ldots, P_{m_1+m_2})$. By translation invariance in space \mathbf{x} and time τ,

$$\hat{G}_{mr}(t \mid \underline{P}) = \delta_{\Lambda^*}(p_1 + \ldots + p_m, 0) I_{mr}(t \mid \underline{P}) \quad (4.122)$$

with a totally antisymmetric function $I_{mr}(t \mid \underline{P})$ of $\underline{P} \in \Gamma^m$ that satisfies

$$(1, \ldots, 1) \cdot \nabla I_{mr}(t \mid \underline{P}) = 0. \quad (4.123)$$

A priori, (4.122) only implies the existence of a function $\tilde{I}_{mr}(t \mid \underline{P})$, defined for those $\underline{P} = (P_1, \ldots, P_m)$ for which $p_1 + \ldots + p_m = 0$. However, since

$$H = \{p_1, \ldots, p_m : p_1 + \ldots + p_m = 0\}. \quad (4.124)$$

is a linear subspace of $(\Lambda^*)^m$, one can simply extend \tilde{I} to a function on all space by defining $I = \tilde{I} \circ \Pi_H$ with Π_H the projection to the subspace H. Since Π_H is symmetric in all its arguments, I is totally antisymmetric.

The product of the two deltas in (4.121) can be combined to cancel the delta in the relation between \hat{G} and I, and to remove the integration over k_1. Thus k_1 is fixed as

$$k_1 = -(k_2 + \ldots + k_i + p_1 + \ldots + p_{m_1-i+1}) \quad (4.125)$$

and the RGE in Fourier space is $\frac{\partial}{\partial t} I_{mr}(t \mid \underline{P}) = \frac{1}{2} A_m \hat{Q}_{mr}(t \mid \underline{P})$, with

$$
\begin{aligned}
\hat{Q}_{mr}(t \mid \underline{P}) \;=\;& \frac{1}{2} A_m \int d\hat{k}_{mr} \int dK_2 \dots dK_i \sum_{\sigma_1, j_1} (-\dot{D}_t(K_1)) \\
& \prod_{s=2}^{i} \bar{D}_t(K_s) \, I_{m_1 r_1}(t \mid \underline{P}^{(1)}, K_1, \dots, K_i) \\
& I_{m_2 r_2}(t \mid \sim K_i, \dots, \sim K_1, \underline{P}^{(2)})
\end{aligned}
\tag{4.126}
$$

with $K_1 = (k_1, \sigma_1, j_1)$, where k_1 is given by (4.125).

4.4 Power Counting for Skeletons

We now proceed to analyze the field theory by the RGE in perturbation theory. To this end, we take the thermodynamic limit. The existence of the thermodynamic limit, as well as that of the limit $n_\tau \to \infty$, is shown in Appendix B.7. There is no problem with singularities as long as the temperature is positive; taking $L \to \infty$ simplifies some of the technical arguments because the discrete momentum sums of the finite volume are replaced by integrals.

4.4.1 Bounds for the Infinite-Volume Propagator

In the thermodynamic limit, the spatial part \mathbf{p} of momentum becomes continuous, $\mathbf{p} \in \mathcal{B}$, and the set of Matsubara frequencies becomes $\mathbb{M} = \{\frac{\pi}{\beta}(2n+1) : n \in \mathbb{Z}\}$. It is convenient to take $(p_0, \mathbf{p}) \in \mathbb{R} \times \mathcal{B}$ and put all the β-dependence into the integrand. To this end, we define the step function $\omega_\beta : \mathbb{R} \to \mathbb{M}$ by

$$
\omega_\beta(p_0) = \frac{\pi}{\beta}(2n+1) \quad \text{if } p_0 \in (\frac{2\pi}{\beta}n, \frac{2\pi}{\beta}(n+1)].
\tag{4.127}
$$

For any continuous and integrable function f,

$$
\int_{\mathbb{R}} \frac{dp_0}{2\pi} \, f(\omega_\beta(p_0)) = \frac{1}{\beta} \sum_{\omega \in \mathbb{M}} f(\omega).
\tag{4.128}
$$

Moreover,

$$
\sup_{p_0 \in \mathbb{R}} |\omega_\beta(p_0) - p_0| = \frac{\pi}{\beta} \quad \text{and} \quad \inf_{p_0 \in \mathbb{R}} |\omega_\beta(p_0)| = \frac{\pi}{\beta},
\tag{4.129}
$$

$|\omega_\beta(p_0)| \geq \frac{p_0}{2}$, and $\omega_\beta(-p_0) = -\omega_\beta(p_0)$ holds Lebesgue-almost everywhere, so that in integrals like (4.128), ω_β can be treated as an antisymmetric function. With this, the propagator now reads $D_t(p) = \mathcal{C}_t(\omega_\beta(p_0), E(\mathbf{p}))$ with

$$
\mathcal{C}_t(x, y) = \frac{1}{ix - y} \chi_1(\epsilon_t^{-2}(x^2 + y^2))
\tag{4.130}
$$

(and $\epsilon_t = \epsilon_0 \, e^{-t}$).

In infinite volume, the restriction on the spatial part \mathbf{p} of momentum improves the bounds on the propagator over the finite-volume ones used in Appendix B.7.

Lemma 4.6. D_t *is bounded*, C^∞ *in* t *and* C^{k_0} *in* \mathbf{p}. *If* $t > \log \frac{\beta \epsilon_0}{\pi}$, *then* $D_t(p) = 0$ *for all* $p \in \mathbb{R} \times B$. *For all multiindices* $\alpha \in \mathbb{N}_0^d$ *with* $|\alpha| \le k_0$, *there is a constant* $B_\alpha > 0$ *such that*

$$\left| \partial^\alpha \dot{D}_t(p) \right| \le B_\alpha \epsilon_t^{-1-|\alpha|} \, \mathbb{1} \left(|i\omega_\beta(p_0) - E(\mathbf{p})| \le \epsilon_t \right) \tag{4.131}$$

$$\le B_\alpha \epsilon_t^{-1-|\alpha|} \, \mathbb{1} \left(|\omega_\beta(p_0)| \le \epsilon_t \right) \, \mathbb{1} \left(|E(\mathbf{p})| \le \epsilon_t \right) .$$

$B_0 = 4$. *For* $t \le \log \frac{\beta \epsilon_0}{\pi}$,

$$\int_\mathbb{R} \frac{dp_0}{2\pi} \left| \partial^\alpha \dot{D}_t(p) \right| \le B_\alpha \epsilon_t^{-|\alpha|} \, \mathbb{1} \left(|E(\mathbf{p})| \le \epsilon_t \right) , \tag{4.132}$$

$$\int_{\mathbb{R} \times B} \frac{d^{d+1} p}{(2\pi)^{d+1}} \left| \partial^\alpha \dot{D}_t(p) \right| \le 2 J_1 B_\alpha \epsilon_t^{1-|\alpha|}, \tag{4.133}$$

and

$$\int_{\mathbb{R} \times B} \frac{d^{d+1} p}{(2\pi)^{d+1}} |D_t(p)| \le 8 J_1 \epsilon_t. \tag{4.134}$$

Proof. For $D_t(p)$ to be nonzero, $|\omega_\beta(p_0)| \le \epsilon_t$ must hold. By (4.129), $D_t = 0$ for $t > \log \frac{\beta \epsilon_0}{\pi}$. Since

$$\dot{D}_t(p) = -2\epsilon_t^{-2}(i\omega_\beta(p_0) + E(\mathbf{p})) \, \chi_1'(\epsilon_t^{-2}(\omega_\beta(p_0)^2 + E(\mathbf{p})^2)) \tag{4.135}$$

and $\|\chi_1'\|_\infty \le 2$, $|\dot{D}_t(p)| \le \frac{4}{\epsilon_t} \, \mathbb{1} \left(|i\omega_\beta(p_0) - E(\mathbf{p})| \le \epsilon_t \right) .$ Thus $B_0 = 4$. Derivatives with respect to \mathbf{p} can act on $E(\mathbf{p})$ or on χ_1', in which case they produce factors bounded by $\epsilon_t^{-1} |\nabla E(\mathbf{p})|$, so there is B_α such that (4.131) holds. Inserting (4.131) into the integral in (4.132) gives

$$\int_\mathbb{R} \frac{dp_0}{2\pi} \left| \partial^\alpha \dot{D}_t(p) \right| \le B_\alpha \epsilon_t^{-1-|\alpha|} \, \mathbb{1} \left(|E(\mathbf{p})| \le \epsilon_t \right) \, M_\beta \tag{4.136}$$

with

$$M_\beta = \frac{1}{\beta} \left| \{ n \in \mathbb{Z} : |2n+1| \le \frac{\beta \epsilon_t}{\pi} \} \right| \le \frac{2}{\pi} \epsilon_t, \tag{4.137}$$

which proves (4.132). This gives for the integral in (4.133)

$$\int_{\mathbb{R} \times B} \frac{d^{d+1} p}{(2\pi)^{d+1}} \left| \partial^\alpha \dot{D}_t(p) \right| \le B_\alpha \epsilon_t^{-|\alpha|} \int_B \frac{d^d \mathbf{p}}{(2\pi)^d} \, \mathbb{1} \left(|E(\mathbf{p})| \le \epsilon_t \right) . \tag{4.138}$$

For $t \geq 0$, $\epsilon_t \leq \epsilon_0$, so by a change of coordinates in the integral,

$$\int_B \frac{d^d p}{(2\pi)^d} \, \mathbb{1}(|E(\mathbf{p})| \leq \epsilon_t) \leq \int_{-\epsilon_t}^{\epsilon_t} d\rho \int d\theta J(\rho, \theta) = 2J_1 \epsilon_t \qquad (4.139)$$

which implies (4.133). (4.134) follows by integration over t. ∎

The geometrical interpretation of the integral in (4.139) is that of the volume of a d-dimensional neighbourhood $\mathcal{R}(\epsilon_t) = \{\mathbf{p} \in \mathcal{B} : |E(\mathbf{p})| \leq \epsilon_t\}$ of thickness ϵ_t of the Fermi surface, and the power counting bound is obtained by weighting the smallness of that volume – as well as that of the according 'volume' M_β of the frequencies, (4.137) – against the largeness of the function itself in that region. Careful estimates of more complicated volumes will be important for the analysis of the model.

4.4.2 Sup Norm Estimates

We first prove a power counting bound in the sup norm $|\cdot|_0$, where, for $F_m : (\Gamma^*)^m \to \mathbb{C}$,

$$|F_m|_0 = \sup_{\underline{P} \in (\Gamma^*)^m} |F(\underline{P})|. \qquad (4.140)$$

Obviously, $|A_m F_m|_0 \leq |F_m|_0$. By Lemma 4.6,

$$|Q_{mr}(t)|_0 \leq 4 \int d\hat{\kappa}_{mr} \|\bar{D}_t\|_1^{i-1} \|\dot{D}_t\|_\infty |I_{m_1 r_1}(t)|_0 \, |I_{m_2 r_2}(t)|_0$$

$$\leq 128 J_1 \int d\hat{\kappa}_{mr} (8 J_1 \epsilon_t)^{i-2} |I_{m_1 r_1}(t)|_0 \, |I_{m_2 r_2}(t)|_0. \qquad (4.141)$$

Integrating the RGE, we have

$$|I_{mr}(t)|_0 \leq |I_{mr}(0)|_0 + \frac{1}{2} \left| \left| \int_0^t ds \, A_m Q_{mr}(s) \right| \right|_0$$

$$\leq |I_{mr}(0)|_0 + \frac{1}{2} \int_0^t ds \, |Q_{mr}(s)|_0. \qquad (4.142)$$

We shall now give bounds for the scaling behaviour of the $I_{mr}(t)$, i.e. of their behaviour as functions of t. It is instructive to look at the case $r = 2$ first. Assuming that only a first order four-fermion interaction is present at $t = 0$, we have $|I_{mr}(0)|_0 = \delta_{r1} \delta_{m4} v$ with some constant $v > 0$, whcih is finite because the interaction is bounded in momentum space, i.e. short range in position space). The restriction $r_1 + r_2 = 2$ in $\int d\hat{\kappa}_{mr}$ implies $r_1 = r_2 = 1$, so

$$|I_{m2}(t)|_0 \le 64 J_1 \int\limits_0^t ds \int d\hat{\kappa}_{mr} (8 J_1 \epsilon_t)^{i-2} v^2 \delta_{m_1 4} \delta_{m_2 4}. \tag{4.143}$$

The restriction $m_1 + m_2 = m + 2i$ in $\int d\hat{\kappa}_{mr}$ leaves only the possibilities $m = 6$ and $i = 1$, $m = 4$ and $i = 2$, and $m = 2$ and $i = 3$. For $m \ge 8$, $I_{m2}(t) = 0$ for all t (in other words, $\bar{m}(2) = 6$). The nontrivial cases correspond to the graphs shown in Fig. 3.9. Inserting all the numerical constants in the definition of $\int d\hat{\kappa}_{mr}$ and doing the integrals over s, we get

$$
\begin{aligned}
|I_{62}(t)|_0 &\le 2 (8v)^2 \epsilon_0^{-1} \int\limits_0^t ds\, e^s \le 2 (8v)^2 \epsilon_0^{-1} e^t \\[2mm]
|I_{42}(t)|_0 &\le (96v)^2 \int\limits_0^t ds = (96v)^2 t \\[2mm]
|I_{22}(t)|_0 &\le (384 J_1 v)^2 \epsilon_0 \int\limits_0^t ds\, e^{-s} \le (384 J_1 v)^2 \epsilon_0.
\end{aligned}
\tag{4.144}
$$

All three bounds give the correct behaviour of $I_{m2}(t)$ as a function of t; this can be seen for $m = 6$ from the explicit expression $|I_{62}(t)| = |C_t(p)|$ (where p is some combination of the external momenta); for $m \le 4$, we shall discuss it in detail below.

A further analysis of the tree graph ($i = 1$) contributions to I_{mr} suggests that for $m \ge 6$, $I_{mr} \sim e^{\frac{t}{2}(m-4)}$, that $I_{4r} \sim t^\alpha$, and $I_{2r} \sim \text{const}$. However, these bounds hold only for the skeleton functions, defined as follows.

Definition 4.7. *The skeleton functions $\tilde{I}_{mr}(t)$ are defined as the solutions of the skeleton RGE $\frac{\partial}{\partial t} \tilde{I}_{mr}(t) = \tilde{Q}_{mr}(t)$, where in $\tilde{Q}_{mr}(t)$, $\int d\hat{\kappa}_{mr}$ is replaced by the weighted sum $\int d\tilde{\kappa}_{mr}$ in which all terms with $m_1 = 2$ or $m_2 = 2$ are left out.*

In the graphical expansion obtained by iterating the RG equation for the skeleton functions, the two-legged insertions that would require renormalization do not occur.

Theorem 4.8. *Assume that $|I_{mr}(0)|_0 \le a_{mr}^{(0)}$ and let $\tilde{I}_{mr}(t)$ be the solution of the skeleton RGE. Then, for all m and r there are constants a_{mr} such that for all $t \ge 0$*

$$
\left| \tilde{I}_{mr}(t) \right|_0 \le \begin{cases} a_{mr}(1+t)^{r-1} e^{\frac{t}{2}(m-4)} & \text{if } m \ge 4 \\ a_{2r} & \text{if } m = 2. \end{cases}
\tag{4.145}
$$

$\tilde{I}_{2r}(t)$ converges as $t \to \infty$, uniformly in β.

Proof. Induction on r, with (4.145) as the inductive hypothesis. For $r = 1$, there is nothing to prove because $\tilde{Q}_{m1} = 0$. Let $r \geq 2$. The inductive hypothesis applies to both factors $|\tilde{I}_{m_k r_k}|_0$ on the RHS of the RGE. Since $m_1 = 2$ and $m_2 = 2$ are excluded from the sum in $\int d\tilde{\kappa}_{mr}$, we can insert the upper line of (4.145) for both factors. Because $m_1 + m_2 = m + 2i$, the exponent becomes independent of $i, m_1,$ and m_2, and we get

$$|Q_{mr}(t)|_0 \leq \alpha_{mr}(1+t)^{r-2}e^{\frac{t}{2}(m-4)} \tag{4.146}$$

with

$$\alpha_{mr} = \frac{1}{J_1\epsilon_0^2}\int d\hat{\kappa}_{mr}(8J_1\epsilon_0)^i a_{m_1 r_1} a_{m_2 r_2}. \tag{4.147}$$

All that remains is to use $|\tilde{I}_{mr}(t)|_0 \leq |\tilde{I}_{mr}(0)|_0 + \frac{1}{2}\int_0^t ds\, |\tilde{Q}_{mr}(s)|_0$ and to do the s-integral. For $m \geq 6$, $m - 4 \geq 0$, so

$$\int_0^t ds\, |\tilde{Q}_{mr}(s)|_0 \leq \alpha_{mr}\int_0^t ds\,(1+s)^{r-2}e^{\frac{s}{2}(m-4)}$$

$$\leq \frac{2\alpha_{mr}}{m-4}(1+t)^{r-2}e^{\frac{t}{2}(m-4)}. \tag{4.148}$$

The induction step for $m \geq 6$ is thus completed by noting that $1 \leq (1+t)^{r-2} \leq (1+t)^{r-1}$ and defining a_{mr} recursively as $a_{mr} = a_{mr}^{(0)} + \alpha_{mr}/(m-4)$. If $m = 4$,

$$\int_0^t ds\, |\tilde{Q}_{4r}(s)|_0 \leq \alpha_{4r}\int_0^t ds\,(1+s)^{r-2} \leq \frac{\alpha_{4r}}{r-1}(1+t)^{r-1}. \tag{4.149}$$

With the definition $a_{4r} = a_{4r}^{(0)} + \alpha_{4r}/2(r-1)$, (4.145) holds for $m = 4$. If $m = 2$,

$$\int_0^t ds\, |\tilde{Q}_{2r}(s)|_0 \leq \alpha_{2r}\int_0^t ds\,(1+s)^{r-2}e^{-s}$$

$$\leq \alpha_{2r}e\int_0^t ds\,(1+s)^{r-2}e^{-(1+s)} \tag{4.150}$$

$$\leq \alpha_{2r}e\int_0^\infty d\sigma\, \sigma^{r-2}e^{-\sigma} \leq \alpha_{2r}e(r-2)!$$

and the appropriate recursive definition of a_{2r} finishes the inductive step for $m = 2$ as well. In particular, $\tilde{I}_{2r}(t)$ is bounded uniformly in β. By (4.146), $|\frac{\partial}{\partial t}\tilde{I}_{2r}(t)|_0 \leq |Q_{2r}(t)|_0 \to 0$ as $t \to \infty$, so $\tilde{I}_{2r}(t)$ converges as $t \to \infty$. ∎

No attempt was made in the above proof to optimize the recursion for the coefficients, which would be necessary for a more careful discussion of the r dependence of the a_{mr}. However, the last bound (4.150) shows the potential danger of the powers of t: a factor t^r produces an $r!$ in the coefficients. If the coefficients grow factorially, the expansion in λ is divergent for all $\lambda \neq 0$, and thus makes sense only as a formal power series. As one sees from $r = 2$ and by inspection of the above proof, these factors come from the four-point function. We shall investigate in more detail when and why these factors arise, and as we shall see, they are no artefacts – their occurrence is linked to physically important phenomena.

If $m_k = 2$ had not been left out on the right hand side of the RGE, the induction step would have failed because $e^{\frac{t}{2}(m-4)} = e^{-t}$ for $m = 2$, but the two-point function is only bounded by a t-independent constant, which is e^t larger. Note that the derivative with respect to t of the two-point function vanishes as e^{-t}, but the function itself does not because the integral starts at zero. Renormalization will change this boundary condition. Theorem 4.8 implies that for getting perturbation expansion termwise finite, it suffices to renormalize the two-point function.

The above bounds only imply the convergence of the skeleton two-point function, but not of the m-point correlation functions for $m \geq 4$. Convergence of the latter cannot hold in sup norm because these functions indeed have singularities. Convergence can, however, be shown in an L^1-norm, which implies that they are finite almost everywhere in momentum space. We do this in the next section.

4.4.3 Estimates in L^1 Norm

To give bounds that have a convergent behaviour also for the m-point functions with $m \geq 6$, we introduce a norm in which part of the P are integrated, and the supremum is taken over the remaining ones. For $s \in \{1, \ldots, m\}$ and a function $F_m : (\Gamma^*)^m \to \mathbb{C}$, let

$$|F_m|_s = \max_{\substack{S \subset \{1,\ldots,m\} \\ |S|=s}} \sup_{(P_q)_{q \notin S}} \int \prod_{q \in S} dP_q \, \delta(p_1 + \ldots + p_m) |F_m(P_1, \ldots, P_m)|.$$

$$(4.151)$$

Because of the maximum over subsets, $|A_m F_m|_s = |F_m|_s$. Conversely, if F_m satisfies $|F_m(\underline{P}_\pi)| = |F_m(\underline{P})|$, where $\underline{P}_\pi = (P_{\pi(1)}, \ldots, P_{\pi(m)})$ for $\underline{P} = (P_1, \ldots, P_m)$, then

$$|F_m|_s = \sup_{P_{s+1},\ldots,P_m} \int dP_1 \ldots dP_s \, \delta(p_1 + \ldots + p_m) |F_m(\underline{P})|. \qquad (4.152)$$

For $s = 1$, the norm is bounded by the sup norm,

$$|F_m|_1 \leq 4 \sup_{\underline{P}} |F_m(\underline{P})| = 4|I_{mr}|_0 \qquad (4.153)$$

(the factor 4 comes from the sum over σ_1 and j_1).

To estimate the right hand side of the RGE, we reinstall the δ functions, to get

$$\delta(p_1 + \ldots + p_m)Q_{mr}(t \mid \underline{P})$$

$$= \int d\hat{\kappa}_{mr} \int dK_1 \ldots dK_i \dot{\bar{D}}_t(K_1) \prod_{j=2}^{i} \bar{D}_t(K_j) \tag{4.154}$$

$$\hat{G}_{m_1r_1}(t \mid \underline{P}^{(1)}, \underline{K})\hat{G}_{m_2r_2}(t \mid \sim \underline{K}, \underline{P}^{(2)}).$$

Here $\int d\hat{\kappa}_{mr} = \int d\kappa_{mr}i\, i!$, and

$$\hat{G}_{mr}(t \mid \underline{P}) = \delta(p_1 + \ldots + p_m)I_{mr}(t \mid \underline{P}). \tag{4.155}$$

In the maximum over sets $S \subset \{1, \ldots, m\}$, three cases can arise:

1. $(P_q)_{q \in S}$ is a subsequence of $\underline{P}^{(1)}$. This can happen only if $s \leq m_1 - 1$ because $i \geq 1$.
2. same as (i), with 1 replaced by 2.
3. s_1 of the variables $(P_q)_{q \in S}$ are arguments of $I_{m_1r_1}$ and $s_2 = s - s_1$ of them are arguments of $I_{m_2r_2}$. Again, because $i \geq 1$, this can happen only if $s_1 \leq m_1 - 1$ and $s_2 \leq m_2 - 1$.

In case (i), we use the momentum conservation δ function in $\hat{G}_{m_2r_2}$ to fix k_1 in terms of the other momenta, and bound $\frac{\partial}{\partial t}\bar{D}_t(K_1)$ by its sup norm and $I_{m_2r_2}$ by $|I_{m_2r_2}|_0$. By the antisymmetry of $I_{m_1r_1}$, its arguments can be permuted such that $(P_q)_{q \in S}$ come first. Integrate over $(P_q)_{q \in S}$. The only factor depending on $(P_q)_{q \in S}$ is now $I_{m_1r_1}$. Taking a sup over all remaining momenta in $I_{m_1r_1}$ gives a contribution

$$T_1(s) = \mathbf{1}(s < m_1) \int d\hat{\kappa}_{mr} \|\bar{D}_t\|_1^{i-1} 4\|\dot{\bar{D}}_t\|_\infty |I_{m_1r_1}(t)|_s\ |I_{m_2r_2}(t)|_0 \tag{4.156}$$

where $\|\bar{D}_t\|_1 = \int dK\, |\bar{D}_t(K)|$.

Case (ii) is similar, and by the symmetry of $\int d\hat{\kappa}_{mr}$ under exchange $1 \leftrightarrow 2$, the contribution is again bounded by $T_1(s)$.

In case (iii), taking suprema after integrating over $(P_q)_{q \in S}$, we get

$$T_2(s_1, s_2) = \mathbf{1}(s_1 < m_1 \wedge s_2 < m_2) \tag{4.157}$$

$$\int d\hat{\kappa}_{mr} \|\bar{D}_t\|_1^{i-1} \|\dot{\bar{D}}_t\|_1 |I_{m_1r_1}(t)|_{s_1}\ |I_{m_2r_2}(t)|_{s_2}.$$

The crucial difference to (4.156) is that both delta functions are used up in the definition of $|\cdot|_{s_k}$ and that therefore K_1 is also integrated. Thus the L^1-norm of the differentiated propagator appears instead of the sup norm. We thus have

$$|A_m Q_{mr}(t)|_s \leq 2T_1(s) + \max_{\substack{s_1, s_2 \geq 1 \\ s_1 + s_2 = s}} T_2(s_1, s_2). \qquad (4.158)$$

By Lemma 4.6, and since $B_0 = 4$,

$$T_1(s) \leq \mathbb{1}(s < m_1) \int d\hat{\kappa}_{mr} \, (8J_1 \epsilon_t)^{i-1} \frac{16}{\epsilon_t} |I_{m_1 r_1}(t)|_s \, |I_{m_2 r_2}(t)|_0 \qquad (4.159)$$

and

$$T_2(s_1, s_2) \leq \mathbb{1}(s_1 < m_1 \wedge s_2 < m_2) \qquad (4.160)$$
$$\int d\hat{\kappa}_{mr} \, (8J_1 \epsilon_t)^i \, |I_{m_1 r_1}(t)|_{s_1} \, |I_{m_2 r_2}(t)|_{s_2}.$$

Thus

$$\left| \tfrac{\partial}{\partial t} I_{mr}(t) \right|_s \leq \int d\hat{\kappa}_{mr} (8J_1 \epsilon_0)^i e^{-ti} \max\{\theta_1(s,t), \theta_2(s,t)\} \qquad (4.161)$$

with

$$\theta_1(s,t) = \mathbb{1}(s < m_1) \, \tfrac{1}{J_1 \epsilon_0^2} e^{2t} |I_{m_1 r_1}(t)|_s \, |I_{m_2 r_2}(t)|_0 \qquad (4.162)$$
$$\theta_2(s,t) = \max_{\substack{s_1, s_2 \geq 1 \\ s_1 + s_2 = s}} \mathbb{1}(s_1 < m_1 \wedge s_2 < m_2) \, |I_{m_1 r_1}(t)|_{s_1} \, |I_{m_2 r_2}(t)|_{s_2}.$$

Theorem 4.9. *Let $|I_{mr}(0)|_s \leq b_{mrs}^{(0)}$ for all $m, r,$ and s. For any $0 < \alpha < 1$, there are sequences $(b_{mrs})_{m,r,s}$ and c_{mr} (depending on α) such that for all $t \geq 0$,*

$$\left| \tilde{I}_{mr}(t) \right|_s \leq \begin{cases} b_{mrs}(1 + t)^{r-1} \, e^{\frac{t}{2}(m - s - \alpha)} & \text{if } s < m \\ c_{mr} & \text{if } s = m. \end{cases} \qquad (4.163)$$

$\tilde{I}_{mr}(t)$ converges in $|\cdot|_m$ to a function \tilde{I}_{mr} as $t \to \infty$, uniformly in β.

Proof. Induction on r, with (4.163) as the inductive hypothesis. For $r = 1$, the statement follows from our hypotheses on the initial interaction $I_{mr}(0)$ because $Q_{m1}(t) = 0$ for all t. Let $r \geq 2$. The inductive hypothesis applies to $I_{m_1 r_1}$ and $I_{m_2 r_2}$. Let $1 \leq s \leq m$. By the inductive hypothesis and Theorem 4.8, and because $m_1 < s$ and $m_1 \geq 4$ and $m_2 \geq 4$,

$$e^{-ti}\theta_1(s,t) \leq \tfrac{1}{J_1 \epsilon_0^2} b_{m_1 r_1 s} a_{m_2 r_2} (1 + t)^{r_1 - 1 + r_2 - 1} e^{\frac{t}{2}(m_1 - s - \alpha + m_2 - 4)}$$
$$\leq \tfrac{1}{J_1 \epsilon_0^2} b_{m_1 r_1 s} a_{m_2 r_2} (1 + t)^{r-2} e^{\frac{t}{2}(m - s - \alpha)}. \qquad (4.164)$$

Similarly, $s_1 < m_1$ and $s_2 < m_2$ and the inductive hypothesis imply that

$$e^{-ti}\theta_1(s,t) \leq \max_{\substack{s_1, s_2 \geq 1 \\ s_1 + s_2 = s}} b_{m_1 r_1 s_1} b_{m_2 r_2 s_2} (1 + t)^{r-2} e^{\frac{t}{2}(m_1 - s_1 - \alpha + m_2 - s_2 - \alpha) - ti}$$
$$\leq \max_{\substack{s_1, s_2 \geq 1 \\ s_1 + s_2 = s}} b_{m_1 r_1 s_1} b_{m_2 r_2 s_2} (1 + t)^{r-2} e^{\frac{t}{2}(m - s - \alpha)}. \qquad (4.165)$$

Thus

$$\left|\frac{\partial}{\partial t}I_{mr}(t)\right|_s \le e^{\frac{t}{2}(m-s-\alpha)}(1+t)^{r-2}\int d\tilde{\kappa}_{mr}(8J_1\epsilon_0)^i\mathcal{M}_{mrs} \qquad (4.166)$$

with

$$\mathcal{M}_{mrs} = \max\{\tfrac{1}{J_1\epsilon_0^2}b_{m_1r_1s}a_{m_2r_2}, \max_{\substack{s_1,s_2\ge 1\\ s_1+s_2=s}} b_{m_1r_1s_1}b_{m_2r_2s_2}\}, \qquad (4.167)$$

and therefore

$$|I_{mr}(t)|_s \le b_{mrs}^{(0)}+\int d\tilde{\kappa}_{mr}(8J_1\epsilon_0)^i\mathcal{M}_{mrs}\int_0^t dt' e^{\frac{t'}{2}(m-s-\alpha)}(1+t')^{r-2}. \qquad (4.168)$$

If $s < m$, then $m - s - \alpha \ge 1 - \alpha > 0$, so the integral over t' is bounded by $\frac{2}{\alpha}(1+t)^{r-2}e^{\frac{t}{2}(m-s-\alpha)}$. Since this quantity is at least one, the induction step is completed for $m < s$ by defining

$$b_{mrs} = b_{mrs}^{(0)} + \int d\tilde{\kappa}_{mr}(8J_1\epsilon_0)^i\mathcal{M}_{mrs}. \qquad (4.169)$$

If $s = m$, $m - s - \alpha = -\alpha < 0$, so the integral over t' is

$$\int_0^t dt'(1+t')^{r-2}e^{-\frac{\alpha}{2}t'} \le e^{\frac{\alpha}{2}}\int_0^t dt'(1+t')^{r-2}e^{-\frac{\alpha}{2}(1+t')} \qquad (4.170)$$

$$\le e^{\frac{\alpha}{2}}\int_0^\infty d\tau\, \tau^{r-2}e^{\tau\frac{\alpha}{2}} = \left(\tfrac{2}{\alpha}\right)^{r-1}e^{\frac{\alpha}{2}}(r-2)!$$

and the induction step is completed by defining

$$c_{mr} = b_{mrm}^{(0)} + \int d\tilde{\kappa}_{mr}(8J_1\epsilon_0)^i\mathcal{M}_{mrm}\left(\tfrac{2}{\alpha}\right)^{r-1}e^{\frac{\alpha}{2}}(r-2)!. \qquad (4.171)$$

For finite β, convergence is trivial because $\frac{\partial I_{mr}}{\partial t} = 0$ for $t > \log\frac{\beta\epsilon_0}{2}$. For $\beta = \infty$, convergence follows from (4.166) because for $s = m$, the right hand side of (4.166) vanishes as $t \to \infty$. ∎

4.5 The Four-Point Function

In this section, we do some explicit calculations for a truncation of the RGDE. This truncation corresponds to the ladder approximation to the four-point function. The justification of this truncation will be discussed in Sect. 4.6.

4.5.1 Motivation

We have seen in the last section that the four-point function produced factors

$$t^r \leq (\log(\beta\epsilon_0))^r \tag{4.172}$$

in the estimates for all Green functions. The powers t^r produced r factorials in the constants in the bounds for the Green functions, for instance in (4.150). If instead, one uses (4.172) to avoid the r factorial in (4.150), one gets bounds that depend on β. It is very important to find out if these bounds are saturated: suppose that for some P,

$$\left| \tilde{I}_{2,r}(t|P) \right|_0 \propto a_{2,r} \tag{4.173}$$

where $a_{2,r} \propto r!\, \text{const}^r$, then the series $\sum_r \lambda^r \tilde{I}_{2,r}(t|P)$ diverges. If, alternatively,

$$\left| \tilde{I}_{2,r}(t|P) \right|_0 \propto \tilde{a}_{2,r} \left(\log(\beta\epsilon_0) \right)^r, \tag{4.174}$$

with $\tilde{a}_{2,r} \leq \text{const}^r$, the series $\sum_r \lambda^r \tilde{I}_{2,r}(t|P)$ converges at most in a temperature-dependent region

$$|\lambda \log(\beta\epsilon_0)| < \text{const} , \tag{4.175}$$

and a similar statement holds for all other Green functions.

Both of the above statements are really true for the class of models considered here. The expansion for the skeleton Green functions does not converge uniformly in the temperature, but it converges in a region given by the condition (4.175). The above bounds on the skeleton functions already gave a hint that such behaviour can hold. To really prove this, one needs to control the combinatorics. Such a proof is possible using a technique invented in [55], but it is beyond the scope of this text.

On the other hand, the above bounds do not imply that the $\log \beta$ is really there. In this and the next section, we motivate why the logarithm actually occurs. In this section, we study a truncation of the right hand side of the differential equation where all terms with number of internal lines $i \geq 3$ are left out. This leads to a differential equation whose solution is a resummation of the particle-particle and particle-hole ladders. These graphs are sometimes called the *parquet graphs*, so we shall call the four-point function obtained by this truncation the parquet four-point function. Under the assumptions we made on the Fermi surface, in particular because of the positive curvature, the particle-particle ladders dominate this flow, and they have the above logarithmic behaviour when the sum of the two ingoing momenta adds up to zero.

The terms with $i \geq 3$ will be treated in the next section. We show that the solution to the truncation of the RGE where the $i = 2$ terms are left out in the equation for the four-point function is bounded uniformly in the temperature,

that is, there are no factors t^r or $(\log \beta)^r$ in that case. Thus indeed, the ladder term $i = 2$ in the RGE is the only cause of high powers of $\log \beta$. The "overlapping loop" method used to prove this applies very generally and will afterwards be of use for bounding derivatives of the selfenergy as well.

4.5.2 The Parquet Four-Point Function

We now discuss the truncation of the flow obtained by keeping only the $i = 2$ term on the RHS of the RGE for the skeleton four-point function. In Q_{4r}, $m_1 + m_2 = 4 + 2i$, so $i \geq \frac{1}{2}(m_1 + m_2 - 4) \geq 2$ since the skeleton condition has removed $m_1 = 2$ and $m_2 = 2$. Let $Q_{4,r,2}$ be the $i = 2$ term in this sum; it corresponds to the 'bubble' graph drawn in Fig. 4.1.

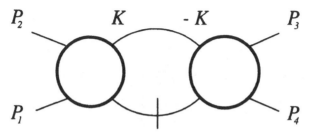

Fig. 4.1. The graph corresponding to $Q_{4,r,2}$

More explicitly, $Q_{4,r,2}$ is given by

$$Q_{4,r,2}(t \mid P_1, \ldots, P_4) =$$
$$\frac{1}{2} A_4 \left[144 \sum_{r_1+r_2=r} \sum_{i_1,\sigma_1,i_2,\sigma_2} \int dk \, \bar{D}_t(K) \, \dot{D}_t(K') \right. \tag{4.176}$$
$$\left. \tilde{I}_{4,r_1}(t \mid P_1, P_2, K', K) \, \tilde{I}_{4,r_2}(t \mid \sim K, \sim K', P_3, P_4) \right]$$

with $K = (k, i_1, \sigma_1)$ and $K' = (-p_1 - p_2 - k, i_2, \sigma_2)$. Let

$$Q_{4,r,\geq 3}(t \mid \underline{P}) = Q_{4,r}(t \mid \underline{P}) - Q_{4,r,2}(t \mid \underline{P}) \tag{4.177}$$

be the contribution from all terms where at least $i \geq 3$ lines connect the two vertices in $Q_{4,r}$.

Although the graph in Fig. 4.1 looks like a bubble graph, the iterative solution to the according differential equation

$$\frac{\partial}{\partial t} I_{4,r}(t \mid \underline{P}) = \frac{1}{2} A_4 Q_{4,r,2}(t \mid \underline{P}) \tag{4.178}$$

generates not just the ladder or the RPA graphs, but many more arising from the combinations of the two types. We call the truncated flow given by (4.178) the parquet flow.

To make the above statement about graphs clearer, we now switch to the more conventional representation in terms of ψ and $\bar{\psi}$ (we also resum the perturbation series over r and denote $F = \sum_r \lambda^r I_{4,r}$, where $I_{4,r}$ is the solution to the truncated RGE (4.178)). Since the j index in $P = (p, \sigma, j)$ fixes whether ψ or $\bar{\psi}$ appears, the four-point function with two incoming fermions and two outgoing ones can be defined by

$$f(t \mid (p_1, \sigma_1), (p_2, \sigma_2), (p_3, \sigma_3), (p_4, \sigma_4)) \tag{4.179}$$
$$= F(t \mid (-p_1, \sigma_1, 1), (-p_2, \sigma_2, 1), (p_3, \sigma_3, 2), (p_4, \sigma_4, 2)).$$

The minus signs in front of p_1 and p_2 mean that the momenta are now counted positive when they are outgoing, so that in f, momentum conservation reads $p_1 + p_2 = p_3 + p_4$. The function f in (4.179) fixes F uniquely because of the antisymmetry $A_4 F = F$. A natural graphical representation of f is drawn in Fig. 4.2.

Fig. 4.2. The vertex corresponding to f

In terms of f, (4.178) reads, with $\xi_k = (p_k, \sigma_k)$,

$$\frac{\partial f}{\partial t}(t \mid \xi_1, \ldots, \xi_4) = \Phi_{pp}(t \mid \xi_1, \ldots, \xi_4) \tag{4.180}$$
$$- \Phi_{ph}(t \mid \xi_1, \xi_2, \xi_3, \xi_4) + \Phi_{ph}(t \mid \xi_1, \xi_2, \xi_4, \xi_3),$$

where

$$\Phi_{pp}(t \mid \xi_1, \ldots, \xi_4) = 12 \int dk \left[\frac{\partial}{\partial t} (D_t(k) D_t(p_1 + p_2 - k)) \right]$$
$$\sum_{\tau_1, \tau_2} f(t \mid \xi_1, \xi_2, (p_1 + p_2 - k, \tau_2), (k, \tau_1)) \tag{4.181}$$
$$f(t \mid (k, \tau_1), (p_1 + p_2 - k, \tau_2), \xi_3, \xi_4)$$

and

$$\Phi_{ph}(t \mid \xi_1, \xi_2, \xi_3, \xi_4) = 24 \int dk \left[\frac{\partial}{\partial t} (D_t(k) D_t(p_2 - p_3 + k)) \right]$$
$$\sum_{\tau_1, \tau_2} f(t \mid (k, \tau_1), \xi_1, \xi_3, (p_2 - p_3 + k, \tau_2)) \tag{4.182}$$
$$f(t \mid \xi_2, (p_2 - p_3 + k, \tau_2), (k, \tau_1), \xi_4)$$

and D_t is the scaled propagator $D_t = C_t$, with C_t given by (4.130). There are two particle-hole terms because the function f has to be antisymmetric under exchange of ξ_1 and ξ_2, and under exchange of ξ_3 and ξ_4. The difference of the two Φ_{ph} functions has these properties. (4.180) has the graphical interpretation shown in Fig. 4.3.

Fig. 4.3. The graphs in the parquet flow equation

We now discuss more explicitly the particle-particle and particle-hole flow and show how the particle-particle flow leads to the restriction $|\lambda| \log \beta <$ const and why the particle-hole flow does not lead to singularities if the Fermi surface has positive curvature. The mixing of the two flows will be discussed after that.

4.5.3 The One-Loop Volume Bound

The right hand side of the parquet equation contains the functions

$$b_t^{(\pm)}(q) = \tfrac{\partial}{\partial t} \int dk\, D_t(k) D_t(q \pm k). \tag{4.183}$$

Let

$$B_t^{(\pm)}(q) = - \int_0^t ds\, b_s^{(\pm)}(q), \tag{4.184}$$

then

$$B_t^{(\pm)}(q) = \int dk\, (D_0(k) D_0(q \pm k) - D_t(k) D_t(q \pm k)) = B_t^{(\pm)}(-q). \tag{4.185}$$

$B_t^{(-)}$ is called the particle-particle bubble, and $B_t^{(+)}$ the particle-hole bubble. The power counting applied in Sect. 4.4 gives the bound

$$|b_t^{(\pm)}(q)| \le 2\|\dot{D}_t\|_\infty \|D_t\|_1 \le 2J_1 B_0^2. \tag{4.186}$$

This bound is independent of q. A better bound can be obtained as follows. We use that D vanishes for $t > \log(\beta\epsilon_0$ to write

$$D_t(k) = \int_t^{\log(\beta\epsilon_0)} dt'\, \dot{D}_{t'}(k), \tag{4.187}$$

to get

$$|b^{(\pm)}(q)| \leq 2 \int\limits_{t}^{\log(\beta\epsilon_0)} dt' \int dk \, |\dot{D}_{t'}(k) \, \dot{D}_t(q \pm k)|. \tag{4.188}$$

We now use the pointwise bound (4.131) of Lemma 4.6, bound the indicator function $\mathbf{1}(|q_0 \pm \omega_\beta(k_0)| \leq \epsilon_{t'}) \leq 1$, and bound the frequency sum with the restriction $|\omega| \leq \epsilon_{t'}$ by $\epsilon_{t'}$ (see (4.137)), to get

$$|b^{(\pm)}(q)| \leq 2B_0^2 \epsilon_t^{-1} \int\limits_{t}^{\log(\beta\epsilon_0)} dt' \, \mathcal{V}_1^{(\pm)}(\mathbf{q}, t, t') \tag{4.189}$$

with

$$\mathcal{V}_1^{(\pm)}(\mathbf{q}, t, t') = \int\limits_{B} \frac{dk}{(2\pi)^d} \, \mathbf{1}(|E(\mathbf{k})| \leq \epsilon_{t'}) \, \mathbf{1}(|E(\mathbf{q} \pm \mathbf{k})| \leq \epsilon_t) \, . \tag{4.190}$$

Let \mathcal{R}_{ϵ_t} be the neighbourhood of the Fermi surface defined by

$$\mathcal{R}(\epsilon_t) = \{\mathbf{p} \in B : |E(\mathbf{p})| \leq \epsilon_t\}, \tag{4.191}$$

then $\mathcal{V}_1^{(\pm)}(\mathbf{q}, t, t')$ is the intersection volume

$$\mathcal{V}_1^{(\pm)}(\mathbf{q}, t, t') = (2\pi)^{-d} \, \mathrm{vol}_d \, (\mathcal{R}(\epsilon_{t'}) \cap (\mathcal{R}(\epsilon_t) \mp \mathbf{q})) \tag{4.192}$$

of $\mathcal{R}_{\epsilon_{t'}}$ and the translate of $\mathcal{R}(\epsilon_t)$ by $\mp\mathbf{q}$. By the symmetry $E(-\mathbf{p}) = E(\mathbf{p})$, $\mathcal{V}_1^{(v)}(\mathbf{q}, t, t')$ is independent of $v \in \{-1, 1\}$.

Lemma 4.10. *Let $\|\mathbf{q}\|_B$ be the distance of \mathbf{q} to 0 on the torus B, and let $t' \geq t$. Then there is a constant $Q_0 > 0$ such that for all $\mathbf{q} \in B$,*

$$\mathcal{V}_1^{(\pm)}(\mathbf{q}, t, t') \leq Q_0 \, \epsilon_{t'} \begin{cases} 1 & \text{if } \|\mathbf{q}\|_B \leq \sqrt{\epsilon_t} \\ \sqrt{\epsilon_t} & \text{otherwise.} \end{cases} \tag{4.193}$$

The geometric idea in the proof of this lemma is that, for the class of Fermi surfaces considered here, the intersection of the Fermi surface and its translate by a vector \mathbf{q} is transversal for almost all \mathbf{q}, in which case the intersection volume is bounded by the product of the thicknesses of the neighbourhoods, $\epsilon_t \epsilon_{t'} \leq \epsilon_{t'} \sqrt{\epsilon_t}$. . For $\mathbf{q} = 0$, both neighbourhoods lie on top of each other, and no gain is possible. For $\mathbf{q} \in 2S$ the intersection is tangential, but the curvature of S provides a gain $\sqrt{\epsilon_t}$. The various cases are drawn in Fig. 4.4.

We prove Lemma 4.10 for a spherical Fermi surface in Appendix B.8.1, and also discuss the generalization to nonspherical Fermi surfaces there.

We now turn back to the analysis of the four-point function. The volume bound implies the following bound.

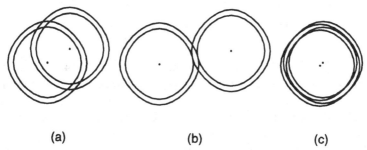

Fig. 4.4. The intersections of a shell around the Fermi surface with its translate by a vector **q**. (a) transversal (b) tangential, with **q** near to $2S$ (c) $\mathbf{q} \approx 0$

Corollary 4.11. *Let $B_t^{(\pm)}(q)$ be as in (4.184). For all $q_0 \in \mathbb{R}$, all $\mathbf{q} \in \mathcal{B}$, and all $t \geq 0$,*

$$B_t^{(\pm)}(q_0, \mathbf{q}) \leq 2B_0^2 Q_O \left(2 + \log(\tfrac{\epsilon_0}{\|\mathbf{q}\|_B^2})\right). \tag{4.194}$$

Proof. By (4.189) and Lemma 4.10,

$$|b_t^{(\pm)}(q)| \leq 2B_0^2 Q_O \, \epsilon_t^{-1} \int\limits_t^{\log(\beta\epsilon_0)} dt' \, \epsilon_{t'} \begin{cases} 1 & \text{if } \|\mathbf{q}\|_B \leq \sqrt{\epsilon_t} \\ \sqrt{\epsilon_t} & \text{otherwise.} \end{cases} \tag{4.195}$$

The integral over t' is bounded by ϵ_t, so

$$\left|b_t^{(\pm)}(q)\right| \leq 2B_0^2 Q_O \begin{cases} 1 & \text{if } \|\mathbf{q}\|_B \leq \sqrt{\epsilon_t} \\ \sqrt{\epsilon_t} & \text{otherwise.} \end{cases} \tag{4.196}$$

To get a bound for $B_t^{(\pm)}$, we now have to integrate the right hand side of (4.196) over t. For $t \leq \log(\tfrac{\epsilon_0}{\|\mathbf{q}\|_B^2})$, the bound for the integrand is constant, so the integral gives this constant times a factor $\log(\tfrac{\epsilon_0}{\|\mathbf{q}\|_B^2})$. For $t \geq \log(\tfrac{\epsilon_0}{\|\mathbf{q}\|_B^2})$, the integral is bounded by a constant times $\int_0^\infty dt\, e^{-t/2} \leq 2$. ∎

Thus the only place where $B^{(\pm)}(q)$ can diverge as $\beta \to \infty$ is $\mathbf{q} = 0$. We now investigate if that is the case.

4.5.4 The Particle–Particle Flow

Let us look first at the modification of (4.178) where the Φ_{ph} terms on the right hand side are dropped, i.e. $\frac{\partial f}{\partial t} = \Phi_{pp}$. For an initial four-point function $f(0)$ that is local in position space, one can solve this equation explicitly. That is, the solution with

$$f(0|\xi_1, \ldots, \xi_4) = -\lambda_0 \varepsilon_{\sigma_1 \sigma_2} \varepsilon_{\sigma_3 \sigma_4}, \tag{4.197}$$

with some initial coupling constant λ_0, is

$$f(t|\xi_1,\ldots,\xi_4) = -\frac{\lambda_0}{1 + 24\lambda_0 B_t^{(-)}(p_1 + p_2)}\varepsilon_{\sigma_1\sigma_2}\varepsilon_{\sigma_3\sigma_4}. \qquad (4.198)$$

where $B_t^{(-)}(q)$ is the particle-particle bubble defined in (4.184). The minus sign in the initial condition appears because we wrote the Boltzmann factor as $e^{\mathcal{G}}$ and the initial Boltzmann factor as $e^{-\lambda \mathcal{V}}$. Thus $\lambda_0 > 0$ corresponds to a repulsive interaction, and $\lambda_0 < 0$ to an attractive one. The ε symbols make f antisymmetric under exchange of ξ_1 and ξ_2, and ξ_3 and ξ_4. The factor 24 and the extra sign are due to $\sum_{\tau_1,\tau_2}\varepsilon_{\tau_1\tau_2}\varepsilon_{\tau_2\tau_1} = -2$. The solution is so simple because the momentum dependence of $f(t|\xi_1,\ldots,\xi_4)$ is only via the sum of the ingoing momenta $p_1 + p_2$. Thus integrating this truncated equation amounts to a resummation of the particle-particle ladders. For $t \to \infty$, $D_t(k) \to 0$, so

$$B^{(-)}(q) = \lim_{t\to\infty} B_t^{(-)}(q) = \int dk\, D_0(k)D_0(q-k). \qquad (4.199)$$

For $T > 0$, $B^{(-)}(0)$ is finite, but proportional to $\log\beta$: up to a term coming from the ultraviolet cutoff that is bounded uniformly in the temperature, we have, by the symmetry $E(-\mathbf{k}) = E(\mathbf{k})$,

$$\begin{aligned}
B^{(-)}(0) &= \tfrac{1}{\beta}\sum_{\omega\in M}\int_{|E(\mathbf{k})|\leq\epsilon_0} dk\, \frac{1}{i\omega - E(\mathbf{k})}\,\frac{1}{-i\omega - E(-\mathbf{k})} \\
&= \int_{-\epsilon_0}^{\epsilon_0} dE\, N(E)\,\tfrac{1}{\beta}\sum_{\omega\in M}\frac{1}{\omega^2 + E^2}
\end{aligned} \qquad (4.200)$$

with the density of states $N(E)$ as in (4.77). Because $\sum\omega^{-2}$ converges we can put in a factor $e^{i\epsilon\omega}$ and get , by (4.68),

$$\begin{aligned}
\tfrac{1}{\beta}\sum_{\omega\in M}\frac{1}{\omega^2 + E^2} &= \lim_{\epsilon\downarrow 0}\tfrac{1}{\beta}\sum_{\omega\in M}\frac{e^{i\epsilon\omega}}{\omega^2 + E^2} \\
&= \lim_{\epsilon\downarrow 0}\tfrac{1}{\beta}\sum_{\omega\in M}\frac{1}{-2E}\left(\frac{e^{i\epsilon\omega}}{i\omega - E} + \frac{e^{i\epsilon\omega}}{-i\omega - E}\right) \\
&= \frac{1}{-2E}\left(\frac{1}{1 + e^{\beta E}} - \frac{1}{1 + e^{-\beta E}}\right) \\
&= \frac{1}{2E}\tanh\frac{\beta E}{2},
\end{aligned} \qquad (4.201)$$

so

$$B^{(-)}(0) = \int_{-\epsilon_0}^{\epsilon_0}\frac{dE}{E}\tanh\frac{\beta E}{2}\, N(E) = \int_{-\frac{\beta\epsilon_0}{2}}^{\frac{\beta\epsilon_0}{2}}\frac{dx}{2x}\tanh x\, N(\tfrac{2x}{\beta}). \qquad (4.202)$$

As shown in Sect. 4.2.6, $N(E)$ is C^{k_0-1} around zero, so a Taylor expansion gives for large β

$$B^{(-)}(0) = \frac{1}{2}N(0)\log(\frac{\beta\epsilon_0}{2}) + O(1) \qquad (4.203)$$

(if $k_0 \geq 3$ then one can use Taylor expansion to second order; the first order term then cancels because the integrand is odd, and one gets $O(\frac{1}{\beta})$ instead of $O(1)$). $N(0)$ is the density of states at the Fermi surface.

For a repulsive interaction ($\lambda_0 > 0$), the function f remains bounded and vanishes for $\beta \to \infty$ at $p_1 + p_2 = 0$ ("asymptotic freedom": a running coupling constant defined by

$$\lambda(t) = -\frac{\lambda_0}{1 + 24\lambda_0 B_t^{(-)}(0)} \qquad (4.204)$$

decreases to 0 as $t \to \infty$.

For an attractive interaction ($\lambda_0 < 0$), $\lambda(t)$ has a singularity at a finite value of t (the BCS case: the effective coupling constant grows as the energy scale is lowered), and f becomes singular if β gets large. However, f remains finite and analytic in λ_0 as long as $|\lambda_0|N(0)\log(\beta\epsilon_0)$ is small enough (this statement was proved in the last section for the skeleton four-point function, not just the ladder resummation). If one fixes λ_0 and increases β, a singularity indicating the onset of superconductivity occurs at a temperature $T_0 = 1/\beta_0$ of order

$$T_0 \propto \epsilon_0 e^{-\frac{1}{24|\lambda|N(0)}}. \qquad (4.205)$$

It is natural that ϵ_0 sets the scale in this equation: the above considerations are valid only when the particle-particle ladders dominate the RG flow. If the Fermi surface has positive curvature, it is possible to choose ϵ_0 such that the ladders dominate the flow for energies below ϵ_0. The role of the curvature for this will become clear when the volume improvement from overlapping loops is shown; it will be discussed further in Sect. 4.8.

The coupling constant λ_0 is not the original coupling constant, but the effective four-point coupling at scale ϵ_0. It is a regular, in fact, analytic, function of the original coupling constant λ, if $|\lambda|$ is small enough (dependent on ϵ_0) but it is of course not true in general that this effective interaction is local, i.e. a constant in momentum space. This can be taken into account in the following way. Let

$$\phi_{\underline{\sigma}}(t|Q, k, l) = f\left(t|(-k + \tfrac{Q}{2}, \sigma_1), (k + \tfrac{Q}{2}, \sigma_2), (l + \tfrac{Q}{2}, \sigma_3), (-l + \tfrac{Q}{2}, \sigma_4)\right),$$

$$\qquad (4.206)$$

then

$$\frac{\partial\phi_{\underline{\sigma}}}{\partial t}(t \mid Q, k, l) = 12\int dr\, [\tfrac{\partial}{\partial t}(D_t(-r + \tfrac{Q}{2})D_t(r + \tfrac{Q}{2}))] \qquad (4.207)$$

$$\sum_{\tau_1,\tau_2} \phi_{\sigma_1\sigma_2\tau_1\tau_2}(t \mid Q, k, r)\phi_{\tau_2\tau_1\sigma_3\sigma_4}(t \mid Q, r, l).$$

Note that the sum of the ingoing momenta Q now plays the role of a parameter in this function. Because $Q = 0$ is the singular case, we now restrict to $Q = 0$. The flow for $\phi_{\underline{\sigma}}$ has the initial condition

$$\phi_{\underline{\sigma}}(0 \mid 0, k, l) = v_{\underline{\sigma}}(k, l) \tag{4.208}$$

where v is a C^2 function obeying t-independent bounds, since it is the initial condition. The particle-particle bubble in (4.207) is independent of the momenta k and l. Thus derivatives with respect to k and l do not act on the propagators D_t and therefore do not deteriorate the scaling behaviour. Writing the integration momentum as $r = (\omega, \pi(\rho, \theta))$, we may thus do a Taylor expansion in ρ without influencing the t-dependence. The terms proportional to ρ have power counting that is a factor e^{-t} better. Denoting

$$w(t \mid \sigma_1, \sigma_2, \theta ; \sigma_3, \sigma_4, \theta') = \phi_{\underline{\sigma}}(t \mid 0, \pi(0, \theta), \pi(0, \theta')), \tag{4.209}$$

we thus have, up to terms that are multiplied by an extra factor e^{-t},

$$\frac{\partial w}{\partial t}(t \mid \sigma_1, \sigma_2, \theta ; \sigma_3, \sigma_4, \theta') =$$
$$12 \int d\omega \, d\rho \frac{\partial}{\partial t} \left(\frac{\chi_1(\epsilon_t^{-2}(\omega^2 + \rho^2))^2}{\omega^2 + \rho^2} \right) \int d\theta'' J(0, \theta'') \tag{4.210}$$
$$\sum_{\tau_1, \tau_2} w(t \mid \sigma_1, \sigma_2, \theta ; \tau_1, \tau_2, \theta'') w(t \mid \tau_2, \tau_1, \theta'' ; \sigma_3, \sigma_4, \theta').$$

We can now interpret [44] the function w as the kernel of an operator acting on the weighted L^2 space $L^2(\{-1, 1\}^2 \times S^{d-1})$ where the weight is given by the local density of states $J(0, \theta)$. An explicit construction of an ONB for spaces associated to Fermi surfaces (also called Fermi surface harmonics) was given by Allen [49]. An expansion in an ONB of eigenfunctions of w, with t-dependent coefficients $g_k(t)$,

$$w(t \mid \sigma_1, \sigma_2, \theta ; \sigma_3, \sigma_4, \theta') = \sum_{k=1}^{\infty} g_k(t) \Psi_k(\sigma_1, \sigma_2, \theta) \overline{\Psi_k(\sigma_3, \sigma_4, \theta')}, \tag{4.211}$$

then gives, because of orthogonality, the uncoupled system of equations

$$\frac{\partial g_k(t)}{\partial t} = 12 \dot{b}_t^{(-)}(0) g_k(t)^2. \tag{4.212}$$

Thus, in the ladder approximation, the coupling constants associated to the various symmetry channels (which generalize angular momentum channels) flow independently, and all according to the law

$$g_k(t) = \frac{g_k(0)}{1 + 12 g_k(0) B_t^{(-)}(0)}. \tag{4.213}$$

Any coupling that starts out negative $(g_k(0) < 0)$ will eventually grow, but one can expect that the one that has largest absolute value will determine the symmetry of the superconducting order parameter.

The full flow does not lead to a decoupling of the different angular momentum sectors (even here, we had to do a Taylor expansion and omit some terms to get this decoupling property). Kohn and Luttinger observed, by a second-order calculation [50], that even if the initial interaction is purely repulsive, the effective action at scale ϵ_0, which determines the initial condition (4.208), must have $g_k(0) < 0$ for some (in general very large) k. Feldman, Knörrer, Sinclair and Trubowitz [51] have shown by a very careful analysis that although in $d = 2$ dimensions, the Kohn–Luttinger effect is not there in second order, there is a similar third order effect that drives the full RG flow to an $l = 1$ superconducting state [51]. The analysis of [51, 52] goes far beyond the original proposal [50]. In particular, the RG flow, and not just the Bethe-Salpeter kernel, is studied in [52].

4.5.5 The Particle–Hole Flow

The particle-hole flow is obtained by leaving out the Φ_{pp} term on the right hand side of (4.178). Similarly to the particle-particle case, this now resums the particle-hole bubbles

$$B_t^{(+)}(q) = \int dk \ (D_0(k)D_0(q+k) - D_t(k)D_t(q+k)) . \tag{4.214}$$

In contrast to the particle particle case, this function is bounded uniformly in the temperature even in the limit $t \to \infty$. This follows for $|q|$ away from zero from Corollary 4.11 and for $|q|$ near zero from the following theorem, which is based on an interpolation formula for Fermi surfaces.

Theorem 4.12. *Let* $\mathbf{q} \in \mathcal{B}$, $q_0 \in \frac{2\pi}{\beta}\mathbb{Z}$, *and*

$$B^{(+)}(q) = \int \frac{d^d\mathbf{p}}{(2\pi)^d} \frac{1}{\beta} \sum_{n \in \mathbb{Z}} \frac{1}{i\omega_n - E(\mathbf{p})} \frac{1}{i(\omega_n + q_0) - E(\mathbf{p+q})} . \tag{4.215}$$

There is $r > 0$ *and* $K > 0$ *such that for all* $|\mathbf{q}| < r$ *and all* q_0

$$|B^{(+)}(q)| \le K \tag{4.216}$$

r is determined by the nearest van Hove singularity (i.e. the nearest point where $\nabla E(\mathbf{p}) = 0$).

Proof. Let $\chi_\beta(x) = (1 + e^{\beta x})^{-1}$ be the Fermi distribution and $\delta_\beta(x) = -\frac{\partial}{\partial x}\chi_\beta(x)$. The function δ_β is an approximate unity: $\int_{\mathbb{R}} dx \ \delta_\beta(x) = 1$, and $\delta_\beta \to \delta$ as $\beta \to \infty$.

Let $\mathbf{q} \neq 0$. Using a Taylor expansion, $B^{(+)}(q)$ can be rewritten as

$$
\begin{aligned}
B^{(+)}(q) &= \int \frac{d^d\mathbf{p}}{(2\pi)^d} \frac{1}{iq_0 + E(\mathbf{p}) - E(\mathbf{p}+\mathbf{q})} \\
&\qquad \frac{1}{\beta} \sum_{n \in \mathbb{Z}} \left(\frac{1}{i\omega_n - E(\mathbf{p})} - \frac{1}{i\omega_n - E(\mathbf{p}+\mathbf{q})} \right) \\
&= \int \frac{d^d\mathbf{p}}{(2\pi)^d} \frac{\chi_\beta(E(\mathbf{p})) - \chi_\beta(E(\mathbf{p}+\mathbf{q}))}{iq_0 + E(\mathbf{p}) - E(\mathbf{p}+\mathbf{q})} \\
&= \int \frac{d^d\mathbf{p}}{(2\pi)^d} \frac{E(\mathbf{p}) - E(\mathbf{p}+\mathbf{q})}{iq_0 + E(\mathbf{p}) - E(\mathbf{p}+\mathbf{q})} \int_0^1 dt\, \delta_\beta(E(\mathbf{p}, \mathbf{q}, t))
\end{aligned}
\tag{4.217}
$$

where

$$
E(\mathbf{p}, \mathbf{q}, t) = (1 - t)E(\mathbf{p}) + tE(\mathbf{p} + \mathbf{q}).
\tag{4.218}
$$

Because

$$
\nabla E(\mathbf{p}, \mathbf{q}, t) = \nabla E(\mathbf{p}) + tE''(\mathbf{p} + \vartheta \mathbf{q}) \cdot \mathbf{q}
\tag{4.219}
$$

with some $\vartheta \in [0, 1]$ (here E'' denotes the matrix of second partial derivatives of E), and because $|\nabla E(\mathbf{p})| \geq g_0$ on the compact set S and the second derivative of E is bounded, there is $r > 0$ such that for all $|\mathbf{q}| < r$, $|\nabla E(\mathbf{p}, \mathbf{q}, t)| \geq g_0/2$. Thus the *interpolating Fermi surface* $S_t = \{\mathbf{p} : E(\mathbf{p}, \mathbf{q}, t) = 0\}$ is still a submanifold for all $t \in [0, 1]$ (it is, in our case, even positively curved). We can therefore introduce a radial coordinate $\rho = E(\mathbf{p}, \mathbf{q}, t)$ and a tangential coordinate $\theta \in S^{d-1}$ in the standard way [45], with a bounded Jacobian \tilde{J}, and get

$$
\left| B^{(+)}(q) \right| \leq \int_0^1 dt \int d\rho\, d\theta\, \tilde{J}(t, \rho, \theta) \delta_\beta(\rho) \leq |J|_0 \int d\theta.
\tag{4.220}
$$

If $\mathbf{q} = 0$ and $q_0 = 0$, use

$$
\frac{1}{\beta} \sum_\omega \frac{1}{(i\omega - E)^2} = \frac{\partial}{\partial E} \chi_\beta(E) = -\delta_\beta(E)
\tag{4.221}
$$

and an analogue of (4.220) to see that for $\beta \to \infty$,

$$
B^{(+)}(0) = -\int d\theta\, J(0, \theta) = -N(0)
\tag{4.222}
$$

is just the density of states at the Fermi energy, which is finite by our assumptions on E. If $\mathbf{q} = 0$, but $q_0 \neq 0$, a similar calculation shows that $B^{(+)}(q_0, 0) = 0$. ∎

Equation (4.217) also implies that at $q_0 = 0$ the particle-hole bubble is a C^{k_0} function of \mathbf{q} for $|\mathbf{q}| < r$ if the dispersion relation E is C^{k_0}.

In Theorem 4.12, we have left out the ultraviolet cutoff on D_0. This is justified because in the integrand for the difference, at least one of the propagators is cut off away from the Fermi surface, hence bounded, and thus the integral converges because $\int dp |D_0(p)|$ is finite.

In summary, the finiteness of the particle-hole ladders holds under two conditions: there is no van Hove singularity on the Fermi surface, and the Fermi surface has positive curvature. If one of these conditions fails, the particle-hole ladders can have singularities. A relevant example for that is the half-filled Hubbard model, where singularities occur (in $d = 2$) on the two lines $p_1 = p_2$ and $p_1 = -p_2$ (here $\mathbf{p} = (p_1, p_2)$). Near to, but not exactly at, half-filling, these singularities are replaced by large values of the function, which may be important in the RG flow, although they do not introduce true instabilities at very low energy scales for small values of the coupling constant.

4.5.6 The Combined Flow

An iteration of the full parquet equation (4.178) also generates terms that mix the particle-particle and particle-hole bubbles. However, these terms all contain overlapping loops (defined in Sect. 4.6), and therefore they affect the leading low-energy behaviour, which is given by the particle-particle ladders, only by exponentially small terms.

The particle-particle singularity is more generic in that it occurs at any value of the filling factor, whereas the particle-hole singularity occurs only for specific filling values (e.g. half-filling in the Hubbard model).

4.6 Improved Power Counting

In the last section, we have seen that the ladder approximation to the flow of the four-point function has a logarithmic growth when the sum of the ingoing momenta equals zero. This by itself does not prove the occurrence of such a singularity in the full four-point function because there may be other contributions that cancel the singularity in the ladders. In this section we shall prove, among other things, that if the ladder term is left out, the four-point function is bounded uniformly in the temperature. Thus, although in the full flow, the ladder term and the others are coupled, such cancellations do not occur.

4.6.1 Overlapping Loops

We now proceed to discuss the bounds that are at the heart of the analysis of perturbation theory to all orders. They apply to contributions from graphs with *overlapping loops*. Examples for such graphs are given in Fig. 4.5. The

first one is a contribution to the four-point function, the second one to the selfenergy. The third one is a graph that occurs when the integral equation corresponding to the parquet RGE is iterated, and when the particle-particle and particle-hole flows mix. All these graphs have at least two loops, and two different loops have one line in common. This line is drawn heavy in the figure.

Fig. 4.5. Examples for overlapping graphs

To motivate the bounds that follow, we discuss the two-legged graph in detail. If associate a vertex function that is identical to one to each of the two vertices, we get the second order contribution to the skeleton selfenergy in the Hubbard model with on-site interaction, given by the integral

$$\dot{\Sigma}_2(q) = \int dk_1 dk_2 \, D_t(k_1) D_t(k_2) \dot{D}_t(q + v_1 k_1 + v_2 k_2) \tag{4.223}$$

with $v_i \in \{-1, 1\}$. The property that a linear combination of two integration momenta, like the $q + v_1 k_1 + v_2 k_2$ in (4.223), appears in the argument of a propagator, is a defining property of graphs with overlapping loops.

A purely graphical definition of the property of a graph to be overlapping can be given [45]. If a spanning tree T for a graph G is fixed, the loop generated by a line $l_1 \notin T$ consists of l_1 and the closed nonselfintersecting path connecting one endpoint of l_1 to the other one. If there are two lines $l_1, l_2 \notin T$, the graph is called overlapping if the loops generated by l_1 and l_2 contain a common line of T. Whether a graph is overlapping or not does not depend on which spanning tree one chooses to check this property [45].

Although the graphical definition and interpretation just given will not be necessary for understanding how the bounds work in the RG differential equation, they allow to decide rather easily if there are gains in power counting also when one does the analysis with discrete renormalization steps. One can give a complete characterization of nonoverlapping Feynman graphs with two and four external legs [45].

The power counting of Sect. 4.4 gave

$$\left| \left(\frac{\partial}{\partial q} \right)^\alpha \dot{\Sigma}_2(q) \right| \le \|\partial^\alpha \dot{D}_t\|_\infty \|D_t\|_1^2 \le B_\alpha (8J_1)^2 \epsilon_t^{1-|\alpha|}. \tag{4.224}$$

For $\alpha = 0$, this leaves a decay $\epsilon_t \propto e^{-t}$, so that $\Sigma_2(q) = \int_0^t ds \, \dot{\Sigma}_2$ is finite. For $\alpha \ge 1$, however, the integral grows with β and diverges for zero temperature,

thus indicating that already the first derivative of the selfenergy is divergent at zero temperature, if the momentum is put on the Fermi surface. This is the true behaviour in one dimension, but not for curved Fermi surfaces for $d \geq 2$:

Proposition 4.13. *[45, 46] For the class of models obeying the assumptions in Sect. 4.2.6, there is a constant (independent of temperature) such that*

$$\left| \left(\tfrac{\partial}{\partial q} \right)^\alpha \dot{\Sigma}_2(q) \right| \leq \text{const } \epsilon_t^{2-|\alpha|} \begin{cases} 1+t & \text{if } d = 2 \\ 1 & \text{if } d \geq 3. \end{cases} \tag{4.225}$$

Thus the first derivative of Σ_2 is bounded uniformly in β and the second derivative grows at most as a power of $\log \beta$.

Here and in the following, derivatives with respect to p_0 are understood as a difference operation $\frac{\beta}{2\pi}(f(p_0 + \frac{2\pi}{\beta}) - f(p_0))$. The RGE actually defines the Green functions for (almost all) real values of p_0, not just the discrete set of Matsubara frequencies, so the effect of such a difference can be bounded in terms of a derivative by Taylor expansion. This changes at most constants, so we shall not write this out explicitly in the proofs.

The proof of this proposition will use the volume improvement bound from overlapping loops. It is a generalization of the one-loop volume bound given in Lemma 4.10, but it has much stronger consequences because it is uniform in the external momentum. It applies to all contributions to the right hand side of the RGE with number of internal lines i at least three. Using this bound we shall generalize Proposition 4.13 to the full skeleton selfenergy.

Proof. We write $k_i = (\omega_i, \mathbf{k}_i)$ for the integration variables, use again (4.187), (4.131), and the inequality $\mathbb{1}(|q_0 + v_1 \omega_1 + v_2 \omega_2| \leq \epsilon_t) \leq 1$, and do the frequency sums over ω_1 and ω_2. This gives the bound

$$\left| \left(\tfrac{\partial}{\partial q} \right)^\alpha \dot{\Sigma}_2(q) \right| \leq B_0^2 B_\alpha \, \epsilon_t^{-1-|\alpha|} \int\limits_t^{\log(\beta\epsilon_0)} dt_1 \int\limits_t^{\log(\beta\epsilon_0)} dt_2 \, V_2(\mathbf{q}, t, t_1, t_2) \tag{4.226}$$

with

$$V_2(\mathbf{q}, t, t_1, t_2) = \max_{v_1, v_2 = \pm 1} \int\limits_{\mathcal{R}(\epsilon_{t_1})} \frac{d\mathbf{k}_1}{(2\pi)^d} \int\limits_{\mathcal{R}(\epsilon_{t_2})} \frac{d\mathbf{k}_2}{(2\pi)^d}$$
$$\prime \quad \mathbb{1}\left(|E(\mathbf{q} + v_1 \mathbf{k}_1 + v_2 \mathbf{k}_2)| \leq \epsilon_t \right). \tag{4.227}$$

and $\mathcal{R}(\epsilon)$ given in (4.191). The proposition now follows from the Theorem 4.14 in the next section, by integration over t_1 and t_2 in (4.226). ∎

4.6.2 Volume Improvement from Overlapping Loops

Theorem 4.14. *Under the assumptions stated in Sect. 4.2.6, there is a constant Q_T such that for all $q \in B$ and all t_1, t_2, t with $t_1 \geq t$ and $t_2 \geq t$,*

$$V_2(\mathbf{q}, t, t_1, t_2) \leq Q_T \epsilon_{t_1} \epsilon_{t_2} \epsilon_t \begin{cases} 1 + t & \text{if } d = 2 \\ 1 & \text{if } d \geq 3. \end{cases} \tag{4.228}$$

The constant Q_T depends on the minimal curvature of the Fermi surface. The naive bound for $V_2(\mathbf{q}, t, t_1, t_2)$, obtained by dropping the indicator function in (4.227), is

$$V_2(\mathbf{q}, t, t_1, t_2) \leq \text{const } \epsilon_{t_1} \epsilon_{t_2}, \tag{4.229}$$

so, up to the factor t in two dimensions, the Lemma implies an improvement by a full power ϵ_t. We call $\epsilon_t \propto e^{-t}$ the *volume improvement factor*.

The most important feature of this bound is that the improvement is uniform in the external momentum \mathbf{q}. It does therefore not matter if \mathbf{q} is near to the Fermi surface (as would be the case in the selfenergy) or if \mathbf{q} is any linear combination of momenta, as is the case for the four-point function and the higher m-point functions.

Volume improvement from overlapping loops was first proven in [45]. The analogue of Theorem 4.14 proven in [45] gave an improvement factor $e^{-\delta t}$ with some $\delta > 0$ under much weaker assumptions on the Fermi surface (it need not be convex, not even connected, and the curvature may vanish on subsets of positive codimension of \mathcal{S}, e.g., in $d = 3$ on a curve on \mathcal{S}). On the other hand, the t-behaviour stated in Theorem 4.14 is optimal, that is, one cannot gain more than one power of ϵ_t from an overlap of two loops.

Theorem 4.14 was first proven in [46] in the form stated in Appendix B.8.2. We prove it for spherical Fermi surfaces in Appendix B.8.2. The proof for the nonspherical case, including surfaces that are not invariant under $\mathbf{p} \to -\mathbf{p}$, is in Appendix B of [46]. It does not use Lemma 4.10: in fact, in absence of the symmetry $\mathbf{p} \to -\mathbf{p}$, $V_1^{(+)}$ and $V_1^{(-)}$ behave differently: an analogue of Lemma 4.10 holds for $V_1^{(+)}$, but only a weaker bound holds for $V_1^{(-)}$ (see Appendix C of [46]).

The existence of a volume improvement by a factor $e^{-t/2}$, as well as its uniformity in \mathbf{q}, can be seen in a very simple way by the following argument [48] using Lemma 4.10. By definition (4.227),

$$V_2(\mathbf{q}, t, t_1, t_2) = \max_{v_1, v_2 = \pm 1} \int_{\mathcal{R}(\epsilon_{t_1})} \frac{d\mathbf{k}_1}{(2\pi)^d} V_1^{(v_2)}(\mathbf{q} + v_1 \mathbf{k}_1, t, t_2). \tag{4.230}$$

By Lemma 4.10,

$$V_1^{(\pm)}(\mathbf{p}, t, t_2) \leq Q_0 \epsilon_{t_2} \begin{cases} 1 & \text{if } \|\mathbf{p}\|_B \leq \sqrt{\epsilon_t} \\ \sqrt{\epsilon_t} & \text{otherwise.} \end{cases} \tag{4.231}$$

Inserting this bound into (4.230), we get

$$\mathcal{V}_2(\mathbf{q}, t, t_1, t_2) \leq Q_0 \epsilon_{t_2} (\sqrt{\epsilon_t} \, \text{vol} \, \mathcal{R}(\epsilon_{t_1})$$
$$+ \quad \text{vol} \, (\mathcal{R}(\epsilon_{t_1}) \cap B(-v_1 \mathbf{q}, \sqrt{\epsilon_t})) \qquad (4.232)$$

with $B(\mathbf{q}, r) = \{\mathbf{p} \in \mathcal{B} : \|\mathbf{p} - \mathbf{q}\|_{\mathcal{B}} \leq r\}$. We have vol $\mathcal{R}(\epsilon_{t_1}) \leq$ const ϵ_{t_1}. The set $B(-v_1 \mathbf{q}, \sqrt{\epsilon_t}) \cap S$ is either empty or a piece of the Fermi surface with diameter $\approx 2\sqrt{\epsilon_{t_1}}$. Thus $\mathcal{R}(\epsilon_{t_1}) \cap B(-v_1 \mathbf{q}, \sqrt{\epsilon_t})$ is a subset of the cylinder of height $2\epsilon_{t_1}$ above that region, and therefore

$$\text{vol}(\mathcal{R}(\epsilon_{t_1}) \cap B(-v_1 \mathbf{q}, \sqrt{\epsilon_t}) \leq \text{const } \epsilon_{t_1} \, \epsilon_t^{(d-1)/2}. \qquad (4.233)$$

Thus, for all $d \geq 2$,

$$\mathcal{V}_2(\mathbf{q}, t, t_1, t_2) \leq \text{const } \epsilon_{t_1} \, \epsilon_{t_2} \, \sqrt{\epsilon_t}. \qquad (4.234)$$

In the following three subsections, we prove stronger ('improved') power counting bounds for the skeleton functions. The proofs in these subsections are taken from [53].

4.6.3 Volume Improvement in the RGE

We now use the volume improvement of Theorem 4.14 to bound those contributions to the right hand side of the RGE that have $i \geq 3$. We also want to bound derivatives. The derivative applied to Q_{mr} gives

$$\partial^\alpha Q_{mr}(t \mid \underline{P}) = \int d\kappa_{mr}(m_1, r_1, m_2, r_2, i) i \, i! \mathcal{X}_{\alpha,i}(t \mid \underline{P}), \qquad (4.235)$$

where, after a change of variables $k_j \to (-1)^{l_j} k_j$,

$$\mathcal{X}_{\alpha,i}(t \mid \underline{P}) = \sum_{\sigma \in \{-1,1\}^i} \sum_{l \in \{1,2\}^i} \sum_{\underline{\alpha}} \frac{\alpha!}{\alpha_0! \alpha_1! \alpha_2!}$$

$$\int_{\mathbb{R} \times \mathcal{B}} dk_2 \dots \int_{\mathbb{R} \times \mathcal{B}} dk_i \, \partial^{\alpha_0} \dot{D}_t(k_1) \prod_{j=2}^{i} D_t(k_j) \qquad (4.236)$$

$$\partial^{\alpha_1} I_{m_1 r_1}(t \mid \underline{P}^{(1)}, K) \, \partial^{\alpha_2} I_{m_2 r_2}(t \mid \sim K, \underline{P}^{(2)})$$

with $K_j = (k_j, \sigma_j, l_j)$, the sum over $\underline{\alpha}$ running over all triples $(\alpha_0, \alpha_1, \alpha_2)$ with $\alpha_0 + \alpha_1 + \alpha_2 = \alpha$, and $k_1 = -\sum_{j=2}^{i}(-1)^{l_j} k_j - (p_1 + \dots + p_{m_1-i+1})$. $\mathcal{X}_{\alpha,i}$ depends also on $(m_1, r_1, m_2, r_2, m, r)$. Application of $|\cdot|_0$ gives

$$|\mathcal{X}_{\alpha,i}(t)|_0 \leq 4^i \sum_{\underline{\alpha}} \frac{\alpha!}{\alpha_0! \alpha_1! \alpha_2!} \, |\partial^{\alpha_1} I_{m_1 r_1}(t)|_0 \, |\partial^{\alpha_2} I_{m_2 r_2}(t)|_0 \, \mathcal{Y}_{\alpha_0, i}(t) \qquad (4.237)$$

with

$$\mathcal{Y}_{\alpha,i}(t) = \sup_{\substack{Q \in \mathbb{M} \times \mathcal{B} \\ v_i \in \{-1,1\}}} \int \prod_{j=2}^{i} dk_j \, |D_t(k_j)| \, |\partial^\alpha \dot{D}_t(\sum_{j=2}^{i} v_j k_j + Q)|. \qquad (4.238)$$

Lemma 4.15. $\mathcal{Y}_{\alpha,i}(t) = 0$ *for* $t > \log \frac{\beta \epsilon_0}{\pi}$, *and*

$$\mathcal{Y}_{\alpha,i}(t) \leq (8J_1)^{i-1} B_\alpha \, \epsilon_t^{i-2-|\alpha|}. \tag{4.239}$$

If $i \geq 3$,

$$\mathcal{Y}_{\alpha,i}(t) \leq K^{(1)} \frac{1}{2}(1+t) \, \epsilon_t^{i-1-|\alpha|}. \tag{4.240}$$

with $K^{(1)} = B_0^2 B_\alpha (8J_1)^{i-3} Q_T$.

Proof. By (4.131), $\mathcal{Y}_{\alpha,i}(t) \leq B_\alpha \epsilon_t^{-1-|\alpha|} \|D_t(k)\|_1^{i-1}$, so (4.239) holds by (4.133). For $i \geq 3$, the integral (4.238) for $\mathcal{Y}_{\alpha,i}(t)$ contains the subintegral $\mathbb{S}(\sum_{j=4}^i v_j k_j + Q)$, with

$$\mathbb{S}(q) = \int dk_2 \int dk_3 |D_t(k_2) D_t(k_3) \dot{D}_t(v_2 k_2 + v_3 k_3 + q)| \tag{4.241}$$

which we recognize as the function $\dot{\Sigma}_2$ (see (4.223)). Thus by Proposition 4.13, $|\mathbb{S}(q)| \leq B_0^2 B_\alpha Q_T \epsilon_t^{2-|\alpha|}$. Inserting this bound, the remaining $i-3$ integrations are independent and contribute a factor $\|D_t(k)\|_1^{i-3}$. Thus (4.240) holds. ■

Lemma 4.16. *For all* i,

$$|\mathcal{X}_{\alpha,i}(t)|_0 \leq 4^i (8J_1)^{i-1} \sum_\alpha \frac{\alpha!}{\alpha_0! \alpha_1! \alpha_2!} |\partial^{\alpha_1} I_{m_1 r_1}(t)|_0 \, |\partial^{\alpha_2} I_{m_2 r_2}(t)|_0$$

$$B_{\alpha_0} \epsilon_t^{i-2-|\alpha_0|}. \tag{4.242}$$

For $i \geq 3$

$$|\mathcal{X}_{\alpha,i}(t)|_0 \leq 4^i (8J_1)^{i-1} \sum_\alpha \frac{\alpha!}{\alpha_0! \alpha_1! \alpha_2!} |\partial^{\alpha_1} I_{m_1 r_1}(t)|_0 \, |\partial^{\alpha_2} I_{m_2 r_2}(t)|_0$$

$$B_{\alpha_0} K^{(1)} \epsilon_t^{i-1-|\alpha_0|} \begin{cases} \frac{1}{2}(1+t) & \text{if } d = 2 \\ 1 & \text{if } d \geq 3. \end{cases} \tag{4.243}$$

In the next sections we use Lemma 4.16 to prove bounds on the non-ladder skeleton functions and on the derivatives of the selfenergy.

4.6.4 Bounds on the Non-Ladder Skeletons

We define the non-ladder skeleton functions as the solutions to the skeleton RGE where the ladder contributions to the four-point function are left out. That is, on the right hand side of the equation for $m = 4$, $m_1 = 4$ and $m_2 = 4$ are left out.

Definition 4.17. *The functions $\tilde{I}_{m,r}^{(N)}$, obtained from the truncation of the skeleton RGE, in which the equation for the four-point function is given by*

$$\frac{\partial}{\partial t}\tilde{I}_{4,r}^{(N)}(t) = \frac{1}{2}A_4 Q_{4,r,\geq 3}(t) \tag{4.244}$$

(i.e., where $m_1 = 4$ and $m_2 = 4$ are left out in the sum $\int d\tilde{\kappa}_{4,r}$) are the non-ladder skeleton functions.

The next theorem [53] states that the non-ladder skeleton four-point function is bounded and that the non-ladder skeleton selfenergy is C^1 uniformly in β.

Theorem 4.18. *For all $r \geq 1$, the non-ladder r^{th} order skeleton selfenergy $\tilde{I}_{2,r}^{(N)}(t)$ converges for $t \to \infty$ to a C^2 function. There are constants L_r and L_r', independent of β, such that*

$$\left|\partial^\alpha \tilde{I}_{2,r}^{(N)}(t)\right|_0 \leq L_r \begin{cases} 1 & \text{if } |\alpha| \leq 1 \\ (\log \beta\epsilon_0)^2 & \text{if } |\alpha| = 2 \text{ and } d = 2 \\ \log \beta\epsilon_0 & \text{if } |\alpha| = 2 \text{ and } d \geq 3. \end{cases} \tag{4.245}$$

and

$$\left|\partial^\alpha \tilde{I}_{4,r}^{(N)}(t)\right|_0 \leq L_r' \begin{cases} 1 & \text{if } |\alpha| = 0 \\ \log \beta\epsilon_0 & \text{if } |\alpha| = 1 \\ \beta\epsilon_0 & \text{if } |\alpha| = 2. \end{cases} \tag{4.246}$$

Thus the removal of the ladder contributions to the four-point function indeed makes $|\tilde{I}_{4,r}^{(N)}|_0$ uniformly bounded in β. Theorem 4.18 has an extension to the $m \geq 6$-point functions with bounds in the L^1 norm.

The following theorem implies Theorem 4.18. Note that all bounds in this Theorem are independent of β.

Theorem 4.19. *Let α be a multiindex with $|\alpha| \leq 2$. For all m, r, let $I_{mr}(0)$ be C^2 in $(\mathbf{p}_1, \ldots, \mathbf{p}_m)$, and assume that there are $K_{mr}^{(0)} \geq 0$ such that $|\partial^\alpha I_{mr}(0)| \leq K_{mr}^{(0)}$, with $K_{mr}^{(0)} = 0$ for $m > 2r + 2$, and $K_{m1}^{(0)} = \delta_{m4} v$, where $v > 0$. Let $\tilde{I}_{mr}^{(N)}(t)$ be the Green functions generated by the non-ladder skeleton RGE, as given in Definition 4.17, with initial values $I_{mr}(0)$. Let $K_{m1} = K_{m1}^{(0)}$ and for $r \geq 2$, let*

$$K_{mr} = K_{mr}^{(0)} + \frac{1}{m}M_1 \int d\tilde{\kappa}_{mr} i!(32J_1)^{i-1} K_{m_1 r_1} K_{m_2 r_2} \tag{4.247}$$

where $M_1 = 240BK^{(1)}$, with $B = \max_\alpha B_\alpha$. Then $K_{mr} = 0$ if $m > 2r + 2$, and for all $t \geq 0$ and all m, r

$$\left|\partial^\alpha \tilde{Q}_{mr}^{(N)}(t)\right|_0 \leq 2K_{mr} \begin{cases} \epsilon_t^{2-\frac{m}{2}-|\alpha|} & \text{if } m \geq 6 \\ \epsilon_t^{1-|\alpha|}\frac{1+t}{2} & \text{if } m = 4 \\ \epsilon_t^{2-|\alpha|}\left(\frac{1+t}{2}\right)^{\delta_{d,2}} & \text{if } m = 2. \end{cases} \tag{4.248}$$

Moreover, for $m \geq 6$,

$$\left|\partial^\alpha \tilde{I}_{mr}^{(N)}(t)\right|_0 \leq K_{mr} \epsilon_t^{2-\frac{m}{2}-|\alpha|}, \tag{4.249}$$

for $m = 4$

$$\left|\partial^\alpha \tilde{I}_{mr}^{(N)}(t)\right|_0 \leq K_{4r} \begin{cases} 1 & \text{if } \alpha = 0 \\ \left(\frac{1+t}{2}\right)^2 & \text{if } |\alpha| = 1 \\ \epsilon_t^{-1}\frac{1+t}{2} & \text{if } |\alpha| = 2, \end{cases} \tag{4.250}$$

and for $m = 2$

$$\left|\partial^\alpha \tilde{I}_{mr}^{(N)}(t)\right|_0 \leq K_{2r} \begin{cases} 1 & \text{if } |\alpha| \leq 1 \\ \left(\frac{1+t}{2}\right)^2 & \text{if } |\alpha| = 2 \text{ and } d = 2 \\ \frac{1+t}{2} & \text{if } |\alpha| = 2 \text{ and } d = 3. \end{cases} \tag{4.251}$$

Proof. Induction in r, with the statement of the Theorem as the inductive hypothesis. $r = 1$ is trivial because $\tilde{Q}_{m1}^{(N)} = 0$ and because the statement holds for $I_{m1}(0)$. Let $r \geq 2$ and the statement hold for all $r' < r$. The inductive hypothesis applies to both factors $\tilde{I}_{m_k r_k}^{(N)}$ in $\tilde{Q}_{mr}^{(N)}$. For $m_k = 4$, it implies that

$$\left|\tilde{I}_{m_k r_k}^{(N)}(t)\right|_0 \leq K_{m_k r_k} \epsilon_t^{-|\alpha|} = K_{m_k r_k} \epsilon_0^{-|\alpha|} e^{t(|\alpha|+\frac{m_k}{2}-2)} \tag{4.252}$$

for all $t \geq 0$. Recall (4.242) and (4.243), and

$$\left|\partial^\alpha A_m \tilde{Q}_{mr}^{(N)}(t)\right|_0 = \left|A_m \partial^\alpha \tilde{Q}_{mr}^{(N)}(t)\right|_0 \leq \left|\partial^\alpha \tilde{Q}_{mr}^{(N)}(t)\right|_0$$
$$\leq \int d\tilde{\kappa}_{mr}^{(N)} i \, i! |\mathcal{X}_{\alpha,i}(t)|_0. \tag{4.253}$$

Let $m \geq 6$. By (4.242) and $m_1 + m_2 = m + 2i$, the t-dependent factors in $\mathcal{X}_{\alpha,i}(t)$ are

$$e^{t(\frac{m_1}{2}-2+|\alpha_1|)} e^{t(\frac{m_2}{2}-2+|\alpha_2|)} e^{t(2-i+|\alpha_0|)} = e^{t(\frac{m}{2}-2+|\alpha|)}. \tag{4.254}$$

Since $\frac{m}{2} - 2 + |\alpha| \geq \frac{m}{2} - 2 \geq 1$, integrating the RGE gives

$$\left|\partial^\alpha \tilde{I}_{mr}^{(N)}(t)\right|_0 \leq \left|\partial^\alpha \tilde{I}_{mr}^{(N)}(0)\right|_0 + \frac{1}{2}\int_0^t ds \left|\partial^\alpha \tilde{Q}_{mr}^{(N)}(s)\right|_0$$
$$\leq K_{mr}^{(0)} + \frac{1}{m-4}\nu_{mr}\epsilon_t^{2-\frac{m}{2}-|\alpha|} \tag{4.255}$$

where

$$\nu_{mr} = 4B \int d\tilde{\kappa}_{mr} i \, i! (32 J_1)^{i-1} \sum_\alpha \frac{\alpha!}{\alpha_0! \alpha_1! \alpha_2!} K_{m_1 r_1} K_{m_2 r_2}. \tag{4.256}$$

For $m \geq 6$, $m - 4 \geq \frac{m}{3}$. Moreover, $\sum \frac{\alpha!}{\alpha_0! \alpha_1! \alpha_2!} \leq 3^{|\alpha|}$, so (4.248) and (4.249) follow.

Let $m = 4$. One of m_1 and m_2 must be at least six because the ladder part is left out in $\tilde{I}_{mr}^{(N)}$. Thus $i = \frac{1}{2}(m_1 + m_2 - 4) \geq 3$. By (4.243), there is an extra small factor $\epsilon_t = \epsilon_0 e^{-t}$, so

$$\left| \partial^\alpha \tilde{Q}_{4,r}^{(N)}(t) \right|_0 \leq \frac{1+t}{2} \epsilon_t^{1-|\alpha|} K^{(1)} \nu_{mr}, \tag{4.257}$$

which proves (4.248). (4.250) follows by integration.

Let $m = 2$. The case $m_1 = m_2 = 2$ is excluded since this is the skeleton RG. Thus $i = \frac{1}{2}(m_1 + m_2 - 2) \geq 3$, and (4.243) implies

$$\left| \partial^\alpha \tilde{Q}_{2,r}^{(N)}(t) \right|_0 \leq \left(\frac{1+t}{2} \right)^{\delta_{d,2}} \epsilon_t^{1-|\alpha|} K^{(1)} \nu_{2r}. \tag{4.258}$$

Thus (4.248) holds, and (4.251) follows by integration. ∎

Theorem 4.18 follows from Theorem 4.19 by the following argument. Let $|\alpha| \leq 1$. By (4.251), $|\partial^\alpha \tilde{I}_{2r}^{(N)}(t)| \leq K_{2,r}$. By (4.248), $|\partial^\alpha \frac{\partial}{\partial t} \tilde{I}_{2r}^{(N)}(t)| \leq K_{2,r}\epsilon_0 e^{-t} \frac{1+t}{2} \to 0$ as $t \to \infty$. Thus the limit $t \to \infty$ of $\tilde{I}_{2,r}^{(N)}(t)$ exists and is a C^1 function of $(\mathbf{p}_1, \mathbf{p}_2)$. All constants are uniform in β. The second derivative of $\tilde{Q}_{2,r}^{(N)}$ is $O(t)$ in $d = 2$ and $O(1)$ in $d = 3$. Since $\tilde{Q}_{2,r}^{(N)} = 0$ for $t > \log \beta \epsilon_0$, the integral for $\tilde{I}_{2,r}^{(N)}$ over t runs only up to $\log \beta \epsilon_0$, which gives the stated dependence on $\log \beta \epsilon_0$.

A look back at the proof shows that no factorials from powers of t were generated in the recursion for the coefficients (because there are no powers of t that grow with r).

4.6.5 The Derivatives of the Skeleton Selfenergy

If the ladder contribution is kept in the RGE, powers of $\log \beta \epsilon_0$ appear in the Green functions. We now prove [53] that if $|\lambda| \log \beta \epsilon_0$ is bounded by a constant, then the derivatives of the selfenergy are bounded independently of β.

Theorem 4.20. *The skeleton functions $\tilde{I}_{2,r}$ and $\tilde{I}_{4,r}$ converge for $t \to \infty$. There are constants $\tilde{L}_{m,r}$ such that for all β,*

$$\left| \partial^\alpha \tilde{I}_{m,r}(t) \right|_0 \leq \tilde{L}_{m,r} (\log \beta \epsilon_0)^r \tag{4.259}$$

for $m = 2$, $|\alpha| \leq 2$, and $m = 4$, $|\alpha| = 0$.

Theorem 4.20 is a consequence of the following, more detailed, theorem.

Theorem 4.21. *Let α be a multiindex with $|\alpha| \leq 2$. For all m, r, let $I_{mr}(0)$ be C^2 in $(\mathbf{p}_1, \ldots, \mathbf{p}_m)$, and assume that there are $K_{mr}^{(0)} \geq 0$ such that $|\partial^\alpha I_{mr}(0)| \leq K_{mr}^{(0)}$, with $K_{mr}^{(0)} = 0$ for $m > 2r + 2$, and $K_{m1}^{(0)} = \delta_{m4} v$, where $v > 0$. Let $\tilde{I}_{mr}(t)$ be the Green functions generated by the skeleton RGE, as given in Definition 4.7, with initial values $I_{mr}(0)$. Let $K_{m1} = K_{m1}^{(0)}$ and for $r \geq 2$, let K_{mr} be given by (4.247). Then for $m \geq 4$*

$$\left| \partial^\alpha \tilde{Q}_{mr}(t) \right|_0 \leq 2 K_{mr} \left(\log \beta \epsilon_0 \right)^{r-2} \epsilon_t^{2 - \frac{m}{2} - |\alpha|} \tag{4.260}$$

and

$$\left| \partial^\alpha \tilde{I}_{mr}(t) \right|_0 \leq K_{mr} \left(\log \beta \epsilon_0 \right)^{r-1} \epsilon_t^{2 - \frac{m}{2} - |\alpha|}. \tag{4.261}$$

For $m = 2$,

$$\left| \partial^\alpha \tilde{Q}_{2r}(t) \right|_0 \leq 2 K_{2r} \left(\log \beta \epsilon_0 \right)^{r - 2 + \delta_{d,2}} \epsilon_t^{2 - |\alpha|} \tag{4.262}$$

and

$$\left| \partial^\alpha \tilde{I}_{2r}(t) \right|_0 \leq K_{2r} \left(\log \beta \epsilon_0 \right)^{r-2} \begin{cases} 1 & \text{if } |\alpha| \leq 1 \\ (\log \beta \epsilon_0)^{1 + \delta_{d,2}} & \text{if } |\alpha| = 2. \end{cases} \tag{4.263}$$

Proof. The proof is by induction in r, with the statement of the theorem as the inductive hypothesis. It is similar to the proof of Theorem 4.19, with only a few changes. Note that $m_1 = 2$ and $m_2 = 2$ never appear on the right hand side of the RGE because of the skeleton truncation in Definition 4.7. For $m \geq 4$, use (4.242); this gives

$$\left| \partial^\alpha \tilde{Q}_{mr}(t) \right|_0 \leq 9 v_{mr} \left(\log \beta \epsilon_0 \right)^{r-2} \epsilon_t^{2 - \frac{m}{2} - |\alpha|}. \tag{4.264}$$

For $m \geq 6$, the scale integral is as in the proof of Theorem 4.19. For $m = 4$, the scale integral is now $\int_0^t ds = t \leq \log \beta \epsilon_0$. This produces the powers of $\log \beta \epsilon_0$ upon iteration. For $m \geq 2$, use (4.243); this gives $K^{(1)} \epsilon_t^{2 - |\alpha|} (\frac{1}{2}(1 + t))^{\delta_{d,2}}$ instead of $\epsilon_t^{2 - \frac{m}{2} - |\alpha|}$ in (4.264). The theorem now follows by integration over t, recalling that the upper integration limit is at most $\log \beta \epsilon_0$. ∎

Theorem 4.21 implies Theorem 4.20 as follows. Convergence of the selfenergy follows for $|\alpha| \leq 1$ as in the proof of Theorem 4.18. For $m = 4$, the function is bounded uniformly in t. For $|\alpha| = 2$, convergence at $\beta > 0$ holds because $\frac{\partial}{\partial t} \tilde{I}_{4r}(t) = 0$ for $t > \log \beta \epsilon_0$.

Corollary 4.22. *For $R \geq 1$, let $\tilde{I}_2^{(R)} = \lim_{t \to \infty} \sum_{r=1}^R \tilde{I}_{2,r}(t) \lambda^r$. Then for all $R \geq 1$, there are constants γ_R, independent of β, such that for all $|\alpha| \leq 2$, and all λ with $|\lambda| \log \beta \epsilon_0 < $ const , the skeleton selfenergy satisfies*

$$|\partial^\alpha \tilde{I}_2^{(R)}|_0 \leq \gamma_R. \tag{4.265}$$

It is expected that $\gamma_R \leq$ const r, so that the skeleton selfenergy is analytic in λ as long as the above temperature-dependent restriction holds. The present treatment of the combinatorics is far from showing anything of the kind, but a proof should be possible by combining the method of [55] with the bounds from overlapping loops [45, 46, 47, 53].

4.7 Renormalization Subtractions

We now discuss what happens when the restriction to skeletons is removed, that is, when two-legged insertions are allowed.

4.7.1 Motivation; the Counterterm

The two-point function in momentum space, $I_{2,r}(t \mid P_1, P_2)$, is for $P_k = (p_k, \sigma_k, j_k)$ of the form

$$I_{2,r}(t \mid P_1, P_2) = \varepsilon_{j_1 j_2} \delta_{\sigma_1 \sigma_2} \, \mathbb{I}_{2,r}(t|(-1)^j p_1) \qquad (4.266)$$

(this holds by antisymmetry and because $\Pi_H I_{2,r} = I_{2,r}$, where Π_H projects to the subset $p_1 = -p_2$). Accordingly, we write the right hand side of the RGE for $m = 2$ as

$$\tfrac{\partial}{\partial t} \mathbb{I}_{2,r}(t|p) = \frac{1}{2} \mathbb{Q}_{2,r}(t|p) \qquad (4.267)$$

with a function $\mathbb{Q}_{2,r}$ that is defined in terms of $Q_{2,r}$ in the same way as $\mathbb{I}_{2,r}$ is determined by $I_{2,r}$. Also, in the neighbourhood of the Fermi surface where the flow takes place, $p = (\omega, \mathbf{p}) = (\omega, \pi(\rho, \theta))$, with ρ and θ the radial and tangential coordinates around S.

The skeleton m-point functions had the simple power counting behaviour $e^{\frac{t}{2}(m-4)}$; this bound iterated through the RGE in an almost trivial way *because on the right hand side of the skeleton RGE, the $m = 2$ contributions were left out.* For $m = 2$, $e^{\frac{t}{2}(m-4)} = e^{-t}$, but we have already seen that the skeleton two-point function is not $O(e^{-t})$, but $O(1)$. Thus putting in the two-legged functions leads to a failure of the bounds. On the other hand, the derivative $\frac{\partial}{\partial t} \mathbb{I}_{2,2} = \frac{1}{2} \mathbb{Q}_{2,2}$ was of order e^{-t}, so one can once again attempt to solve the problem by a change of boundary conditions. The goal will be to choose an initial condition for the flow such that

$$\lim_{t \to \infty} \mathbb{I}_{2,r}(t|0, \pi(0, \theta)) = 0 \qquad \text{for all } \theta \in S^{d-1}. \qquad (4.268)$$

with the idea that, if $\mathbb{Q}_{2,r}$ remains of order e^{-t}, one can then integrate the RGE as

$$\mathbb{I}_{2,r}(t|(0, \pi(0, \theta))) = -\frac{1}{2} \int_t^\infty ds \, \mathbb{Q}_{2,r}(s|0, \pi(0, \theta)) \qquad (4.269)$$

which implies that $\mathbb{I}_{2,r}(t|(0, \pi(0, \theta)))$ is of order e^{-t}, and hence gives hope that the bounds would not be upset when the skeleton equation is generalized to the full RGE.

This strategy works [45, 46, 47], but this is not at all trivial. From a technical point of view, one has to be careful because in order to get from the Fermi surface $(0, \pi(0, \theta))$ to general values of p, one has to do Taylor expansion, and taking derivatives potentially generates singularities. From a conceptual point of view, one has to justify the change of boundary conditions. This is an important point, which itself generates further technical complications.

One might think of bypassing the above discussion about Taylor expansion by requiring that $\mathbb{I}_{2,r}(t|p) \to 0$ as $t \to \infty$ for all $p \in M \times B$. Indeed, there is no unique way of imposing the boundary condition, so one has some freedom. However, both from a conceptual and from a technical point of view, there are significant differences between different boundary conditions. The boundary condition for $t \to \infty$ is really achieved by an appropriate *initial* condition $\mathbb{I}_{2,r}(0|p) = \mathbb{I}_{2,r}^{(0)}(p)$. This initial condition changes the original model to one where the quadratic part of the original action, $\int dp\, \bar{\psi}(p)(i\omega - E(\mathbf{p}))\psi(p)$, is changed to

$$\int dp\, \bar{\psi}(p)(i\omega - E(\mathbf{p}) - \mathbb{I}_{2,r}^{(0)}(\omega, \mathbf{p}))\psi(p). \tag{4.270}$$

This change has to be interpreted correctly (and justified, see below). For instance, it makes a difference whether the extra term really depends on ω or not. If it does, then the quadratic part of the initial action will in general not correspond to a time-independent Hamiltonian, because the ω-dependence of the action is fixed as $i\omega$ in the functional integral we derived.

We shall therefore require that the initial condition is given by a function K that depends only on \mathbf{p}, and, to fix K uniquely, we furthermore require that $K(\mathbf{p}) = K(\pi(\rho, \theta))$ depends only on the angular variable θ.

4.7.2 Full Amputation

The singular behaviour of derivatives makes it necessary to transform from \mathbb{I}_2 to another function. The best way of getting optimal bounds is to transform to the one-particle irreducible functions. For the purposes of the present discussion, we shall, however, not need to Legendre transform, but just transform from $\mathbb{I}_2 = \sum_{r \geq 1} \lambda^r \mathbb{I}_{2,r}$ to an analogue of the full selfenergy and amputate full propagators from the m-point functions for $m \geq 4$, as follows. Let

$$\Sigma(t|p) = \frac{\mathbb{I}_2(t|p)}{1 + C_t(p)\mathbb{I}_2(t|p)}. \tag{4.271}$$

This equation, as all following ones, is to be read as a formal power series in λ, i.e. $\Sigma = \sum_p \mathbb{I}_2(C_t \mathbb{I}_2)^p$. Since \mathbb{I}_2 is $O(\lambda)$, Σ is a well-defined formal power series in λ. The inverse relation is

$$\mathbb{I}_2(t|p) = \frac{\Sigma(t|p)}{1 - C_t(p)\Sigma(t|p)}. \tag{4.272}$$

This transformation is motivated by the relation

$$\mathbb{I}_2(\infty) = C^{-1}(G_2 - C)C^{-1} \tag{4.273}$$

relating \mathbb{I}_2 to the full unamputated two-point function G_2. In this equation, C gets subtracted because \mathbb{I}_2 contains only terms of order λ^r with $r \geq 1$, whereas G_2 contains also the free term, and the factors C^{-1} are there because \mathbb{I}_2 is an amputated function.

The derivative with respect to t gives

$$\dot{\Sigma}(t) = (1 - C_t\Sigma(t))^2 \left(\dot{\mathbb{I}}_2(t) - \mathbb{I}_2(t)\dot{C}_t\mathbb{I}_2(t)\right). \tag{4.274}$$

The term $\mathbb{I}_2(t)\dot{C}_t\mathbb{I}_2(t)$ is the $i = 1$ term in the RGE for \mathbb{I}_2, so we see that this term drops out of the RGE for Σ.

Similarly, we define the fully amputated m-point functions for $m \geq 4$ by setting

$$A(t|P) = \left(1 - C_t((-1)^j p)\Sigma(t|(-1)^j p)\right)^{-1} \tag{4.275}$$

and defining

$$W_{mr}(t|P_1, \ldots, P_m) = I_{mr}(t|P_1, \ldots, P_m) \prod_{a=1}^{m} A(t|P_a). \tag{4.276}$$

When inserting this into the RGE, on the internal lines, $D_t(p)$ gets replaced by $D_t(p)(1 - C_t(p)\Sigma(t|p))^{-2}$, and $\dot{D}_t(p)$ gets replaced by $\dot{D}_t(p)(1 - C_t(p)\Sigma(t|p))^{-2}$, and the external amputations cancel out. The $i = 1$ term $I_m\dot{C}_t I_2$ gets replaced by $W_m C_t \dot{\Sigma}(t)(1 - C_t\Sigma(t))^{-1}$. Thus the RGE for Σ and the W_m reads

$$\frac{\partial}{\partial t}\Sigma_r(t|p) = L_{2,r}^{(2)}(t|p) + \tilde{Q}_{2,r}(t|p) \tag{4.277}$$

$$\frac{\partial}{\partial t}W_{mr}(t|\underline{P}) = L_{mr}^{(1)}(t|\underline{P}) + L_{mr}^{(2)}(t|\underline{P}) + \tilde{Q}_{mr}(t|\underline{P})$$

with

$$L_{m,r}^{(1)}(t|\underline{P}) = \left[W_m(t|\underline{P})C_t((-1)^{j_m}p_m)\dot{\Sigma}(t|(-1)^{j_m}p_m)A(t|P_m)\right]_r$$

$$L_{mr}^{(2)}(t|\underline{P}) = \left[\int dk\, W_{m+2}(t|P_1, .., P_{m-2}, (k, 1, 2), (-k, 1, 1))\right.$$
$$\left.\dot{D}_t(k)D_t(k)\frac{\Sigma(t|k)}{(1 - C_t\Sigma)^3(k)}\right]_r \tag{4.278}$$

$$\tilde{Q}_{mr}(t|\underline{P}) = \left[\sum_{\underline{\sigma},j}\prod_{q=2}^{i} dk_q \frac{D_t(k_q)}{(1 - C_t\Sigma(t))^2(k_q)}\frac{\dot{D}_t(k_1)}{(1 - C_t\Sigma(t))^2(k_1)}\right.$$
$$\left. W_{m_1}(t|\underline{P}^{(1)}, \underline{K})W_{m_2}(t| \sim \underline{K}, \underline{P}^{(2)})\right]_r.$$

Apart from the replacements on the internal lines, \tilde{Q}_{mr} is of the same form as the right hand side of the skeleton RGE. The notation $[\,\cdot\,]_r$ means that the coefficient of λ^r of the formal power series is taken.

4.7.3 Bounds for a Truncation

In the following, we consider the truncation of the equations (4.277), where for $m = 2$ and $m = 4$, the "generalized tadpole" terms $L_{m,r}^{(2)}$ are left out. The only reason for this is to keep the following proofs and theorems as simple as possible; it will be discussed at the end how one can take these terms into account.

Again, we also distinguish between the cases where the ladder contribution $i = 2$ to $\tilde{Q}_{4,r}$ is removed from (non-ladder) or left in the RGE. As before, the difference in the result is that powers of $\log \beta\epsilon_0$ occur when the ladder term is taken into account.

The recursion for the constants will be similar to the ones we have done up to now. We therefore do not keep track of these constants any more but use the following shorthand. We say that

$$W_{mr}(t|\underline{P}) = \mathcal{O}\left(f(t, \beta)\right) \tag{4.279}$$

if there is a constant c_{mr}, independent of β and t, such that

$$|W_{mr}(t|\underline{P})| \le c_{mr} f(t, \beta). \tag{4.280}$$

Let the nonladder equation be the RGE (4.277) with $L_{m,r}^{(2)} = 0$ for $m = 2, 4$ and without the ladder term ($i = 2$ in \tilde{Q}_{4r}. Also, for $m \ge 4$ and $r \ge 1$ let $W_{m,r}^{(0)}$ be bounded C^{k_0} functions on $(\mathbb{R} \times B)^m$ satisfying $\Pi_H W_{m,r}^{(0)} = W_{m,r}^{(0)}$, and for $p = (\omega, \pi(\rho, \theta)) \in \mathbf{M} \times B$ let

$$\Pi(p) = (0, \pi(0, \theta)). \tag{4.281}$$

The following theorem is an analogue of Theorem 4.8 for the full Green functions. It implies finiteness of the two- and four-point function. An analogue of Theorem 4.9, which implies finiteness of all the m-point functions with $m \ge 6$ as well, can be proven similarly, by using the norms $|\,\cdot\,|_s$.

Theorem 4.23. *For all $r \ge 1$, there are functions $K_r \in C^{k_0}(B, \mathbb{R})$ satisfying*

$$\frac{\partial}{\partial \rho} K_r(\pi(\rho, \theta)) = 0 \tag{4.282}$$

such that the solutions Σ_r and W_{mr} of the non-ladder equations with the initial condition

$$\begin{aligned} \Sigma_r(0|p) &= K_r(p) \\ W_{mr}(0|\underline{P}) &= W_{m,r}^{(0)}(\underline{P}) \end{aligned} \tag{4.283}$$

are C^{k_0} in \underline{p} and satisfy for all α with $|\alpha| \le 1$

$(1)_r$ $\partial^\alpha \dot{\Sigma}_r(t|p) = \partial^\alpha \dot{\tilde{Q}}_{2,r}(t|p) = \mathcal{O}\left((1+t)e^{-t(2-|\alpha|)}\right)$

$(2)_r$ $\partial^\alpha K_r(\mathbf{p}) = \mathcal{O}(1)$

$(3)_r$ $\partial^\alpha \dot{\Sigma}_r(t|\Pi(p)) = \mathcal{O}\left((1+t)e^{-t(2-|\alpha|)}\right)$

$(4)_r$ $\partial^\alpha \Sigma(t|p) = \mathcal{O}(1)$

$(5)_r$ $\Sigma_r(t|p) = \mathcal{O}\left(e^{-t} + |p - \Pi(p)|\right)$

$(6)_r$ $\Sigma_r(t|p)C_t(p) = \mathcal{O}(1)$ if $p \in$ supp D_t

$(7)_r$ $\partial^\alpha W_{4,r}(t|\underline{P}) = \mathcal{O}\left((1+t)^\alpha\right)$

$(8)_r$ $\partial^\alpha W_{m,r}(t|\underline{P}) = \mathcal{O}\left(e^{t(\frac{m}{2}-2+|\alpha|)}\right).$

Thus, as $t \to \infty$, for all $r \geq 1$, the selfenergy $\Sigma_r(t)$ and the non-ladder four-point function $W_{4,r}(t)$ converge to functions that are uniformly bounded in β.

Proof. The proof is by induction on r, with (4.282), $(1)_r$–$(8)_r$, and the hypothesis

$$\begin{aligned} \Sigma(t|-p_0, \mathbf{p}) &= \overline{\Sigma(t|p_0, \mathbf{p})} \\ W_{mr}(t|R\underline{P}) &= \overline{W_{mr}(t|\underline{P})}, \end{aligned} \qquad (4.284)$$

where $R\underline{P}$ changes all $(p_i)_0$ to $-(p_i)_0$, as the inductive hypothesis. We do the case $r = 2$ explicitly, to show the procedure for construction of K_r before going into the more complicated argument for general r. Note that the right hand side of the equation for Σ_r and $W_{m,r}$ contains only $\Sigma_{r'}$ and $W_{m,r'}$ with $r' < r$, thus the inductive hypothesis applies to all factors in there.

Let $r = 1$. Set $K_1(\mathbf{p}) = 0$ for all \mathbf{p}. Then $\Sigma_1(t|p) = 0$ for all $t \geq 0$. Thus $(1)_1$–$(6)_1$ hold trivially. Since $\tilde{Q}_{m1} = 0$ for all m, $(7)_1$ and $(8)_1$ hold by the properties of $W_{m,1}^{(0)}$.

Let $r = 2$. The $\Sigma_{r'}(t)$ occurring on the right hand side of (4.277) must have $r' = 1$, so all of them vanish. Thus $\Sigma_2(t|p)$ is the second order skeleton selfenergy, and hence the results of Sect. 4.6 imply that $(1)_2$ holds for $\partial^\alpha \dot{\Sigma}_2(t|p) = \partial^\alpha \dot{\tilde{Q}}_{22}(t|p)$. In the integral for $\dot{\Sigma}_2(t|-p_0, \mathbf{p})$, change variables from k_0 to $-k_0$ for every loop momentum k. Because $D_t(-p_0, \mathbf{p}) = \overline{D_t(p_0, \mathbf{p})}$, and because the same holds for the potential \hat{v} by the hypotheses stated in Sect. 4.2.6, it follows that

$$\dot{\Sigma}_2(t|-p_0, \mathbf{p}) = \overline{\dot{\Sigma}_2(t|p_0, \mathbf{p})} \qquad (4.285)$$

so (4.284) holds by integration over t. Set

$$K_2(\mathbf{p}) = -\int_0^{\log \beta \epsilon_0} dt\, \tilde{Q}_{22}(t|\Pi(p)), \qquad (4.286)$$

then $K_2(\mathbf{p}) \in \mathbb{R}$ by (4.285), and (4.282) holds for K_2 because $\frac{\partial}{\partial p}\Pi(p) = \frac{\partial}{\partial p}\pi(0, \theta) = 0$. Also, $(1)_2$ implies $(2)_2$ because $\int_0^\infty dt\, \partial^\alpha Q_{22}(t)$ converges for $|\alpha| \leq 1$. Integration of the RGE gives

$$\Sigma_2(t|p) = K_2(p) + \int_0^t ds \, Q_{22}(s|p).$$ (4.287)

Thus, by definition of K_2,

$$\Sigma(t|\Pi(p)) = -\int_t^\infty ds \, Q_{22}(s|\Pi(p))$$ (4.288)

which, together with $(1)_2$, implies $(3)_2$. (4.287), $(1)_2$, and $(2)_2$ imply $(4)_2$. Now $(5)_2$ follows by Taylor expansion in ρ and ω, and by $(3)_2$ with $\alpha = 1$. On supp D_t, $|p - \Pi(p)| \leq$ const e^{-t}, and $\|C_t\|_\infty \leq$ const e^t, so $(6)_2$ holds. In second order, $L^{(1)}$ and $L^{(2)}$ are zero because $\Sigma_1 = 0$, so $(7)_2$ and $(8)_2$ follow from the results of Sect. 4.4, and (4.284) follows again by the properties of D_t and \hat{v}.

Let $r \geq 3$. The main difference to the $r = 2$ case is that $L_{m,r}^{(1)}$ occurs for $m \geq 4$, $L_{m,r}^{(2)}$ occurs for $m \geq 6$, and that $1 - C_t \Sigma(t)$ is now $\neq 1$ because the $O(r')$, $r' < r$ selfenergy will in general be nonvanishing. Whenever a factor $(1 - C_t \Sigma(t))^{-1}$ appears, we expand it in a formal power series $\sum_n (C_t \Sigma(t))^n$. As mentioned, since Σ is $O(\lambda)$, all terms on the right hand side are of order $< r$ and so the inductive hypothesis applies to them.

Proof of $(1)_r$. Let $\alpha = 0$. By $(6)_{<r}$, $C_t \Sigma(t) = \mathcal{O}(1)$, so, using $(8)_{<r}$, the estimate reduces to the skeleton case, since $(8)_{<r}$ is identical to the estimate for skeleton functions. Let $|\alpha| = 1$. The only difference to the skeleton case now comes from the term where ∂^α acts on $C_t(k_1)\Sigma(t|k_1)$. Because $|\partial C_t(k_1)| \leq$ const e^{2t}, and because $k_1 \in$ supp D_t, $(5)_{<r}$ implies that

$$\partial\big(C_t(k_1)\Sigma(t|k_1)\big) = \mathcal{O}\left(e^t\right),$$ (4.289)

which is a factor e^t larger than the bound for $C_t \Sigma(t)$ itself. This is the same factor that occurs when a derivative acts on $\dot{D}_t(k_1)$. Thus the skeleton proof applies, and $(1)_r$ holds. Let

$$K_r(p) = -\int_0^\infty dt \, \tilde{Q}_{2,r}(t|\Pi(p)).$$ (4.290)

Then $(2)_r$ and (4.282) hold by $(1)_r$. The same sign-flipping argument as for $r = 2$, and the inductive hypothesis (4.284) for $r' < r$ imply (4.284) for r and the reality of K_r. $(3)_r$ and $(4)_r$ now follow from $(1)_r$ in exactly the same way as $(3)_2$ and $(4)_2$ followed from $(1)_2$, and $(5)_r$ and $(6)_r$ follow from $(3)_r$ and $(4)_r$ by Taylor expansion.

Again, (4.284) follows for W_{mr} by flipping signs in the integrals. To prove $(7)_r$, we have to bound $\mathbb{L}_{4,r}^{(1)}$ and $\tilde{Q}_{4,r}$. The bound for $\tilde{Q}_{4,r}$ is the same as in the non-ladder skeleton case by (4.289). By $(1)_r$,

$$\partial^\alpha \left(C_t(k) \dot\Sigma(t|k) \right) = \mathcal{O}\left(e^{t(|\alpha|-1)}(1+t) \right) \tag{4.291}$$

is $\mathcal{O}(1)$ for $\alpha = 0$ and $\mathcal{O}(1 = t)$ for $|\alpha| = 1$ so $(7)_r$ follows directly from $(7)_{<r}$ and the product rule. Similarly, $(8)_r$ follows by the skeleton case, which applies to \tilde{Q}_{mr} by (4.289), and by $(8)_{<r}$ and the inequality

$$\int dk\, \dot{D}_t(k) D_t(k) \Sigma(t|k)(C_t(k)\Sigma(t|k))^p = \mathcal{O}\left(e^{-t} \right) \tag{4.292}$$

which follows from the properties of D_t, from $(5)_{<r}$, and from $(6)_{<r}$. ∎

If the ladder term $i = 2$ is included for $m = 4$, it produces powers of $\log \beta\epsilon_0$ in the same way as it did in the skelton case. Once one allows for such logarithms, one can also bound the second derivative of Σ. There is an analogue of Theorem (4.23) where $(7)_r$ is dropped, a factor $(\log \beta\epsilon_0)^{r-1}$ appears everywhere on the RHS, and where the second derivative of the selfenergy is bounded as in Theorem 4.21 and Corollary 4.22. In other words, if $|\lambda| \log \beta\epsilon_0 < \text{const}$, the selfenergy satisfies

$$\left| \partial^\alpha \Sigma^{(R)} \right|_0 \leq \gamma_R \tag{4.293}$$

for $|\alpha| \leq 2$, with constants γ_R independent of β, and the four-point function is bounded.

The generalized tadpole term $L_{mr}^{(2)}$ is a technical complication for two reasons. First, it leads to one-particle reducible contributions to Σ (this is why we called Σ only an analogue of the selfenergy), and second, it leads to graphs that are nonoverlapping. Both these complications can be dealt with - the first one by doing a Legendre transform, the second one by a more careful analysis, using techniques of [45]. Instead of going through these proofs, we turn back to the question of what adding K means. The discussion in the next section is based on the results of [45, 46, 47].

4.7.4 The Meaning of K

In the last section, we renormalized the perturbation expansion by changing the initial condition by a formal power series

$$K(\lambda, E, \mathbf{p}) = \sum_{r=1}^{\infty} \lambda^r K_r(\mathbf{p}). \tag{4.294}$$

Thus, the initial Hamiltonian now has a λ-dependent free part

$$H_0'(\lambda) - \mu N = \sum_\sigma \int d\mathbf{k}\, (E(\mathbf{k}) + K(\lambda, E, \mathbf{k})) a_{\mathbf{k},\sigma}^* a_{\mathbf{k},\sigma}. \tag{4.295}$$

K is a formal power series in λ, and a functional of E. The goal, and the effect, of this was to make the selfenergy vanish on the Fermi surface (see

$(5)_r$ of Theorem 4.23: for $t \to \infty$, $\Sigma_r \propto |p - \Pi(p)|$). This means that the Fermi surface S does not change with λ because Σ_r vanishes on S for all r; thus S is the Fermi surface of the interacting system. It is this property that removes the infrared divergences, which came from expanding around the propagator with the singularity at the wrong place. The renormalized expansion still amounts to expanding the full propagator $(i\omega - E - \Sigma)^{-1}$ in terms of the free propagator $(i\omega - E)^{-1}$, but the powers of $\frac{\Sigma}{i\omega - E}$ arising in this expansion are now bounded because Σ vanishes on S and because its first derivative is bounded.

There remains the question how to justify putting in the counterterm function K. In the jellium case, where one has rotational symmetry, K does not depend on \mathbf{k} and thus is a correction to μ. In the case without rotational symmetry, things are not as simple. In order to have a fixed, that is, λ-independent, starting Hamiltonian, we have to require that

$$E(\mathbf{k}) + K(\lambda, E, \mathbf{k}) = e(\mathbf{k}) \tag{4.296}$$

is a fixed dispersion relation (e.g. that of the Hubbard model). In other words, once one does renormalization subtractions, the dispersion relation in the propagator is not the original dispersion relation of the model any more, but it has to be obtained from the latter by solving an implicit function equation. To solve such an equation, we must now regard λ as a real parameter, not just a formal variable, to have $\Phi(E) = E + K(\lambda, E)$ a real-valued function. With the above bounds, we can do this if we restrict to contributions up to order R, i.e., take $K^{(R)} = \sum_{r=1}^{R} \lambda^r K_r$ instead of K. $K^{(R)}$ is a polynomial in λ, so it can be made arbitrarily small by restricting to $|\lambda| \leq \lambda_R$. It is therefore natural to try an iteration to solve (4.296). To do this, one first needs a set of E that is mapped to itself by the map $E \mapsto \Phi^{(R)}(E) = E + K^{(R)}(\lambda, E)$. This is the reason why we were careful about finding bounds for derivatives of K. The above results imply that for $|\alpha| = 2$ and $|\lambda| \log \beta \epsilon_0 \leq \text{const}$, $|\partial^\alpha K^{(R)}| \leq \gamma_R$. Thus, for every E satisfying the assumptions of Sect. 4.2.6 and every $R \geq 1$, there is an open ball B_R around E such that $\Phi^{(R)}(B_R) \subset B_R$, and we can iterate the map in this ball. We have not proven the convergence of the iteration; to do this, one needs more detailed bounds, obtained in [46, 47] by an analysis involving a discrete RG flow (such bounds can, with a good deal more work, be given also using the continuous RG flow developed here). Uniqueness of the solution to (4.296) holds under weaker assumptions [45].

Once one can solve (4.296), one has a clear procedure for doing renormalization. Given the interaction, the dispersion relation e of the model, e.g. the tight-binding dispersion relation, and given any $R \geq 1$, one solves for $E = E^{(R)} = (\Phi^{(R)})^{-1}(e)$. Then $E + K^{(R)} = e$ by construction of E, and one can put E into the propagator by the Gaussian shift formula

$$d\mu_{C_e}(\psi) = \det(1 - C_e K^{(R)}) \, d\mu_{C_E}(\psi) e^{-\int dp \, \bar\psi(p) K^{(R)}(p) \psi(p)} \tag{4.297}$$

(which follows directly from the definition of the Gaussian measure; see Appendix B). This automatically puts the counterterm $K^{(R)}$ into the action,

and therefore the expansion in λ up to order R is now the renormalized expansion. One can interpret the procedure of solving (4.296) as calculating the interaction effects in two steps: first, one determines $K^{(R)}$, which is the part of the selfenergy that changes the Fermi surface. Going from e to E means changing from the Fermi surface of the free system to that of the interacting system. Once this is done, all other effects of the interaction can be calculated at the fixed Fermi surface $\{\mathbf{p} : E(\mathbf{p}) = 0\}$.

It should be clear from the description of the solution of (4.296) that with the bounds we have, one can only show that the solution $E = \Phi^{-1}(e)$ is C^2 in λ and in \mathbf{p}. Thus a solution of (4.296) does not make sense in a formal power series setting, and even if the power series in λ for K converges, the solution of (4.296) cannot be analytic in λ (if it were, the infrared divergent unrenormalized expansion would converge, by the identity theorem).

4.8 Conclusion

At the end of this long, and partly technical, discussion we review some of the arguments and results, to put them into perspective. The discussion in this chapter consisted of some rigorous bounds for the perturbative Green functions, together with an analysis of truncations of the flow that provide heuristic guidelines of how to understand the bounds. Needless to say, this leaves many open problems, a few of which we shall discuss briefly now.

4.8.1 Summary

We have analyzed perturbation theory in the coupling constant λ to all orders for the skeleton Green functions, and also for the ones with properly subtracted two-legged insertions (i.e. with the appropriate boundary condition). The results show that in $d \geq 2$ spatial dimensions, and for Fermi surfaces that have strictly positive curvature, for all $R \geq 1$, the order $\leq R$ renormalized selfenergy $\Sigma^{(R)} = \lim_{t\to\infty} \sum_{r=1}^{R} \Sigma_r(t)\lambda^r$ exists and satisfies

$$\left|\partial^\alpha \Sigma^{(R)}\right|_0 \leq \gamma_R \tag{4.298}$$

for all $|\alpha| \leq 2$, and the order $\leq R$ renormalized four-point function $\tilde{I}_4^{(R)} = \lim_{t\to\infty} \sum_{r=1}^{R} \tilde{I}_{4,r}(t)\lambda^r$ exists and satisfies

$$\left|\tilde{I}_4^{(R)}\right|_0 \leq \gamma_R', \tag{4.299}$$

with γ_R and γ_R' independent of β, provided the condition

$$|\lambda| \log \beta\epsilon_0 < \text{const} \tag{4.300}$$

holds. The higher Green functions are bounded in an L^1 norm. The detailed bounds also imply that if a single term in the RGE for the four-point function, the ladder term ($i = 2$), is left out, then the statements for the four-point function still hold, but without the restriction in (4.300), i.e. for $|\lambda| \leq$ const, uniformly in the temperature. This conclusion comes from the overlapping loop bounds; it applies to rather general Fermi surfaces because it holds whenever any kind of volume improvement is there (see [45] for more precise statements).

One can get a heuristic insight in the behaviour of the full solution by the opposite truncation of the RGE where only the ladder term $i = 2$ is kept in the four-point function. The solution of the RG differential equation then describes a flow of a coupling constant that reproduces the familiar BCS formulas, e.g. the form of the critical temperature (4.205).

Renormalization was done by introducing a counterterm function K that, in general, is momentum dependent. The justification for this requires the solution of a functional equation that relates the free and the interacting Fermi surface . This is a mathematically nontrivial problem even in finite-order perturbation theory, and we have given only a hint at its solution [46, 47].

We have not given any nonperturbative bounds here. Doing so is possible, at least in $d = 2$ spatial dimensions [55]. The nonperturbative analysis of these systems is the subject of ongoing research [56, 61] and goes beyond the scope of this introductory text. On the whole, it is expected that the above perturbative analysis is confirmed in the nonperturbative setting (but a proof of this is difficult). Also, some technical parts of it, in particular the overlapping loop method, are important for nonperturbative studies as well. The continuous RG method also works nonperturbatively for fermionic quantum field theory if the RGE is integrated [57, 63]; the continuous method may work without explicit integration as well [54, 53], but whether it does is at the moment an open problem.

4.8.2 A Fermi Liquid Criterion

The results lead us to formulate a criterion for Fermi liquid behaviour at positive temperatures [53]. This is necessary because Fermi liquid (or non-Fermi liquid) behaviour is observed only above the critical temperature for superconductivity. Because of the Kohn-Luttinger effect, one may expect that every system satisfying the assumptions of Sect. 4.2.6 has such a superconducting transition (if the reflection symmetry is broken, the Cooper instability is suppressed and Fermi liquid behaviour can occur at zero temperature, as shown in [56]). It is not completely straightforward to formulate such a criterion that involves only equilibrium observables and that can distinguish between Fermi liquids and other states of many fermion systems (such as the Luttinger liquid behaviour that occurs in $d = 1$ [59, 60]) because at positive temperatures, there is no discontinuity in the occupation number density which could serve

as a distinction (at $T = 0$ there isn't either, because the energy gap in the superconducting state removes the discontinuity).

A criterion for Fermi liquid behaviour at positive temperature is as follows.

1. The renormalized perturbation expansion in λ for the full Green functions and the counterterm function K converges in a region $\mathcal{R} = \{(\lambda, \beta) : |\lambda| \log \beta < \text{const}\}$.
2. The selfenergy is C^2 in p, and its second derivative is uniformly bounded on compact subsets of \mathcal{R}.
3. The counterterm function K is C^{k_0} in \mathbf{p}, with $k_0 > d$ (where d is the spatial dimension), and there is a unique solution to (4.296).

A slightly different, but equivalent, definition of Fermi liquid behaviour at positive temperature (referring to the skeleton functions instead of the full Green functions) was given in [53]. We conjecture that the above conditions can be verified in $d = 2$ dimensions for the jellium band relation and the Hubbard band relation, in the latter case for a filling that is sufficiently far away from half-filling, by a combination of existing techniques [45, 46, 47, 55]. The jellium model is much easier than the Hubbard model because in that case K is a constant, hence Condition 3 is trivially fulfilled. Condition 3 was proven in [45, 46, 47] in perturbation theory by a rather detailed singularity analysis [46] and an extension of the overlapping loop method [47]. Such an analysis is also possible using the RGDE, but not included here. Condition 3 is needed to invert the map Φ and to have certain summability properties of the propagator.

While a proof that the above version of Fermi liquid behaviour occurs is, even for the jellium model, beyond what can be explained here, it is easy to explain why it is sensible to call systems fulfilling Conditions 1–3 Fermi liquids, and in particular, why the above definition allows to distinguish between Fermi and Luttinger liquids. This distinction is in the selfenergy.

By Condition 2, one can do a first-order Taylor expansion in $p - \Pi(p)$, to get (writing $\Pi(p) = (0, \mathbf{P}(\mathbf{p}))$)

$$\Sigma(p) = p_0 (\partial_0 \Sigma)(0, \mathbf{P}(\mathbf{p})) + (\mathbf{p} - \mathbf{P}(\mathbf{p})) \cdot \nabla \Sigma(0, \mathbf{P}(\mathbf{p})) + \tilde{\Sigma}(p) \qquad (4.301)$$

from which one obtains a finite wave function renormalization

$$Z(\mathbf{p}) = 1 + i(\partial_0 \Sigma)(0, \mathbf{P}(\mathbf{p})) \qquad (4.302)$$

and a finite correction to the Fermi velocity $\nabla E(\mathbf{p})$, and the Taylor remainder $\tilde{\Sigma}(p)$ vanishes quadratically in the distance of the momentum p to its projection $\Pi(p)$ on the Fermi surface.

This property distinguishes Fermi liquids from other possible states of the many-fermion system, such as Luttinger liquids: In one dimension (where 'Luttinger liquid behaviour' has been proven [58, 59, 60]), already the first derivative of the selfenergy does not obey Condition 2. The geometric reason for this is that in one dimension, the Fermi "surface" consists only of two

points. There is no transversal direction and hence, of course, no curvature, and all the volume improvement effects are absent. Higher-dimensional systems with a flat Fermi surface also have singularities in the first derivative of the selfenergy, again because of absence of any volume improvement.

Note that this distinction between the behaviours of the selfenergy in one and more than one dimension can only be made if β is allowed to vary; at fixed β, the requirement that something is bounded independently of β is trivial. This is the reason why a whole range of values of β and λ is included in the above definition.

As discussed, the restriction (4.300) comes from the four-point function, in particular the ladder terms analyzed in Sect. 4.5.4. Thus this restriction is there to avoid the BCS instability: the temperature is above the critical temperature for superconductivity.

Under the above definition, one can justify using renormalized perturbation theory: given a model with dispersion relation e, interaction \hat{v}, and coupling constant λ, and at temperature T such that $|\lambda| \log \beta$ is small enough, solve the inversion equation (4.296) for E and change to E, whose zero set is the interacting Fermi surface, as described in Sect. 4.6. Because λ is fixed, E is now fixed. By condition 1, and because λ is fixed with $|\lambda| \log \beta$ small enough, the renormalized expansion in λ converges.

4.8.3 How the Curvature Sets a Scale

Finally, we revisit the heuristic discussion about the validity of the ladder flow near to half-filling in the Hubbard model in the light of the above volume and power counting bounds. It was mentioned that an important role in the perturbative justification of the ladder approximation by the overlapping loop bound was played by the energy scale at which the flow is started, the 'initial' energy scale ϵ_0. The reason for this is that the volume improvement is curvature dependent, and will lead to effects only below a certain scale, as follows.

For the band relation $E(\mathbf{p}) = -t \sum_{i=1}^{d} \cos k_i - \mu$, the curvature of the Fermi surface vanishes for $\mu = 0$ (half-filling). For small μ, the size r_0 of the neighbourhood will therefore depend on μ and vanish with μ. The initial energy scale ϵ_0 for the flow must be chosen $\leq r_0$. The analysis of Sect. 4.6 showed that the $i \geq 3$ terms have an extra scaling factor $Q_T \epsilon_t$, where the constant Q_T is proportional to $1/\kappa$. Since $\kappa \sim \mu$, we now see that the chemical potential sets a scale ϵ_0, below which the ladders dominate, in a natural way: if $\epsilon_t/\mu > 1$ is large, then the "improvement factor" $Q_T \epsilon_t > 1$, so there is no improvement at all, and thus no justification for restricting to the ladder flow. If ϵ_0 is chosen such that $\epsilon_0/\mu < 1$, the improvement holds for all $t \geq 0$, and then the ladders dominate the flow. The flow of the coupling constants in the various angular momentum sectors is then essentially independent, and one that started out negative will eventually diverge.

Thus the geometrical picture given by the volume bounds is simply that when $\epsilon_t/\mu > 1$, the neighbourhood is so thick that curvature effects do not lead to any improvement; the scaling behaviour is similar to that of a one-dimensional system. Only below scale μ, the curvature becomes visible and its effects single out the ladders in the flow of the four-point function (see Fig. 4.6).

Fig. 4.6. A shell around the Fermi surface intersecting its translate, for the tight-binding dispersion relation near half-filling $\mu \approx 0$. The energy scale of the shell is above the scale set by μ, therefore the shell is so thick that curvature effects do not improve power counting

There are a number of other effects that become important near to half-filling. The effective one-dimensionality suggested by the above considerations also implies that the selfenergy behaves rather differently above and below scale μ, so that in particular a careful analysis of the wave-function renormalization is important for scales above μ. It should be noted that above scale μ, the parquet flow is also not justified as the leading contribution by the overlapping loop bounds. Finding out which approximation is tractable and gives the correct behaviour requires different arguments.

A. Appendix to Chapters 1–3

A.1 A Topology on the Ring of Formal Power Series

In this appendix, we recall briefly some basic facts about formal power series. The ring of formal power series is the set $\mathbb{F} = \mathbb{C}^{\mathbb{N}_0}$ of sequences $S = (s_k)_{k \geq 0}$ with addition defined in the obvious way and multiplication as

$$(SS')_k = \sum_{l=0}^{k} s_l s'_{k-l}. \tag{A.1}$$

\mathbb{F} is a domain of integrity with unit element $(1, 0, 0, \ldots)$ and the map $\imath : \mathbb{C} \to \mathbb{F}$, $a \mapsto (a, 0, 0, \ldots)$ is an injection. The rule for the product is remembered easiest if one associates a formal expression $S(z) = \sum_{k \geq 0} z^k s_k$, to a formal variable z, to the sequence S.

The lower degree of $S \in \mathbb{F}$ is defined as $\omega(S) = \min\{k \geq 0 : s_k \neq 0\}$ for $S \neq 0$, and $\omega(0) = \infty$. Then $\omega(S + S') \geq \min\{\omega(S), \omega(S')\}$, and we can introduce a metric on \mathbb{F} by the definition $d(S, S) = 0$ and for $S \neq S'$,

$$d(S, S') = 2^{-\omega(S-S')}. \tag{A.2}$$

With this, \mathbb{F} becomes a complete metric space.

We call a family $(S_\alpha)_\alpha \in \mathbb{N}$ of formal power series summable if the limit

$$\lim_{n \to \infty} \sum_{\nu=1}^{n} S_\nu \tag{A.3}$$

exists in \mathbb{F}. Then a sequence $(S_\alpha)_\alpha \in \mathbb{N}$ is summable if and only if $\omega(S_\alpha) \to \infty$ as $\alpha \to \infty$. In particular, the substitution of one formal power series in another, $S(S'(z))$, is a well-defined formal power series if and only if $\omega(S') \geq 1$.

A.2 Fourier Transformation

Recall that $\frac{L}{2\varepsilon} \in \mathbb{N}$, so $M = \frac{L}{\varepsilon} \in 2\mathbb{N}$. The dual lattice to

$$\Gamma = \varepsilon \mathbb{Z}^d / L \mathbb{Z}^d = \{x \in \varepsilon \mathbb{Z}^d : -\frac{L}{2} \leq x_k < \frac{L}{2} \quad \forall k \in \{1, \ldots, d\}\} \tag{A.4}$$

is

$$\Gamma^* = \frac{2\pi}{L}\mathbb{Z}^d / \frac{2\pi}{\varepsilon}\mathbb{Z}^d = \{p \in \frac{2\pi}{L}\mathbb{Z}^d : -\frac{\pi}{\varepsilon} \leq p_k < \frac{\pi}{\varepsilon} \quad \forall k \in \{1,\ldots,d\}\}. \quad (A.5)$$

Both lattices have the same number of points

$$|\Gamma| = |\Gamma^*| = M^d = (\frac{L}{\varepsilon})^d. \quad (A.6)$$

We define $p \cdot x = \sum_{x=1}^{d} p_k x_k$,

$$\int_{\Gamma^*} dp\hat{F}(p) = L^{-d} \sum_{p \in \Gamma^*} \hat{F}(p), \quad (A.7)$$

and

$$\delta_{\Gamma^*}(p,q) = L^d \delta_{pq}. \quad (A.8)$$

The Fourier transform of $f : \Gamma \to \mathbb{C}$ is $\hat{f} : \Gamma^* \to \mathbb{C}$, given by

$$\hat{f}(p) = \int_{\Gamma} dx\, e^{-ip\cdot x} f(x). \quad (A.9)$$

The inverse relation is

$$f(x) = \int_{\Gamma^*} dp\, e^{ip\cdot x} \hat{f}(p). \quad (A.10)$$

The Fourier sums have the periodicity properties

$$\hat{f}(p + \frac{2\pi}{\varepsilon}n) = \hat{f}(p) \quad \forall n \in \mathbb{Z}^d \quad (A.11)$$

and

$$f(x + Ln) = f(x) \quad \forall n \in \mathbb{Z}^d. \quad (A.12)$$

We denote the scalar product for functions $\hat{\phi}, \hat{\psi} \in \mathbb{C}^{\Gamma^*}$ by

$$(\hat{\phi}, \hat{\psi})_{\Gamma^*} = \int_{\Gamma^*} dp\, \overline{\hat{\phi}(p)}\hat{\psi}(p) = L^{-d} \sum_{p \in \Gamma^*} \overline{\hat{\phi}(p)}\hat{\psi}(p). \quad (A.13)$$

As mentioned, one can view the transition from Γ to Γ^* as a change of basis to the basis given by $F_p(x) = e^{ip\cdot x}$, so that $\hat{F}_p(q) = \delta_{\Gamma^*}(p,q)$. The orthogonality of this basis, $(\hat{F}_p, \hat{F}_q)_{\Gamma^*} = \delta_{\Gamma^*}(p,q)$, implies

$$(\hat{\phi}, \hat{\psi})_{\Gamma^*} = (\phi, \psi)_{\Gamma} \quad (A.14)$$

so that, with this choice of scalar products, the Fourier transform is indeed a unitary map. Moreover,

$$\int_{\Gamma^*} dp\, e^{ip\cdot x} = \delta_{x0}\, \varepsilon^{-d} = \delta_\Gamma(x,0) \qquad \text{if } x \in \Gamma \tag{A.15}$$

and

$$\int_\Gamma dx\, e^{ip\cdot x} = \delta_{p0}\, L^d = \delta_{\Gamma^*}(p,0) \qquad \text{if } p \in \Gamma^*. \tag{A.16}$$

The matrix representation of a linear operator is in the Fourier basis

$$(A\phi)\hat{}(p) = \int_{\Gamma^*} dp\, \hat{A}(p,q)\hat{\phi}(q) \tag{A.17}$$

with

$$\hat{A}(p,q) = (F_p, AF_q) = \int_\Gamma dx \int_\Gamma dy\, e^{-ip\cdot x} A(x,y)\, e^{iq\cdot y}. \tag{A.18}$$

The inverse formula is

$$A(x,y) = \int_{\Gamma^*} dp \int_{\Gamma^*} dq\, e^{ipx}\, \hat{A}(p,q)\, e^{-iqy}. \tag{A.19}$$

We have chosen our conventions such that the limiting cases $\varepsilon \to 0$ and $L \to \infty$ can be discussed separately without further rescalings. They are

(a) the *thermodynamic limit* $L \to \infty$:

$$\begin{aligned}
\Gamma &= \varepsilon \mathbb{Z}^d, \quad \Gamma^* = \mathbb{R}^d / \frac{2\pi}{\varepsilon} \mathbb{Z}^d =: \mathcal{B} \\
\hat{f}(p) &= \varepsilon^d \sum_{x \in \Gamma} e^{-ipx} f(x) \\
f(x) &= \int_\mathcal{B} \frac{d^d p}{(2\pi)^d} e^{ipx} \hat{f}(p) \\
\varepsilon^d \sum_{x \in \Gamma} e^{ipx} &= \delta(p) \quad \text{for } p \in \Gamma^* \\
\int_\mathcal{B} \frac{d^d p}{(2\pi)^d} e^{ipx} &= \delta_{x0} \cdot \varepsilon^{-d}.
\end{aligned} \tag{A.20}$$

(b) the *continuum limit* $\varepsilon \to 0$:

$$\begin{aligned}
\Gamma &= \mathbb{R}^d / L\mathbb{Z}^d, \quad \Gamma^* = \frac{2\pi}{L} \mathbb{Z}^d \\
\hat{f}(p) &= \int d^d x\, e^{-ipx} f(x) \\
f(x) &= L^{-d} \sum_{p \in \Gamma^*} e^{ipx} \hat{f}(p) \\
L^{-d} \sum_{p \in \Gamma^*} e^{ipx} &= \delta(x) \\
\int dx\, e^{ipx} &= \delta_{p0} L^d \quad \text{if } p \in \Gamma^*.
\end{aligned} \tag{A.21}$$

Note that under our conditions, the limit $\varepsilon \to 0$ at fixed L is the limit of a sequence $\varepsilon_N = \frac{L}{2N}$ for $N \to \infty$.

A.3 Properties of the Boson Propagator

In this Appendix, we prove Lemma 1.10. Write $p = (p_1, k)$, $k \in \mathbb{R}^{d-1}$.

$$1 + p^2 = 1 + k^2 + p_1^2 = (p_1 + i\mu(k))(p_1 - i\mu(k)) \tag{A.22}$$

with $\mu(k) = \sqrt{1 + k^2}$. Since $t > 0$, and the integrand is $O(|k|^{-2})$ as $|k| \to \infty$, we can evaluate the integral using the residue theorem, closing in the upper half plane. This gives

$$
\begin{aligned}
\tilde{C}_d(t) &= \int \frac{d^{d-1}k}{(2\pi)^{d-1}} \frac{e^{-t\mu(k)}}{2\mu(k)} \\
&= \frac{1}{2(2\pi)^{d-1}} |S^{d-2}| \int_0^\infty d\rho\, \rho^{d-2} \frac{e^{-t\mu(\rho)}}{\mu(\rho)}
\end{aligned}
\tag{A.23}
$$

This shows that $\tilde{C}_d(t)$ is finite for all $t > 0$ and that (i) holds. The constant $|S^{d-2}| = 2\pi^{\frac{d-1}{2}} \Gamma(\frac{d-1}{2})$. We call the constant in front of the integral in (A.23) A_d. Since

$$\mu(\rho) = 1 + \int_0^\rho d\sigma \frac{\sigma}{(1+\sigma^2)^{1/2}} \geq 1 + \frac{\rho^2}{2(1+\rho^2)^{1/2}}, \tag{A.24}$$

we obtain

$$\mu(\rho) \geq \begin{cases} 1 + \delta\rho^2 & \text{if } 0 \leq \rho \leq 1 \\ 1 + \delta\rho & \text{if } \rho \geq 1 \end{cases} \tag{A.25}$$

with $\delta = \frac{1}{3}$. Therefore, for $t \geq 1$

$$
\begin{aligned}
\tilde{C}_d(t) &= A_d \left(\int_0^1 d\rho\, \rho^{d-2} \frac{e^{-t\mu(\rho)}}{\mu(\rho)} + \int_1^\infty d\rho\, \rho^{d-2} \frac{e^{-t\mu(\rho)}}{\mu(\rho)} \right) \\
&\leq A_d\, e^{-t} \left(\int_0^1 d\rho\, \rho^{d-2} e^{-t\delta\rho^2} + \int_1^\infty d\rho\, \rho^{d-2} \frac{e^{-t\delta\rho}}{\delta\rho} \right) \\
&\leq A_d\, e^{-t} \left(t^{-\frac{d-1}{2}} \int_0^t dx\, x^{d-2} e^{-\delta x^2} + t^{-d} \frac{1}{\delta} \int_t^\infty dx\, x^{d-1} e^{-\delta x} \right) \\
&\leq A_d e^{-t} t^{-\frac{d-1}{2}} \left(\int_0^\infty dx\, x^{d-2} e^{-\delta x^2} + \frac{1}{\delta} \int_1^\infty dx\, x^{d-1} e^{-\delta x} \right) \\
&\leq \text{const} \cdot e^{-t}\, t^{-\frac{d-1}{2}}
\end{aligned}
\tag{A.26}
$$

($t \geq 1$ was used to replace $\int_t^\infty dx$ by $\int_1^\infty dx$).

For $0 < t \leq 1$, write $\rho = \frac{s}{t}$, $d\rho = \frac{1}{t}ds$, to get

$$\tilde{C}_d(t) = A_d t^{-(d-1)} \int_0^\infty ds \, s^{d-2} \frac{e^{-t\mu(\frac{s}{t})}}{\mu(\frac{s}{t})} \tag{A.27}$$

so

$$t^{d-2}\tilde{C}_d(t) = A_d \int_0^\infty ds \, s^{d-2} \frac{e^{-s\mu(\frac{t}{s})}}{s\mu(\frac{t}{s})} \tag{A.28}$$

(we used $t\mu(\frac{s}{t}) = t\sqrt{1+(\frac{s}{t})^2} = \sqrt{s^2+t^2} = s\sqrt{1+(\frac{t}{s})^2} = s\mu(\frac{t}{s})$).

For $d \geq 3$, simply use $\mu(\frac{t}{s}) \geq 1$ to get

$$t^{d-2}\tilde{C}_d(t) \leq A_d \int_0^\infty ds \, s^{d-3} e^{-s} \tag{A.29}$$

so $\tilde{C}_d(t) \leq A_d\Gamma(d-2)t^{-(d-2)}$. Since $\mu(\frac{t}{s}) \to 1$ as $t \to 0$, this becomes exact as $t \to 0$.

For $d = 2$, write $s = t\mu(\rho) = t(1+\rho^2)^{1/2}$, so $\rho = \sqrt{(\frac{s}{t})^2 - 1}$, $\rho \, d\rho = \frac{2s}{t^2}ds$, then

$$
\begin{aligned}
\tilde{C}_2(t) &= A_2 \int_t^\infty ds \, \frac{2s}{t^2} t \frac{e^{-s}}{s} \frac{1}{\sqrt{(\frac{s}{t})^2 - 1}} \\
&= A_2 \, 2 \int_t^\infty ds \, e^{-s} \frac{1}{\sqrt{s^2 - t^2}} \sim \log t. \tag{A.30}
\end{aligned}
$$

A.4 Wick Reordering for Bosons

It is convenient to derive the component form of the RG in position space and take the Fourier transform afterwards.

The quadratic term on the right hand side of the RG equation is

$$Q_r^{\Lambda\Lambda_0}(\phi) = \sum_{\substack{r_1 \geq 1, r_2 \geq 1 \\ r_1 + r_2 = r}} \sum_{m_1=1}^{\bar{m}(r_1)} \sum_{m_2=1}^{\bar{m}(r_2)} \mathcal{Y}(r_1, m_1, r_2, m_2) \tag{A.31}$$

with

$$\mathcal{Y}(r_1, m_1, r_2, m_2) = \int d\underline{x} \, G_{m_1 r_1}^{\Lambda\Lambda_0}(\underline{x}^{(1)}) \, G_{m_2 r_2}^{\Lambda\Lambda_0}(\underline{x}^{(2)}) \, P(\underline{x}, \phi) \tag{A.32}$$

where $\underline{x}^{(1)} = (x_1, \ldots, x_{m_1})$, $\underline{x}^{(2)} = (x_{m_1+1}, \ldots, x_{m_1+m_2})$, $\underline{x} = (x_1, \ldots, x_{m_1+m_2})$,

$$P(\underline{x}, \phi) = \int dx \, dy \, \dot{C}_{\Lambda\Lambda_0}(x, y) \, \omega_{x,y}(\phi), \tag{A.33}$$

(here $\dot{C}_{\Lambda\Lambda_0} = \frac{\partial}{\partial\Lambda}C_{\Lambda\Lambda_0}$) and

$$\omega_{x,y}(\phi) = \left(\frac{\delta}{\delta\phi(x)}\Omega_{C_{0A}}\left(\prod_{k_1=1}^{m_1}\phi(x_{k_1})\right)\right)\frac{\delta}{\delta\phi(y)}\Omega_{C_{0A}}\left(\prod_{k_2=1}^{m_2}\phi(x_{m_1+k_2})\right).$$

$$(A.34)$$

Let

$$\alpha^{(1)} = \sum_{k=1}^{m_1}\alpha_k\,\delta_\Gamma(x,x_k), \quad \alpha^{(1)} = \sum_{k=1}^{m_2}\alpha_{m_1+k}\,\delta_\Gamma(x,x_{m_1+k}), \qquad (A.35)$$

and $\alpha = \alpha^{(1)} + \alpha^{(2)}$. With this,

$$\begin{aligned}
\frac{\delta}{\delta\phi(x)}\Omega_{C_{0A}}\left(\prod_{k_1=1}^{m_1}\phi(x_{k_1})\right) &= \left[\frac{\delta}{\delta\phi(x)}\prod_{k=1}^{m_1}\frac{\partial}{\partial\alpha_k}\Omega_{C_{0A}}\left(e^{(\alpha^{(1)},\phi)_\Gamma}\right)\right]_{\alpha=0} \\
&= \left[\prod_{k=1}^{m_1}\frac{\partial}{\partial\alpha_k}\frac{\delta}{\delta\phi(x)}e^{(\alpha^{(1)},\phi)_\Gamma-\frac{1}{2}(\alpha^{(1)},C_{0A}\alpha^{(1)})_\Gamma}\right]_{\alpha=0} \\
&= \left[\prod_{k=1}^{m_1}\frac{\partial}{\partial\alpha_k}\alpha^{(1)}(x)e^{(\alpha^{(1)},\phi)_\Gamma-\frac{1}{2}(\alpha^{(1)},C_{0A}\alpha^{(1)})_\Gamma}\right]_{\alpha=0}. \quad (A.36)
\end{aligned}$$

Similarly,

$$\Omega_{C_{0A}}\left(\prod_{k=1}^{m_2}\phi(x_k)\right) = \left[\prod_{k=1}^{m_2}\frac{\partial}{\partial\alpha_{m_1+k}}\alpha^{(2)}(y)e^{(\alpha^{(2)},\phi)_\Gamma-\frac{1}{2}(\alpha^{(2)},C_{0A}\alpha^{(2)})_\Gamma}\right]_{\alpha=0}.$$

$$(A.37)$$

Since $\alpha^{(1)}$ is independent of $\alpha_{m_1+1},\dots,\alpha_{m_1+m_2}$, the derivatives with respect to $\alpha^{(2)}$ can be commuted through so that

$$P(\underline{x}) = \left[\frac{\partial^{m_1+m_2}}{\partial\alpha_1\dots\partial\alpha_{m_1+m_2}}Z(\alpha,\phi)\right]_{\alpha=0} \qquad (A.38)$$

with

$$Z(\alpha,\phi) = (\alpha^{(1)},\,\dot{C}_{A A_0}\,\alpha^{(2)})\,e^{(\alpha,\phi)_\Gamma-\frac{1}{2}(\alpha^{(1)},\,C_{0A}\,\alpha^{(2)})_\Gamma-\frac{1}{2}(\alpha^{(1)},\,C_{0A}\,\alpha^{(2)})_\Gamma}.$$

$$(A.39)$$

Wick ordering of $e^{(\alpha,\phi)_\Gamma}$, the symmetry of $C_{A A_0}$ and C_{0A}, and $\dot{C}_{A A_0} = -\dot{C}_{0A}$, give

$$\begin{aligned}
Z(\alpha,\phi) &= (\alpha^{(1)},\,\dot{C}_{A A_0}\,\alpha^{(2)})\,e^{(\alpha^{(1)},\,C_{0A}\,\alpha^{(2)})_\Gamma}\,\Omega_{C_{0A}}\left(e^{(\alpha,\phi)_\Gamma}\right) \\
&= (\alpha^{(1)},\,(-\dot{C}_{0A})\,\alpha^{(2)})\,e^{(\alpha^{(1)},\,C_{0A}\,\alpha^{(2)})_\Gamma}\,\Omega_{C_{0A}}\left(e^{(\alpha,\phi)_\Gamma}\right) \\
&= \left(-\frac{\partial}{\partial t}\,e^{(\alpha^{(1)},\,C_{0A}\,\alpha^{(2)})_\Gamma}\right)\Omega_{C_{0A}}\left(e^{(\alpha,\phi)_\Gamma}\right). \quad (A.40)
\end{aligned}$$

Let $m_1 + m_2 = \mu$. By Lemma 2.6

$$\left[\frac{\partial^\mu}{\partial\alpha_1\dots\partial\alpha_\mu}Z(\alpha,\phi)\right]_{\alpha=0} = \left[Z(\frac{\partial}{\partial\alpha})\,\alpha_1\dots\alpha_\mu\right]_{\alpha=0}. \qquad (A.41)$$

By definition of $\alpha^{(1)}$ and $\alpha^{(2)}$,

$$\left(\frac{\partial}{\partial \alpha^{(1)}}, C_{0\Lambda} \frac{\partial}{\partial \alpha^{(2)}} \right) = \sum_{k_1=1}^{m_1} \sum_{k_2=m_1+1}^{\mu} \frac{\partial}{\partial \alpha_{k_1}} C_{0\Lambda}(x_{k_1}, x_{k_2}) \frac{\partial}{\partial \alpha_{k_2}}. \qquad (A.42)$$

Every derivative can act only once on $\alpha_1 \ldots \alpha_\mu$ without giving zero because the α's are all different. So

$$\exp \left(C_{0\Lambda} \frac{\partial}{\partial \alpha^{(1)}}, \frac{\partial}{\partial \alpha^{(2)}} \right) \alpha_1 \ldots \alpha_\mu \qquad (A.43)$$

$$= \sum_{L \subset M_1 \times M_2} \prod_{(k_1,k_2) \in L} \left(\frac{\partial}{\partial \alpha_{k_1}} C_{0\Lambda}(x_{k_1}, x_{k_2}) \frac{\partial}{\partial \alpha_{k_2}} \right) \alpha_1 \ldots \alpha_\mu$$

where $M_1 = \{1, \ldots, m_1\}$ and $M_2 = \{m_1+1, \ldots, \mu\}$. $L = \emptyset$ contributes to the sum, but gives the t-independent result $\alpha_1 \ldots \alpha_\mu$, so the t-derivative removes this term in (A.40). Let $\pi_1(k_1, k_2) = k_1$ and $\pi_2(k_1, k_2) = k_2$. For a term given by $L \neq \emptyset$ to be nonzero, $\pi_1|_L$ and $\pi_2|_L$ must be injective, because otherwise a derivative would act twice. Thus the sum can be restricted to the set

$$\mathcal{L} = \{ L \subset M_1 \times M_2 : L \neq \emptyset, \text{ and } \pi_k |_L \text{ injective for } k = 1, 2 \} \qquad (A.44)$$

If $L \in \mathcal{L}$, $\pi_1(L) = \pi_2(L)$. Thus

$$\mathcal{L} = \bigcup_{i=1}^{\min\{m_1,m_2\}} \bigcup_{\substack{B_i \subset M_i \\ |B_1|=|B_2|=i}} \mathcal{L}(B_1, B_2) \qquad (A.45)$$

with

$$\mathcal{L}(B_1, B_2) = \{ L \in \mathcal{L} : \pi_1(L) = B_1 \text{ and } \pi_2(L) = B_2 \}. \qquad (A.46)$$

Let

$$B_1 = \{b_1, \ldots, b_i\}, \qquad 1 \leq b_1 < \ldots b_i \leq m_1,$$
$$B_2 = \{b_{i+1}, \ldots, b_{2i}\}, \qquad m_1 + 1 \leq b_{i+1} < \ldots b_{2i} \leq m_1 + m_2, \quad (A.47)$$

then for any $L \in \mathcal{L}(B_1, B_2)$, there is a unique permutation $\pi \in \mathcal{S}_i$ such that

$$L = \{ (b_k, b_{i+\pi(k)}) : k \in \{1, \ldots, i\} \}. \qquad (A.48)$$

Thus the sum over L splits into a sum over $i \geq 1$, a sum over sequences $b = (b_1, \ldots, b_{2i})$ with

$$1 \leq b_1 < b_2 < \ldots < b_i \leq m_1 < b_{i+1} < \ldots < b_{2i} \leq m_1 + m_2, \qquad (A.49)$$

and a sum over permutations $\pi \in \mathcal{S}_i$, and

$$\exp\left(\frac{\partial}{\partial\alpha^{(1)}}, C_{0A}\frac{\partial}{\partial\alpha^{(2)}}\right)\alpha_1\ldots\alpha_\mu \tag{A.50}$$

$$= \sum_{i=1}^{\min\{m_1,m_2\}} \sum_{\pi\in S_i}\sum_b \prod_{k=1}^i D(x_{b_k}, x_{b_{i+\pi(k)}})\, H(i, b, \pi)$$

with

$$H(i, b, \pi) = \prod_{k=1}^i \left(\frac{\partial}{\partial\alpha_{b_k}}\frac{\partial}{\partial\alpha_{b_{i+\pi(k)}}}\right)\alpha_1\ldots\alpha_\mu. \tag{A.51}$$

The derivatives give (calling $\tilde{b} = \{b_1,\ldots,b_{2i}\}$)

$$H(i, b, \pi) = \prod_{\substack{k=1\\k\notin\tilde b}}^{\mu} \alpha_k . \tag{A.52}$$

The remaining derivatives acting in (A.41) come from the Wick ordered exponential in (A.40). They now act on H. By Lemma 2.6, they give

$$\left[\Omega_{C_{0A}}\left(e^{\left(\frac{\partial}{\partial\alpha},\phi\right)}\right)\prod_{\substack{k=1\\k\notin\tilde b}}^{\mu}\alpha_k\right]_{\alpha=0} = \left[\prod_{\substack{k=1\\k\notin\tilde b}}^{\mu}\frac{\partial}{\partial\alpha_k}\Omega_{C_{0A}}\left(e^{(\alpha,\phi)}\right)\right]_{\alpha=0}$$

$$= \Omega_{C_{0A}}\left(\prod_{\substack{k=1\\k\notin\tilde b}}^{m_1+m_2}\phi(x_k)\right). \tag{A.53}$$

The final step is to rename the integration variables to rewrite the Wick ordered product in the form in which it appears in the result. Before that, we use the permutational symmetry of the $G_{mr}(t \mid x^{(k)})$ to permute its arguments such that x_1,\ldots,x_{b_i} appear as the first i entries. Then the summand does not depend on b and π any more, so the sum over b and π, with the constraint (A.49), gives

$$\sum_{\pi\in S_i}\sum_{\substack{B_1\subset M_1, B_2\subset M_2\\|B_1|=|B_2|=i}} = \binom{m_1}{i}\binom{m_2}{i}i!, \tag{A.54}$$

and \mathcal{Y} becomes

$$\mathcal{Y}(r_1, m_1, r_2, m_2) = \sum_{i=1}^{\min\{m_1,m_2\}}\int dx_1\ldots dx_m\, \Omega_{C_{0A}}\left(\prod_{k=1}^m \phi(x_k)\right)$$

$$Y_{r_1,m_1,r_2,m_2,i}(x_1,\ldots,x_m) \tag{A.55}$$

with

$$Y_{r_1,m_1,r_2,m_2,i}(\underline{x}) = \int dv_1 \dots dv_i \, dw_1 \dots dw_i \left(-\frac{\partial}{\partial t} \prod_{k=1}^{i} C_{0\Lambda}(v_k, w_k) \right)$$
$$G_{m_1 r_1}^{\Lambda\Lambda_0}(v_1, \dots, v_i, \underline{x_1}) \, G_{m_2 r_2}^{\Lambda\Lambda_0}(w_1, \dots, w_i, \underline{x_2}) \quad \text{(A.56)}$$

with $\underline{x_1} = (x_1, \dots, x_{m_1-i})$ and $\underline{x_2} = (x_{m_1-i+1}, \dots, x_m)$. The derivative with respect to t and an additional reordering give another factor i. Let $l = i - 1$, then

$$i \, i! \binom{m_1}{i} \binom{m_2}{i} = m_1 \, m_2 \, l! \binom{m_1 - 1}{l} \binom{m_2 - 1}{l}, \quad \text{(A.57)}$$

which is the combinatorial factor defined in (3.34). The constraint $2i + m = m_1 + m_2$ implies the constraint in (3.33). Performing the Fourier transform and renaming $m_1 = m'$, $m_2 = m''$ etc. gives the RG equation in momentum space.

In the graphical interpretation, most permutations $\pi \neq id$ occurring in the sum give rise to *nonplanar* graphs. If one restricts the sum to planar graphs, only $\pi = id$ and the shifts $\pi(i) = i + k \bmod i$, $k \in \{1, \dots, i-1\}$, remain. Thus only i of the $i!$ permutations give rise to planar graphs. The combinatorial factor corresponding to a 'planarized' theory obtained by leaving out the $l!$ is therefore

$$\tilde{\kappa}_{m_1 m_2 l} = m_1 \, m_2 \binom{m_1 - 1}{l} \binom{m_2 - 1}{l}. \quad \text{(A.58)}$$

A.5 The Lower Bound for the Sunset Graph

In this appendix, we prove (3.104). Since $K'(x) = 0$ unless $x \in (1,4)$,

$$\partial_\Lambda C_{\Lambda\Lambda_0}(q) = -K'(\frac{q^2}{\Lambda^2}) \frac{2}{\Lambda^3} \frac{q^2}{q^2 + m_\phi^2} \quad \text{(A.59)}$$

vanishes unless $\Lambda^2 \leq q^2 \leq 4\Lambda^2$. By a change of variables in the integral

$$I_{\Lambda\Lambda_0}(0) = \int d^3 p \, C_{0\Lambda}(p) \int d^3 q \, C_{0\Lambda}(q - p) \, \partial_\Lambda C_{\Lambda\Lambda_0}(q). \quad \text{(A.60)}$$

The integrand is nonnegative, so $I_{\Lambda\Lambda_0}(0) \geq J(\Lambda)$ where

$$J(\Lambda) = \int_{A(\frac{4}{8}, 4)} d^3 p \, \frac{1}{p^2 + m_\phi^2} \int_{A(\frac{9}{8}\Lambda, \frac{5}{4}\Lambda)} d^3 q \, C_{0\Lambda}(q - p) \, \partial_\Lambda C_{\Lambda\Lambda_0}(q) \quad \text{(A.61)}$$

with

$$A(a, b) = \{p \in \mathbb{R}^3 : a \leq |p| \leq b\}. \quad \text{(A.62)}$$

Since $K'(x) < 0$ for all $x \in (1, 4)$, there is $K_0 > 0$ such that

$$\min\{|K'(x)| : x \in [(\frac{9}{8})^2, (\frac{5}{4})^2]\} \geq K_0. \quad \text{(A.63)}$$

Thus, for $\Lambda \geq m$ and $q \in A(\frac{9}{8}\Lambda, \frac{5}{4}\Lambda)$,

$$\partial_\Lambda C_{\Lambda\Lambda_0}(q) \geq \frac{2K_0}{\Lambda^3} \frac{(\frac{5}{4}\Lambda)^2}{(\frac{9}{8}\Lambda)^2 + m_\phi{}^2} \geq \frac{K_0}{3}\frac{1}{\Lambda^3}. \tag{A.64}$$

Moreover, in the support of the integrand for $J(\Lambda)$,

$$|q - p| \geq ||q| - |p|| \geq \frac{9}{8}\Lambda - \frac{\Lambda}{4} = \frac{7}{8}\Lambda, \tag{A.65}$$

$$|q - p| \leq |q| + |p| \leq \frac{5}{4}\Lambda + \frac{\Lambda}{4} = \frac{3}{2}\Lambda, \tag{A.66}$$

so, with $L_0 = \frac{4}{13}\min\{K(x) : (\frac{7}{8})^2 \leq x \leq \frac{9}{4}\} > 0$,

$$C_{0\Lambda}(q - p) = \frac{1}{(q-p)^2 + m_\phi{}^2} K\left(\frac{(q-p)^2}{m_\phi{}^2}\right) \geq \frac{\frac{13}{4}L_0}{(\frac{3}{2}\Lambda)^2 + m_\phi{}^2} \geq \frac{L_0}{\Lambda^2}. \tag{A.67}$$

Inserting these bounds, we get

$$J(\Lambda) \geq \int\limits_{A(\frac{4}{8},\frac{4}{4})} \frac{\mathrm{d}^3 p}{p^2 + m_\phi{}^2} \frac{L_0}{\Lambda^2}\frac{K_0}{3\Lambda^3} \int\limits_{A(\frac{9}{8}\Lambda,\frac{5}{4}\lambda)} \mathrm{d}^3 q = \tilde{K}\frac{1}{\Lambda^2}\int\limits_{\frac{4}{8}}^{\frac{4}{4}} \frac{\rho^2 \mathrm{d}\rho}{\rho^2 + m_\phi{}^2} \tag{A.68}$$

with

$$\tilde{K} = (\frac{4\pi}{3})^2 K_0 L_0\left((\frac{5}{4})^3 - (\frac{9}{8})^3\right). \tag{A.69}$$

In the last integral, we use $m_\phi{}^2 \leq \Lambda^2 \leq (8\rho)^2$, i.e., $m_\phi{}^2 + \rho^2 \leq 65\rho^2$, to get

$$\int\limits_{\frac{4}{8}}^{\frac{4}{4}} \frac{\rho^2 \mathrm{d}\rho}{\rho^2 + m_\phi{}^2} \geq \frac{1}{65}\int\limits_{\frac{4}{8}}^{\frac{4}{4}} \mathrm{d}\rho = \frac{\Lambda}{520}. \tag{A.70}$$

Thus the lower bound in (3.104) holds with $Q_1 = \frac{\tilde{K}}{520}$.

B. Appendix to Chapter 4

B.1 Fermionic Fock Space

We just give a brief summary here; for details, see [64]. If \mathcal{H} is the one-particle Hilbert space (assumed as separable), the configuration space for an n-particle system is $\mathcal{H}^n = \mathcal{H} \otimes \ldots \otimes \mathcal{H}$ (tensor products are defined, e.g., in [65]). The Fock space \mathcal{F} over \mathcal{H} is defined as

$$\mathcal{F} = \bigoplus_{n=0}^{\infty} \mathcal{H}^n \tag{B.1}$$

with $\mathcal{H}^0 = \mathbb{C}$. elements of \mathcal{F} are sequences $\psi = (\psi^{(0)}, \psi^{(1)}, \ldots)$ with $\psi^{(0)} \in \mathbb{C}$, $\psi^{(1)} \in \mathcal{H}$, etc. The infinite direct sum is a Hilbert space with the natural inner product

$$\langle \varphi \mid \psi \rangle_{\mathcal{F}} = \overline{\varphi^{(0)}} \psi^{(0)} + \sum_{n \geq 1} \langle \varphi^{(n)} \mid \psi^{(n)} \rangle_{\mathcal{H}^n}. \tag{B.2}$$

The set \mathcal{F}_0 of finite vectors, which have only a finite number of $\psi^{(n)}$ nonzero, is dense in \mathcal{F}.

Let P_{\pm} be the linear operator defined by

$$P_{\pm}(f_1 \otimes \ldots \otimes f_n) = \frac{1}{n!} \sum_{\pi \in S_n} (\pm 1)^{\pi} f_{\pi(1)} \otimes \ldots \otimes f_{\pi(n)} \tag{B.3}$$

where $(+1)^{\pi} = 1$ and $(-1)^{\pi} = \varepsilon(\pi)$ is the sign of the permutation π. The P_{\pm} are projectors. The Bose Fock space is $\mathcal{F}_+ = P_+ \mathcal{F}$, the Fermi Fock space is $\mathcal{F}_- = P_- \mathcal{F}$.

For $f \in \mathcal{H}$, the creation and annihilation operators $a^*(f)$ and $a(f)$ are defined on \mathcal{F}_0 as the linear operators satisfying $a(f)\psi^{(0)} = 0$, $a^*(f)\psi^{(0)} = f$, and

$$\begin{aligned} a(f)(f_1 \otimes \ldots \otimes f_n) &= \sqrt{n} \, \langle f \mid f_1 \rangle_{\mathcal{H}} \, f_2 \otimes \ldots \otimes f_n \\ a^*(f)(f_1 \otimes \ldots \otimes f_n) &= \sqrt{n+1} \, f \otimes f_1 \otimes \ldots \otimes f_n. \end{aligned} \tag{B.4}$$

The map $f \mapsto a^*(f)$ is linear and the map $f \mapsto a(f)$ is antilinear. An elementary calculation shows that on \mathcal{H}^n, $\|a(f)\| \leq \sqrt{n}\,\|f\|$ and $\|a^*(f)\| \leq$

$\sqrt{n+1}\|f\|$. The bosonic and fermionic creation and annihilation operators are defined by

$$a_{\pm}^{(*)}(f) = P_{\pm} a^{(*)}(f) P_{\pm}. \tag{B.5}$$

From now on we focus on the fermionic case. The fermionic operators obey the *canonical anticommutation relations* (CAR)

$$\begin{aligned} \{a_-(f), a_-(g)\} &= \{a_-^*(f), a_-^*(g)\} = 0 \\ \{a_-(f), a_-^*(g)\} &= \mathbb{1}\,\langle f \mid g \rangle_{\mathcal{H}}. \end{aligned} \tag{B.6}$$

A consequence of the CAR is that the fermionic creation and annihilation operators are bounded,

$$\|a(f)\| = \|a^*(f)\| = \|f\|_{\mathcal{H}}, \tag{B.7}$$

and thus extend uniquely to bounded operators defined on \mathcal{F} and that, if $(f_\alpha)_{\alpha \in A}$ is an ONB of \mathcal{H}, a ONB of \mathcal{F} is given by the vectors $\Omega = (1, 0, 0, \ldots)$ and $\prod_{\alpha \in S} a_*(f_\alpha)\Omega$, where S runs over all finite subsets of A.

The physical interpretation is that Ω is the vacuum, the state without any particles, and that the n-particle state

$$\prod_{k=1}^{n} a_*(f_{\alpha_k})\Omega = \sqrt{n!}\,P_-(f_{\alpha_1} \otimes \ldots \otimes f_{\alpha_n}) \tag{B.8}$$

has n fermions occupying the states $f_{\alpha_1}, \ldots, f_{\alpha_n}$. The CAR imply in particular that $a_-^*(f)^2 = 0$, i.e. that no state can be occupied twice.

In the text, we only consider the Fermionic Fock space, so we drop the subscript and write $a(f) = a_-(f)$.

B.2 Calculus on Grassmann Algebras

Let \mathcal{N} be a finite set, without loss of generality $\mathcal{N} = \{1, \ldots, N\}$ (this is no loss of generality because the Grassmann Gaussian measure introduced below changes at most by a sign if the elements of \mathcal{N} are permuted). The formal algebraic objects $\psi_1, \ldots \psi_N$ are called Grassmann variables if they obey

$$\{\psi_i, \psi_j\} := \psi_i \psi_j + \psi_j \psi_i = 0 \tag{B.9}$$

for all $i, j \in \Lambda$. The Grassmann algebra $\mathbb{C}\{\psi_1, \ldots, \psi_N\}$ generated by ψ_1, \ldots, ψ_N consists of polynomials

$$g = b(g) + \sum_{n \geq 1} \sum_{i_1, \ldots, i_n \in \Lambda} c_{i_1, \ldots, i_n}(g)\psi_{i_1} \ldots \psi_{i_n} \tag{B.10}$$

where $b, c_{i_1, \ldots, i_n}(g) \in \mathbb{C}$, with addition and multiplication defined as usual for polynomials and by taking account of the anticommutation rules for the generators ψ_1, \ldots, ψ_N, namely, for any permutation π,

$$\psi_{i_{\pi(1)}} \cdots \psi_{i_{\pi(n)}} = \varepsilon(\pi)\psi_{i_1} \cdots \psi_{i_n}. \tag{B.11}$$

The coefficients $c_{i_1,\ldots,i_n}(g)$ in (B.10) are not uniquely fixed by g because the set of all monomials in the generators is not linearly independent by (B.11). However, the set

$$B = \{1\} \cup \{\psi_{i_1} \cdots \psi_{i_n} \mid n \geq 1, i_1, \ldots, i_n \in \Lambda, i_1 < i_2 < \ldots < i_n\} \tag{B.12}$$

is a basis of the vector space $\mathbb{C}\{\psi_1, \ldots, \psi_N\}$, so any element g can be written as

$$
\begin{aligned}
g &= b(g) + \sum_{n \geq 1} \sum_{i_1 < \ldots < i_n \in \Lambda} s_{i_1,\ldots,i_n}(g)\psi_{i_1} \cdots \psi_{i_n} \\
&= b(g) + \sum_{I \subset \Lambda} s_I \psi^I
\end{aligned} \tag{B.13}
$$

and the map $\mathcal{A} \to \mathbb{C}^{2^N}$, $g \mapsto (b, s)$ is a bijection. In other words, we may assume that the coefficients are totally antisymmetric,

$$c_{i_{\pi(1)},\ldots,i_{\pi(n)}}(g) = \varepsilon(\pi)c_{i_1,\ldots,i_n}(g) \tag{B.14}$$

because any other part would cancel by (B.11).

There are plenty of ways to see that $\mathcal{A} = \mathbb{C}\{\psi_1, \ldots, \psi_N\}$, as rather sketchily introduced above, is a well-defined mathematical object, indeed, a \mathbb{C}-algebra. In fact, if $(f_\alpha)_\alpha$ is an ONB of a Hilbert space \mathcal{H}, then the fermionic creation operators $a^*(f_\alpha)$ form a Grassmann algebra, and the same holds for the annihilation operators (if \mathcal{H} has infinite dimension, this Grassmann algebra is infinite-dimensional).

An explicit matrix representation $\rho(\mathcal{A}) \subset M(2^N, \mathbb{C})$ is generated by

$$\rho(\psi_k) = \sigma_3 \otimes \ldots \otimes \sigma_3 \otimes \sigma_+ \otimes \sigma_3 \otimes \ldots \otimes \sigma_3, \tag{B.15}$$

where $\sigma_1, \sigma_2, \sigma_3$ are the Pauli matrices and $\sigma_+ = \sigma_1 + i\sigma_2$ appears as the k-th factor of the product (B.15).

The nilpotency

$$\psi_i^2 = 0 \qquad \forall i \in \Lambda \tag{B.16}$$

implied by the anticommutativity (B.9) has many important consequences: the degree of any element $g \in \mathcal{A}$, defined as the maximum value of n for which a monomial with n generators occurs in (B.10), cannot exceed N. Moreover, viewed as a ring, $\mathbb{C}\{\psi_1, \ldots, \psi_N\}$ is certainly not a domain of integrity since every element without constant term b is a divisor of zero. In the framework of C^*-algebras, the situation may be summarized by the statement that the spectrum of every $g \in \mathcal{A}$ is the set consisting only of the point $b(g)$. If $b(g) \neq 0$, the inverse of g can be calculated as a geometric series which terminates after a finite number of terms because of nilpotency.

Functions on a Grassmann algebra are definable whenever they have some formal or even convergent power series expansion. Let f be a formal power series in some indeterminate z,

$$f = \sum_{k \geq 0} f_k z^k, \tag{B.17}$$

then for all $g \in \mathcal{A}$ with $b(g) = 0$, f(g) can be defined as

$$f(g) = \sum_{k \geq 0} f_k g^k \tag{B.18}$$

since the sum on the right side involves only a finite number of terms. In the case $b(g) \neq 0$, $f(g)$ can be defined similarly by Taylor expansion around $z = b(g)$ if the first N derivatives of f at this point exist.

In analogy to the bosonic, that is, usual, differential and integral calculus, derivatives and integrals with respect to the Grassmann variables are given by special linear mappings. The derivative with respect to ψ_k, acting to the right, is the map

$$\frac{\partial}{\partial \psi_k} : \mathcal{A} \to \mathcal{A} \tag{B.19}$$

determined by its action on elements of B,

$$\frac{\partial}{\partial \psi_k} \psi_{i_1} \ldots \psi_{i_n} = \sum_{l=1}^{n} (-1)^{l-1} \delta_{k,i_l} \psi_{i_1} \ldots \psi_{i_{l-1}} \psi_{i_{l+1}} \ldots \psi_{i_n} \tag{B.20}$$

and the requirement that it be linear. Similarly, the derivative with respect to ψ_k, acting to the left, is the linear map

$$\frac{\overleftarrow{\partial}}{\partial \psi_k} \quad : \quad \mathcal{A} \to \mathcal{A}$$

$$\psi_{i_1} \ldots \psi_{i_n} \frac{\overleftarrow{\partial}}{\partial \psi_k} \quad = \quad \sum_{l=1}^{n} (-1)^{n-l} \delta_{k,i_l} \psi_{i_1} \ldots \psi_{i_{l-1}} \psi_{i_{l+1}} \ldots \psi_{i_n}. \tag{B.21}$$

For all $i, k \in \Lambda$

$$\frac{\partial}{\partial \psi_i} \frac{\partial}{\partial \psi_k} = - \frac{\partial}{\partial \psi_k} \frac{\partial}{\partial \psi_i} \tag{B.22}$$

so that the $\frac{\partial}{\partial \psi}$ themselves generate another Grassmann algebra, and that the same holds for the $\frac{\overleftarrow{\partial}}{\partial \psi}$.

We have seen already when introducing Fock space that changes of basis in the one-particle space \mathcal{H} correspond to linear transformations of the annihilation operators. Since the algebra of operators exists independently of the basis chosen in \mathcal{H}, such transformations do not change the Grassmann algebra generated by them. Similarly, the Grassmann algebra generated by

$\Psi' = A(\psi_1, \ldots, \psi_N)^T$ is the same as that generated by $\Psi = (\psi_1, \ldots, \psi_N)^T$ if $A \in GL(N, \mathbb{C})$. In particular, if ψ_1, \ldots, ψ_N and η_1, \ldots, η_N are Grassmann variables, then $\psi_1 + \eta_1, \ldots \psi_N + \eta_N$ generate the same Grassmann algebra as ψ_1, \ldots, ψ_N and η_1, \ldots, η_N (This is a change of generators $(\psi, \eta)^T \mapsto A(\psi, \eta)^T$ with $A = \begin{pmatrix} 1 & 1 \\ 0 & 1 \end{pmatrix}$).

Under a linear change of generators,

$$\Psi = (\psi_1, \ldots, \psi_N)^T \mapsto \Psi' = A(\psi_1, \ldots, \psi_N)^T, \tag{B.23}$$

with $A \in GL(N, \mathbb{C})$, the derivatives transform as

$$\frac{\partial}{\partial \Psi'} = (A^{-1})^t \frac{\partial}{\partial \Psi}, \tag{B.24}$$

that is, for all $a \in \mathcal{A}$

$$\frac{\partial}{\partial \psi_l'} a = \sum_{k=1}^{N} (A^{-1})_{kl} \frac{\partial}{\partial \psi_k} a. \tag{B.25}$$

Contrary to the bosonic case, the integral with respect to the variable ψ_k is defined to be identical to the derivative,

$$\int d\psi_k\, g = \frac{\partial}{\partial \psi_k} g \qquad \forall g \in \mathcal{A}. \tag{B.26}$$

Multiple integrals are denoted by

$$\int d\psi_{i_1} \ldots d\psi_{i_n} = \int d\psi_{i_1} \int d\psi_{i_2} \ldots \int d\psi_{i_n}, \tag{B.27}$$

the most important case being the integral with respect to all generators

$$g \mapsto \int d\psi_N \ldots d\psi_1 g \tag{B.28}$$

which maps $\mathcal{A} \to \mathbb{C}$.

Lemma B.1. *For all $\alpha \in \mathbb{C}$,*

$$\int d\psi_1\, F(\psi_1 - \alpha \eta_1, \psi_2, \ldots, \psi_N) = \int d\psi_1\, F(\psi_1, \psi_2, \ldots, \psi_N). \tag{B.29}$$

Proof. In every polynomial of the Grassmann algebra, such as F, the variable ψ_1 can appear only linearly by nilpotency. Thus the coefficient of ψ_1 is unchanged by addition of $\alpha \eta_1$. ∎

From now on we assume that we have an even number $N = 2L$ of generators which we label as ψ_1, \ldots, ψ_L and $\bar\psi_1, \ldots, \bar\psi_L$.

Corollary B.2. *Let $A, B \in M_N(\mathbb{C})$. Then*

$$\int \prod_{k=1}^{L} d\bar\psi_k d\psi_k\, f(\bar\psi + A\bar\eta, \psi + B\eta) = \int \prod_{k=1}^{L} d\bar\psi_k d\psi_k\, f(\bar\psi, \psi). \tag{B.30}$$

B.3 Grassmann Gaussian Integrals

For $Q \in M_L(\mathbb{C})$ let

$$\langle \bar{\psi} \mid Q\psi \rangle = \sum_{i,j=1}^{L} \bar{\psi}_i Q_{ij} \psi_j \tag{B.31}$$

and let $D\bar{\psi}D\psi = d\bar{\psi}_1 d\psi_1 \dots d\bar{\psi}_L d\psi_L$.

Lemma B.3.

$$\int D\bar{\psi}D\psi \, e^{-\langle \bar{\psi} \mid Q\psi \rangle} = \det Q. \tag{B.32}$$

Proof. Expanding the exponential, we have

$$I = \int D\bar{\psi}D\psi \, e^{-\langle \bar{\psi} \mid Q\psi \rangle} = \sum_{p=0}^{\infty} \frac{(-1)^p}{p!} \int D\bar{\psi}D\psi \, \langle \bar{\psi} \mid Q\psi \rangle^p. \tag{B.33}$$

$\langle \bar{\psi} \mid Q\psi \rangle^p$ is a monomial of degree p in ψ and p in $\bar{\psi}$. Thus only $p = L$ can contribute to the integral, and

$$
\begin{aligned}
I &= \frac{(-1)^L}{L!} \int \langle \bar{\psi} \mid Q\psi \rangle^L D\bar{\psi}D\psi \\
&= \frac{(-1)^L}{L!} \sum_{\substack{i_1,\dots,i_L \\ j_1,\dots,j_L}} \prod_{k=1}^{L} Q_{i_k j_k} \int D\bar{\psi}D\psi \prod_{k=1}^{L} \bar{\psi}_{i_k} \psi_{j_k}.
\end{aligned}
\tag{B.34}
$$

The anticommutation rules for the Grassmann variables give

$$
\begin{aligned}
\prod_{k=1}^{L} \bar{\psi}_{i_k} \psi_{j_k} &= (-1)^{L(L+1)/2} \prod_{k=1}^{L} \bar{\psi}_{i_k} \prod_{k=1}^{L} \psi_{j_k} \\
&= (-1)^{L(L+1)/2} \varepsilon_{i_1,\dots,i_L} \varepsilon_{j_1,\dots,j_L} \prod_{k=1}^{L} \bar{\psi}_k \prod_{k=1}^{L} \psi_k \\
&= \varepsilon_{i_1,\dots,i_L} \varepsilon_{j_1,\dots,j_L} \prod_{k=1}^{L} \bar{\psi}_k \psi_k.
\end{aligned}
\tag{B.35}
$$

Since $\int d\bar{\psi}_r d\psi_r \bar{\psi}_r \psi_r = -1$, the Grassmann integral gives a factor $(-1)^L$, and

$$
\begin{aligned}
I &= \frac{1}{L!} \sum_{\substack{i_1,\dots,i_L \\ j_1,\dots,j_L}} \prod_{k=1}^{L} Q_{i_k j_k} \\
&= \frac{1}{L!} \sum_{\pi,\sigma \in S_L} \varepsilon(\pi)\varepsilon(\sigma) \prod_{k=1}^{L} Q_{\pi(k)\sigma(k)} \\
&= \sum_{\rho \in S_L} \varepsilon(\rho) \prod_{k=1}^{L} Q_{k,\rho(k)} \frac{1}{L!} \sum_{\pi \in S_L} 1 = \det Q.
\end{aligned}
\tag{B.36}
$$

∎

Definition B.4. *Let $Q \in GL(N,\mathbb{C})$. The Grassmann Gaussian measure with covariance $C = Q^{-1}$ is the linear functional $f \mapsto \langle f \rangle_Q$ on the Grassmann algebra given by*

$$\langle f \rangle_Q = \frac{1}{\det Q} \int \prod_{k=1}^{L} d\bar{\psi}_k d\psi_k \, e^{-\langle \bar{\psi} | Q\psi \rangle} f(\bar{\psi}, \psi). \tag{B.37}$$

We also write this as

$$\langle f \rangle_Q = \int d\mu_C(\bar{\psi}, \psi) f(\bar{\psi}, \psi). \tag{B.38}$$

If f depends on no other Grassmann variables, $\langle f \rangle_Q \in \mathbb{C}$. But f may also depend on other Grassmann variables η, in which case $\langle f \rangle_Q$ is an element of the 'remaining' subalgebra generated by the η.

Remark B.5. Since $f \mapsto \langle f \rangle_Q$ is a linear map on finite-dimensional spaces, it is continuous in f. If f depends differentiably on a parameter λ, derivatives with respect to λ can be taken under the integral sign.

Lemma B.6.

$$\int d\mu_C(\bar{\psi}, \psi) e^{\langle \bar{\eta} | \psi \rangle + \langle \bar{\psi} | \eta \rangle} = e^{\langle \bar{\eta} | C\eta \rangle}. \tag{B.39}$$

Proof. Completing the square and using Corollary B.2, we get

$$\int D\bar{\psi} D\psi e^{-\langle \bar{\psi} | Q\psi \rangle + \langle \bar{\eta} | \psi \rangle + \langle \bar{\psi} | \eta \rangle}$$

$$= e^{\langle \bar{\eta} | C\eta \rangle} \int D\bar{\psi} D\psi e^{-\langle \bar{\psi} - C^T \bar{\eta} | Q(\psi - C\eta) \rangle} \tag{B.40}$$

$$= e^{\langle \bar{\eta} | C\eta \rangle} \int D\bar{\psi} D\psi e^{-\langle \bar{\psi} | Q\psi \rangle}$$

$$= e^{\langle \bar{\eta} | C\eta \rangle} \det Q.$$

∎

Lemma B.7. *If $f = \sum_{I,J \subset \{1,\dots,L\}} \varphi_{IJ}(f) \bar{\psi}^I \psi^J$, then*

$$\langle f \rangle_Q = \sum_{\substack{I,J \subset \{1,\dots,L\} \\ |I| = |J|}} \varphi_{IJ}(f)(-1)^{|I|(|I|-1)/2} \det \Gamma_{JI}(C) \tag{B.41}$$

where $\Gamma_{IJ}(C)$ is the $|I| \times |I|$ matrix with elements $(\Gamma_{IJ}(C))_{kl} = C_{j_k i_l}$ (and $I = \{i_1, \dots, i_L\}$, $J = \{j_1, \dots, j_L\}$, with $i_1 < \dots < i_L$ and $j_1 < \dots < j_L$).

Remark B.8. The $\varphi_{IJ}(f)$ may be Grassmann-algebra valued, as long as they are independent of $\bar{\psi}$ and ψ.

Remark B.9. Setting $\eta = 0$ can be done by putting $\eta_k = \alpha_k \chi_k$, $\alpha_k \in \mathbb{R}$ and χ_k Grassmann variables, and letting $\alpha_k \to 0$. Since all anticommutators of Grassmann generators vanish, the η_k form a Grassmann algebra.

Proof. By linearity of $\langle \cdot \rangle_Q$, it suffices to calculate $\langle \bar{\psi}^I \psi^J \rangle_Q$. By expansion of the exponential we see that $\langle \bar{\psi}^I \psi^J \rangle_Q = 0$ if $|I| \neq |J|$ because only monomials of the same degree L in $\bar{\psi}$ and ψ have a nonzero integral. Let $|I| = |J| = m$. We write

$$\bar{\psi}^I \psi^J = \left[\prod_{k=1}^{m} \left(-\frac{\partial}{\partial \eta_{i_k}} \right) \prod_{k=1}^{m} \frac{\partial}{\partial \bar{\eta}_{j_k}} e^{\langle \bar{\eta} | \psi \rangle + \langle \bar{\psi} | \eta \rangle} \right]_{\eta=0}. \tag{B.42}$$

Exchanging the differentiation with respect to η at $\eta = 0$ and the integral with respect to ψ is allowed since all operations are continuous and they commute. By Lemma B.6,

$$
\begin{aligned}
\langle \bar{\psi}^I \psi^J \rangle_Q &= \left[\prod_{k=1}^{m} \left(-\frac{\partial}{\partial \eta_{i_k}} \right) \prod_{k=1}^{m} \frac{\partial}{\partial \bar{\eta}_{j_k}} e^{\langle \bar{\eta} | C \eta \rangle} \right]_{\eta=0} \\
&= \left[\prod_{k=1}^{m} \left(-\frac{\partial}{\partial \eta_{i_k}} \right) \prod_{l=1}^{m} \sum_{r=1}^{L} C_{j_l r} \eta_r \right]_{\eta=0} \\
&= \sum_{r_1,\ldots,r_m=1}^{L} \prod_{l=1}^{m} C_{j_l r_l} \left[\prod_{k=1}^{m} \left(-\frac{\partial}{\partial \eta_{i_k}} \right) \prod_{l=1}^{m} \eta_{r_l} \right]_{\eta=0} \\
&= \sum_{1 \leq r_1 < \ldots < r_m \leq L} \sum_{\pi \in S_m} \prod_{l=1}^{m} C_{j_l r_{\pi(l)}} \left[\prod_{k=1}^{m} \left(-\frac{\partial}{\partial \eta_{i_k}} \right) \prod_{l=1}^{m} \eta_{r_{\pi(l)}} \right]_{\eta=0} \\
&= \sum_{\pi \in S_m} \varepsilon(\pi) \sum_{1 \leq r_1 < \ldots < r_m \leq L} \prod_{l=1}^{m} C_{j_l r_{\pi(l)}} \prod_{k=1}^{m} \delta_{i_k, r_k} (-1)^{m(m+1)/2} \\
&= (-1)^{m(m+1)/2+m} \det \Gamma_{JI}(C).
\end{aligned}
\tag{B.43}
$$

\blacksquare

Corollary B.10. *If C depends continuously or differentiably on a parameter t, then for all $f \in \mathcal{A}$, $\int d\mu_C(\bar{\psi}, \psi) f(\bar{\psi}, \psi)$ depends continuously or differentiably on t.*

Proof. $\int d\mu_C(\bar{\psi}, \psi) f(\bar{\psi}, \psi)$ is a polynomial in C, hence smooth in C. \blacksquare

B.4 Gram's Inequality; Bounds for Gaussian Integrals

Definition B.11. *Let $q > 0$. For $f = \sum f_{IJ} \bar{\psi}^I \psi^J$, with $f_{IJ} \in \mathbb{C}$, let*

$$\|f\|_q = \sum |f_{IJ}| q^{\frac{1}{2}(|I|+|J|)}. \tag{B.44}$$

Lemma B.12. *For all* $q > 0$, $\| \cdot \|_q$ *is a norm on the Grassmann algebra* $\mathcal{A} = \mathbb{C}\{\bar{\psi}_1, \ldots, \psi_L\}$, *and for all* f *and* $g \in \mathcal{A}$,

$$\|fg\|_q \leq \|f\|_q \, \|g\|_q \, . \tag{B.45}$$

Proof. It is obvious that $\| \cdot \|_q$ is a norm on \mathcal{A}. The product inequality holds because

$$\|fg\|_q = \sum_{\substack{I \cap K = \emptyset \\ J \cap L = \emptyset}} |f_{IJ}g_{KL}| q^{\frac{1}{2}(|I|+|J|+|K|+|L|)} \tag{B.46}$$

and

$$\|f\|_q \, \|g\|_q = \sum_{I,J,K,L} |f_{IJ}||g_{KL}| q^{\frac{1}{2}(|I|+|J|+|K|+|L|)}. \tag{B.47}$$

∎

Lemma B.13. (Gram's inequality) *Let* \mathcal{H} *be a separable Hilbert space, and for* $f = (f_1, \ldots f_n) \in \mathcal{H}^n$ *and* $g = (g_1, \ldots, g_n) \in \mathcal{H}^n$ *let*

$$(f, g) = \det\left(((f_i \mid g_j))_{1 \leq i,j \leq n}\right) \tag{B.48}$$

Then

$$|(f, g)|^2 \leq (f, f)\,(g, g) \tag{B.49}$$

$$|(f, f)| = |\det\left(((f_i \mid f_j))_{i,j}\right)| \leq \prod_{i=1}^{n} \|f_i\|^2, \tag{B.50}$$

and

$$|(f, g)| \leq \prod_{k=1}^{n} \|f_k\| \, \|g_k\| . \tag{B.51}$$

Proof. Let \mathcal{F} be the Fock space over \mathcal{H}, and let $F = \sqrt{n!}P_- f_1 \otimes \ldots \otimes f_n$ and $G = \sqrt{n!}P_- g_1 \otimes \ldots \otimes g_n$ (where P_- is the antisymmetrization projection). Then

$$
\begin{aligned}
\langle F \mid G \rangle_{\mathcal{F}} &= n! \langle P_- f_1 \otimes \ldots \otimes f_n \mid P_- g_1 \otimes \ldots \otimes g_n \rangle \\
&= \frac{1}{n!} \sum_{\pi,\sigma \in S_n} \varepsilon(\pi)\varepsilon(\sigma) \langle f_{\pi(1)} \otimes \ldots \otimes f_{\pi(n)} \mid g_{\sigma(1)} \otimes \ldots \otimes g_{\sigma(n)} \rangle \\
&= \frac{1}{n!} \sum_{\pi,\sigma \in S_n} \varepsilon(\pi)\varepsilon(\sigma) \prod_{k=1}^{n} \langle f_{\pi(k)} \mid g_{\sigma(k)} \rangle \\
&= \frac{1}{n!} \sum_{\pi,\sigma \in S_n} \varepsilon(\pi^{-1} \circ \sigma) \prod_{k=1}^{n} \langle f_k \mid g_{\pi^{-1}\circ\sigma(k)} \rangle \\
&= \sum_{\pi \in S_n} \varepsilon(\pi) \prod_{k=1}^{n} \langle f_k \mid g_{\pi(k)} \rangle \\
&= \det\left((\langle f_k \mid g_l \rangle)\right) . \tag{B.52}
\end{aligned}
$$

Thus (B.49) follows from the Cauchy-Schwarz inequality for the inner product on \mathcal{F}. Consider now the quadratic form on \mathbb{C}^n defined by

$$u \mapsto Q(u) = \left\| \sum_{i=1}^{n} u_i f_i \right\|^2. \tag{B.53}$$

Obviously, $Q(u) \geq 0$ for all $u \in \mathbb{C}^n$. Since $Q(u) = \sum_{i,j} \bar{u}_i \langle f_i \mid f_j \rangle u_j$, we see that the hermitian matrix Φ given by $\Phi_{ij} = \langle f_i \mid f_j \rangle$ satisfies $\Phi \geq 0$, so all eigenvalues of Φ are nonnegative, and $\det \Phi = \langle F \mid F \rangle_{\mathcal{F}} \geq 0$. If f_1, \ldots, f_n are linearly dependent, then, by (B.53), $Q(u) = 0$ for a nonzero u, so Φ has a zero eigenvalue. Then $\det \Phi = 0$ and (B.50) holds trivially. If $f_1, \ldots f_n$ are linearly independent, $Q(u) > 0$ for all $u \neq 0$, and the Gaussian integral

$$(\det \Phi)^{-1} = \int_{\mathbb{C}^n} \prod_{i=1}^{n} \frac{du_i d\bar{u}_i}{2\pi} e^{-\frac{1}{2}Q(u)} \tag{B.54}$$

converges (here for $u = a + ib$, $dud\bar{u} = dadb$). Since $f_1 \neq 0$, we have $\varphi = \langle f_1 \mid f_1 \rangle > 0$, and

$$Q(u) = \varphi |u_1|^2 + (\bar{\alpha} u_1 + \alpha \bar{u}_1) + \tilde{Q}(\tilde{u}) \tag{B.55}$$

where $\tilde{Q}(\tilde{u}) = \sum_{i,j \geq 2} \bar{u}_i \Phi_{ij} u_j$ and $\alpha = \alpha(\tilde{u}) = \sum_{j=2}^{n} \Phi_{1j} u_j$. Thus

$$\begin{aligned} Q(u) &= \varphi \left| u_1 - \frac{\alpha}{\varphi} \right|^2 - \frac{|\alpha|^2}{\varphi} + \tilde{Q}(\tilde{u}) \\ &\leq \varphi \left| u_1 - \frac{\alpha}{\varphi} \right|^2 + \tilde{Q}(\tilde{u}). \end{aligned} \tag{B.56}$$

By Fubini's theorem we can take the u_1-integral first in (B.54) and use (B.56) to get

$$(\det \Phi)^{-1} \geq \int_{\mathbb{C}^n} \prod_{i=2}^{n} \frac{du_i d\bar{u}_i}{2\pi} e^{-\frac{1}{2}\tilde{Q}(\tilde{u})} \int_{\mathbb{C}^n} \frac{du_1 d\bar{u}_1}{2\pi} e^{-\frac{1}{2}\varphi |u_1 - \frac{\alpha}{\varphi}|^2}. \tag{B.57}$$

The u_1-integral gives $\frac{1}{\varphi}$; the integral over \tilde{u} then gives the inverse determinant of the matrix where the first row and column of Φ are removed. Thus

$$\det \left((\langle f_i \mid f_j \rangle)_{1 \leq i,j \leq n} \right) \leq \|f_1\|^2 \det \left((\langle f_i \mid f_j \rangle)_{2 \leq i,j \leq n} \right). \tag{B.58}$$

(B.50) now follows by induction on n, and (B.51) is obvious from (B.49) and (B.50). ∎

Theorem B.14. *Let* $C = Q^{-1}$ *have the representation*

$$C_{ij} = \langle a_i \mid b_j \rangle, \tag{B.59}$$

where a_1, \ldots, a_n *and* b_1, \ldots, b_n *are vectors in a separable Hilbert space* \mathcal{H}, *and for all* $i \in \{1, \ldots, n\}$, *let* $\|a_i\|^2 \leq q$ *and* $\|b_i\|^2 \leq q$. *Then*

$$|\langle f \rangle_Q| \leq \|f\|_q. \tag{B.60}$$

Proof. Let $f = \sum_{I,J} f_{IJ} \bar{\psi}^I \psi^J$. By Lemma B.7 and (B.59),

$$\langle f \rangle_Q = \sum_{|I|=|J|} f_{IJ} \det((\langle a_j \mid b_i \rangle)_{i \in I, j \in J}. \tag{B.61}$$

By Lemma B.13,

$$|\langle f \rangle_Q| \leq \sum_{|I|=|J|} |f_{IJ}| \prod_{i=1}^{|I|} \|a_i\| \|b_i\|$$

$$\leq \sum_{|I|=|J|} |f_{IJ}| q^{\frac{1}{2}(|I|+|J|)} . \tag{B.62}$$

Dropping the constraint that $|I| = |J|$ from the sum, we obtain the bound by $\|f\|_q$. ∎

Remark B.15. This is not the only determinant bound one could use. There is also Hadamard's bound

$$|\det A| \leq \prod_{i=1}^{n} \left(\sum_{j=1}^{n} |A_{ij}|^2 \right)^{1/2} \tag{B.63}$$

or the much cruder bound $|\det A| \leq \prod_{i=1}^{n} (\sum_{j=1}^{n} |A_{ij}|)$.

B.5 Grassmann Integrals for Fock Space Traces

In this appendix, we derive the Grassmann integral representation for the Fock space traces. The presentation follows [66] (except for the convergence proof, which was not given in [66]). Recall that a basis of \mathcal{F} is given by $\Phi_\Delta = \prod_{\xi \in \Delta} a_\xi^* \Omega$. Let \mathcal{A} be the Grassmann algebra generated by $(\bar{\psi}(\xi))_{\xi \in \Gamma}$, and let μ be the linear map $\mu : \mathcal{F} \to \mathcal{A}$, defined by $\mu(\Phi_\Delta) = \bar{\psi}^\Delta = \prod_{\xi \in \Delta} \bar{\psi}(\xi)$. Since μ maps a basis to a basis, it is an isomorphism of \mathcal{F} and \mathcal{A}. If $v \in \mathcal{F}$ has the representation $v = \sum_\Delta v_\Delta \Phi_\Delta$, then $\mu(v) = \sum_\Delta v_\Delta \bar{\psi}^\Delta = v(\bar{\psi})$.

The action of the creation and annihilation operators is

$$\mu(a_\xi^* v) = \bar{\psi}(\xi) v(\bar{\psi}), \quad \mu(a_\xi v) = \frac{\partial}{\partial \bar{\psi}(\xi)} v(\bar{\psi}). \tag{B.64}$$

Since

$$\langle \Phi_\Delta \mid \Phi_{\Delta'} \rangle = \delta_{\Delta,\Delta'} = \left[\left(\frac{\partial}{\partial \bar{\psi}} \right)^{\tilde{\Delta}} \bar{\psi}^{\Delta'} \right]_{\bar{\psi}=0}, \tag{B.65}$$

where $\bar{\psi}^{\tilde{\Delta}} = \prod_{\xi \in \Delta} \bar{\psi}(\xi)$, but with the product taken in the opposite order, the inner product on \mathcal{F} has the representation

$$\langle v \mid w \rangle \quad = \quad \left[\tilde{v} \left(\frac{\partial}{\partial \bar{\psi}} \right) w(\bar{\psi}) \right]_{\bar{\psi}=0} \tag{B.66}$$

where $\tilde{v}(\bar{\psi}) = \sum_{\Delta} \bar{v}_{\Delta} \bar{\psi}^{\Delta}$.

Any linear operator $Q \in \mathcal{L}(\mathcal{F}, \mathcal{F})$ can be written as a polynomial in the creation and annihilation operators. The *normal ordered form* $Q(a^*, a)$ of such a Q is the form where all creation operators are moved to the left of all annihilation operators. Specifically, we define

$$Q(a^*, a) = \sum_{X, Y \subset \Gamma} Q_{XY} \prod_{\xi \in X} a_{\xi}^* \overset{\sim}{\prod_{\eta \in Y}} a_{\eta} \tag{B.67}$$

where the \sim on the product means that it is taken in the opposite order.

The action of a linear operator Q with normal ordered form $Q(a^*, a)$ is transferred to the Grassmann algebra generated by the $\bar{\psi}$ as

$$\mu(Q(a^*, a)v) = A \left(\bar{\psi}, \frac{\partial}{\partial \bar{\psi}} \right) v(\bar{\psi}). \tag{B.68}$$

B.5.1 Delta Functions and Integral Kernels

We now derive an integral representation for the action of the operator A on an element v in the Grassmann algebra, as given in (B.68). For two Grassmann variables $\bar{\psi}$ and $\bar{\psi}'$, the 'Grassmann delta function' is simply given by

$$\delta(\bar{\psi} - \bar{\psi}') = (\bar{\psi}' - \bar{\psi}) \tag{B.69}$$

because for $f(\bar{\psi}) = a + b\bar{\psi}$,

$$\int d\bar{\psi}' \ (\bar{\psi}' - \bar{\psi}) f(\bar{\psi}') = f(\bar{\psi}). \tag{B.70}$$

By an expansion of the exponential and the nilpotency of the Grassmann variables, one sees that

$$\delta(\bar{\psi} - \bar{\psi}') = \int d\psi \ e^{(\bar{\psi} - \bar{\psi}')\psi}. \tag{B.71}$$

One should keep in mind, however, that the thus introduced 'delta function' is an odd element of the Grassmann algebra, so that one has to take care of the order in products. To avoid this, one combines the formulas to

$$\int d\bar{\psi}' \int d\psi \ e^{(\bar{\psi} - \bar{\psi}')\psi} f(\bar{\psi}') = f(\bar{\psi}). \tag{B.72}$$

Now everything is reexpressed in bilinears that commute. The generalization of this formula to finitely many Grassmann variables is as follows.

Lemma B.16. Let $f(\bar{\psi}) \in \mathcal{A}_\Gamma(\bar{\psi})$. Then

$$\int \left(\prod_{x \in \Gamma} d\bar{\psi}'(x)d\psi(x)\right) e^{\sum_{x \in \Gamma}(\bar{\psi}(x)-\bar{\psi}'(x))\psi(x)} f(\bar{\psi}') = f(\bar{\psi}). \qquad (B.73)$$

Proof. f is of the form $f(\bar{\psi}) = \sum_{\Delta \subset \Gamma} f_\Delta \bar{\psi}^\Delta$. By linearity of the integral, and because bilinears commute, the left hand side of (B.73) is

$$\sum_{\Delta \subset \Gamma} f_\Delta \int \prod_{x \in \Gamma} \left(d\bar{\psi}'(x)d\psi(x)e^{(\bar{\psi}(x)-\bar{\psi}'(x))\psi(x)}\right)(\bar{\psi}')^\Delta, \qquad (B.74)$$

and it does not matter in which order the product over $x \in \Gamma$ is taken. Consider the contribution from a fixed Δ, and reorder the product over $x \in \Gamma$ such that the integrals with $x \in \Delta$ are taken first. For $x \in \Delta$,

$$\int d\bar{\psi}'(x)d\psi(x) \; e^{(\bar{\psi}(x)-\bar{\psi}'(x))\psi(x)} (\bar{\psi}')^\Delta$$

$$= \int d\bar{\psi}'(x) \; (\bar{\psi}'(x) - \bar{\psi}(x)) \; (\bar{\psi}')^\Delta$$

$$= \int d\bar{\psi}'(x) \; (-\bar{\psi}(x)) \; (\bar{\psi}')^\Delta \qquad (B.75)$$

$$= \bar{\psi}(x) \int d\bar{\psi}'(x) \; (\bar{\psi}')^\Delta = \bar{\psi}(x) \frac{\partial}{\partial \bar{\psi}'(x)} (\bar{\psi}')^\Delta.$$

Because $\bar{\psi}(x)\frac{\partial}{\partial\bar{\psi}'(x)}$ is bilinear, it commutes with all factors $\bar{\psi}'(x')$, with $x' \ne x$, in the product $(\bar{\psi}')^\Delta$. Thus its effect is just to replace $\bar{\psi}'(x)$ by $\bar{\psi}(x)$. After this replacement has been done for all $x \in \Delta$, the result is $f(\bar{\psi})$, which does not depend on $\bar{\psi}'$ any more, so the remaining integrals leave it unchanged. ∎

Corollary B.17. Let $D_\Gamma \bar{\psi} D_\Gamma \psi = \varepsilon_\Gamma^{-|\Gamma|} \prod_{x \in \Gamma} d\bar{\psi}(x)d\psi(x)$. Then

$$\int D_\Gamma \bar{\psi} D_\Gamma \psi \; e^{(\bar{x}-\bar{\psi},\psi)} f(\bar{\psi}) = f(\bar{x}). \qquad (B.76)$$

Let v and w be in \mathcal{F}_Γ. Then

$$\langle v \mid w \rangle = \int D_\Gamma \bar{\psi} D_\Gamma \psi \; e^{-(\bar{\psi},\psi)} \; \bar{v}(\psi) \; w(\bar{\psi}). \qquad (B.77)$$

Let $A(c^*, c)$ be the normal ordered form of the operator A. Then

$$(Av)(\bar{\psi}) = \int D_\Gamma \bar{\chi} D_\Gamma \chi \; e^{-(\bar{x},x)} \; \tilde{A}(\bar{\psi},\chi) \; v(\bar{x}) \qquad (B.78)$$

where

$$\tilde{A}(\bar{\psi},\chi) = A(\bar{\psi},\chi) \; e^{(\bar{\psi},x)} \qquad (B.79)$$

is the integral kernel of A.

Proof. The extra power of ε^{-1} in $Dr\bar{\psi}Dr\psi$ is there because (ψ, η) is defined as $\varepsilon \sum \psi(x)\eta(x)$. By (B.76)

$$\left(\tfrac{\delta}{\delta\bar{\chi}}\right)^{\Delta} f(\bar{\chi}) = \int Dr\bar{\psi}Dr\psi \; e^{(\bar{\chi}-\bar{\psi},\psi)} \; \psi^{\Delta} \; f(\bar{\psi}), \qquad (B.80)$$

so (B.77) holds by (B.66). Because in the normal ordered form of A, all derivatives are on the right of all factors of $\bar{\psi}$, (B.80) implies (B.78). Note that (B.78) holds only with the normal ordered form of A. ∎

In (B.78), the action of A is expressed as a Grassmann integral. The integral kernel of the product AB is the convolution

$$(AB)^{\sim}(\bar{\psi}, \psi) = \int Dr\bar{\chi}Dr\chi \; e^{-(\bar{\chi},\chi)} \; \tilde{A}(\bar{\psi}, \chi) \; \tilde{B}(\bar{\chi}, \psi). \qquad (B.81)$$

B.5.2 The Formula for the Trace

The trace of an operator A on \mathcal{F}_Γ can be evaluated in the basis Φ_Δ as

$$\text{Tr}\, A = \sum_{\Delta \subset \Gamma} \langle \Phi_\Delta \mid A\,\Phi_\Delta \rangle. \qquad (B.82)$$

Lemma B.18. *If* $A(c^*, c)$ *is the normal ordered form of* A,

$$\text{Tr}\, A = \int Dr\bar{\psi}Dr\psi \; \tilde{A}(\bar{\psi}, \psi) \; e^{(\bar{\psi},\psi)}. \qquad (B.83)$$

Proof. By (B.77) and (B.78), $\text{Tr}\, A = \sum_{\Delta \subset \Gamma} \varepsilon_\Gamma^{|\Delta|} \alpha_\Delta$ with

$$\alpha_\Delta = \int Dr\bar{\theta}Dr\theta Dr\bar{\chi}Dr\chi \; e^{-(\bar{\theta},\theta)-(\bar{\chi},\chi)} \; \theta^{\Delta} \; \tilde{A}(\bar{\theta}, \chi) \; \bar{\chi}^{\Delta}. \qquad (B.84)$$

The only nonvanishing contributions to the integral are those whose degree in $\bar{\theta}$ and θ is the same, and similarly for χ. Let $|\Delta| = d$. Then only terms of A of total degree $2d$ contribute to α_Δ. These terms have even degree, so we may replace $\theta^{\Delta} \; \tilde{A}(\bar{\theta}, \chi) \; \bar{\chi}^{\Delta}$ by $\tilde{A}(\bar{\theta}, \chi) \; \theta^{\Delta} \; \bar{\chi}^{\Delta}$ in the integrand. Because $\theta^{\Delta}\bar{\chi}^{\Delta} = \prod_{x \in \Delta} (-\bar{\chi}(x)\theta(x))$, the sum over Δ gives

$$\sum_{\Delta \subset \Gamma} \varepsilon_\Gamma^{|\Delta|} \prod_{x \in \Delta}(-\bar{\chi}(x)\theta(x)) \;=\; \prod_{x \in \Gamma}(1 - \varepsilon_\Gamma \bar{\chi}(x)\theta(x))$$

$$=\; \prod_{x \in \Gamma} e^{-\varepsilon_\Gamma \bar{\chi}(x)\theta(x)} \;=\; e^{-(\bar{\chi},\theta)}. \qquad (B.85)$$

Renaming the integration variables as $\bar{\psi} = \bar{\theta}$, $\psi = \chi$, $\lambda = \theta$, $\bar{\lambda} = \bar{\chi}$, we get (B.83) because a shift in the integration variable gives

$$\int D_\Gamma \bar\lambda D_\Gamma \lambda \, e^{-(\bar\lambda,\lambda)+(\bar\psi,\lambda)-(\bar\lambda,\psi)}$$

$$= \int D_\Gamma \bar\lambda D_\Gamma \lambda \, e^{-(\bar\lambda-\bar\psi,\lambda+\psi)+(\bar\psi,\psi)}$$

$$= e^{(\bar\psi,\psi)} \int D_\Gamma \bar\lambda D_\Gamma \lambda \, e^{-(\bar\lambda,\lambda)} \tag{B.86}$$

$$= e^{(\bar\psi,\psi)}.$$

∎

The trace of a product of n factors is

$$\operatorname{Tr} \prod_{k=1}^{n-1} A_k \;=\; \int D_\Gamma \bar\chi_0 D_\Gamma \chi_n e^{(\bar\chi_0,\chi_n)} \int \prod_{k=1}^{n-1} D_\Gamma \bar\chi_k D_\Gamma \chi_k$$

$$\prod_{k=1}^{n-1} \left(A_k(\bar\chi_{k-1},\chi_k) e^{-(\bar\chi_k-\bar\chi_{k-1},\chi_k)} \right). \tag{B.87}$$

Renaming the integration variables as $\bar\eta_k = \bar\chi_k$ and $\eta_k = \chi_{k+1}$, we get

$$\operatorname{Tr} \prod_{k=1}^{n-1} A_k = \int \prod_{k=0}^{n-1} (D_\Gamma \bar\eta_k D_\Gamma \eta_k \, A_k(\bar\eta_k,\eta_k))$$

$$\exp\left(\sum_{k=0}^{n-2} (\bar\eta_k - \bar\eta_{k+1}, \eta_k) + (\bar\eta_{n-1} + \bar\eta_0, \eta_{n-1}) \right). \tag{B.88}$$

The form of the exponent shows that the positive sign in the exponent of the integral (B.83) for the trace leads to antiperiodic boundary conditions. Thus we define

$$\bar\eta_n = -\bar\eta_0, \quad \eta_n = -\eta_0, \quad \eta_{-1} = -\eta_{n-1} \tag{B.89}$$

whenever these variables appear in the above sums. The exponent then becomes

$$\sum_{k=0}^{n-1} (\bar\eta_k - \bar\eta_{k+1}, \eta_k) = \sum_{k=0}^{n-1} (\bar\eta_k, \eta_k - \eta_{k-1}). \tag{B.90}$$

B.5.3 The Time Continuum Limit

We have now made all preparations for deriving the Grassmann integral representation for traces of the form $\zeta(A) = \operatorname{Tr} e^{-A}$, for A any linear operator on \mathcal{F}. Since all linear maps between finite-dimensional spaces are continuous, $\zeta(\beta A)$ is an entire function of β, and $\zeta(\beta A) = \lim_{n\to\infty} \tilde\zeta_n(\beta A)$, where

$$\tilde\zeta_n(\beta A) = \operatorname{Tr}\left((1 - \tfrac{\beta}{n} A)^n \right). \tag{B.91}$$

We shall use this to rewrite of the partition function; rewriting the expectation values of observables is similar.

Let $A(a^*, a)$ be the normal ordered form of A. Then, by (B.88),

$$\tilde\zeta_n(\beta A) = \int \prod_{k=0}^{n-1} D r \bar\eta_k D r \eta_k e^{\sum_{k=1}^{n-1}(\bar\eta_k, \eta_k - \eta_{k-1})} \prod_{k=0}^{n-1} \left(1 - \tfrac{\beta}{n} A(\bar\eta_k, \eta_k)\right). \quad (B.92)$$

Note that no extra reordering needs to be done when one uses (B.91). If one uses the Lie product formula , Lemma 1.1, one has to calculate the normal ordered form $(e^{-A})(a^*, a)$ of e^A, which is not the same as $e^{-A(a^*,a)}$. This is not difficult for simple Hamiltonians, but can get arbitrarily complicated for general A.

We would now like to show that one can reexponentiate by writing

$$1 - \tfrac{\beta}{n} A(\bar\eta_k, \eta_k) = e^{-\frac{\beta}{n} A(\bar\eta_k, \eta_k)} - \frac{1}{n^2} B(\bar\eta_k, \eta_k) \quad (B.93)$$

with

$$B(\bar\eta, \eta) = \sum_{p=2}^{\infty} \frac{n^{2-p}}{p!} (-\beta A(\bar\eta, \eta))^p \quad (B.94)$$

and showing that any integral containing a remainder factor B vanishes in the limit $n \to \infty$. The remainders are $O(\frac{1}{n^2})$, but the measure also depends on n and therefore the proof is not trivial; it makes essential use of the Gram estimate.

We define the discrete Euclidean time variable $\tau_k = -\frac{\beta}{2} + \frac{k}{n}\beta$ and denote $\mathbb{T} = \{\tau_k : k \in \{0, \ldots, n-1\}\}$. The lattice spacing in time direction is $\varepsilon_\tau = \tau_{k+1} - \tau_k = \frac{\beta}{n}$. We rename the Grassmann generators as $\psi(\tau_k) = \eta_k$ and $\bar\psi(\tau_k) = \bar\eta_k$, and define $(\partial_\tau \psi)(\tau_k) = \frac{1}{\varepsilon_\tau}(\psi(\tau_k) - \psi(\tau_{k-1}))$. With this, and the notation $D^n\bar\psi D^n\psi = \prod_{k=0}^{n-1} D r \bar\psi(\tau_k) D r \psi(\tau_k)$,

$$\tilde\zeta_n(\beta A) = \int D^n\bar\psi D^n\psi \, e^{\varepsilon_\tau \sum_{\tau \in \mathbb{T}} [(\bar\psi(\tau), (\partial_\tau \psi)(\tau)) - A(\bar\psi(\tau), \psi(\tau))]} \, \mathcal{D}_n \quad (B.95)$$

with

$$\mathcal{D}_n = \prod_{k=0}^{n-1} \left(1 + \tfrac{1}{n^2} R(\bar\psi(\tau_k), \psi(\tau_k))\right) \quad (B.96)$$

and $R(\bar\psi, \psi) = -e^{\varepsilon_\tau A(\bar\psi, \psi)} B(\bar\psi, \psi)$.

Theorem B.19. *Let $A(a^*, a)$ be the normal ordered form of A, and*

$$\zeta_n(\beta A) = \int D^n\bar\psi D^n\psi \, e^{\varepsilon_\tau \sum_{\tau \in \mathbb{T}} ((\bar\psi(\tau), (\partial_\tau \psi)(\tau)) - A(\bar\psi(\tau), \psi(\tau)))}. \quad (B.97)$$

Then $\zeta(\beta A) = \mathrm{Tr}\, e^{-\beta A} = \lim_{n \to \infty} \zeta_n(\beta A)$.

The proof of Theorem B.19 constitutes the remainder of this section.

We first prove that the Gram bound applies to the covariance ∂_τ^{-1}; this also holds for the more general case where a quadratic part of A is included in the measure. Assume that $A(a^*, a) = \sum_{x,y,\sigma} \tilde{E}(x - y) a^*_{x,\sigma} a_{y,\sigma}$, and that the Fourier transform of \tilde{E} is $E(p)$. Then

$$\tilde{\zeta}_n(\beta A) = \int D^n \bar{\psi} D^n \psi \, e^{-\varepsilon_\tau \sum_{\tau \in T} (\bar{\psi}(\tau), \, (-\partial_\tau + \tilde{E}) \psi(\tau))} \quad \mathcal{D}_n$$

$$= \det(\varepsilon_\tau(-\partial_\tau + \tilde{E})) \int d\mu_C(\bar{\psi}, \psi) \, \mathcal{D}_n \qquad (B.98)$$

where $C = (\varepsilon_\tau(-\partial_\tau + \tilde{E}))^{-1}$.

By Fourier transformation, we see that the eigenvalues of $-\partial_\tau + \tilde{E}$ are $-i\hat{\omega} + E(p)$, labelled by $(\omega, p, \sigma) \in \mathbb{M}_n \times \Gamma$, where $i\hat{\omega} = \frac{1}{\varepsilon_\tau}(1 - e^{-i\omega\varepsilon_\tau})$. Thus

$$\det(\varepsilon_\tau(-\partial_\tau + \tilde{E})) = \prod_{(\omega, p, \sigma) \in \mathbb{M}_n \times \Gamma} (1 - \varepsilon_\tau E(p) - e^{-i\omega\varepsilon_\tau}). \qquad (B.99)$$

Now $\omega\varepsilon_\tau = \frac{\pi}{n}(2l + 1)$, so, as l varies, $e^{-i\omega\varepsilon_\tau} = e^{-i\frac{\pi}{n}(2l+1)}$ runs over all roots z_l of -1. Thus

$$\prod_{\omega \in \mathbb{M}_n} (x - e^{-i\omega\varepsilon_\tau}) = x^n + 1, \qquad (B.100)$$

and therefore

$$\det(\varepsilon_\tau(-\partial_\tau + \tilde{E})) = \prod_{(p,\sigma) \in \Gamma^*} \left(1 + (1 - \tfrac{\beta}{n} E(p))^n\right). \qquad (B.101)$$

In the limit $n \to \infty$, this gives the partition function $\prod(1 + e^{-\beta E(p)})$ of the Fermi gas.

The covariance C is the Fourier sum

$$C(\tau, x, \sigma, \tau', x', \sigma') = \frac{1}{\beta} \sum_{\omega \in \mathbb{M}_n} e^{i\omega(\tau - \tau')} L^{-d} \sum_{p \in \Lambda^*} \frac{e^{ip(x - x')}}{-i\hat{\omega} + E(p)} \delta_{\sigma\sigma'}. \qquad (B.102)$$

Lemma B.20. *For all $\omega \in \mathbb{M}_n$, $\omega\varepsilon_\tau \in [-\pi, \pi)$. Let $E \in \mathbb{R}$. For all $n \geq 2\beta|E|$,*

$$|i\hat{\omega} - E| \geq \begin{cases} \frac{n}{2\beta} & \text{if } |\omega\varepsilon_\tau| \geq \frac{\pi}{2} \\ \frac{2}{\pi}|\omega| & \text{if } |\omega\varepsilon_\tau| \leq \frac{\pi}{2}. \end{cases} \qquad (B.103)$$

Proof. For $\omega \in \mathbb{M}_n$, $\omega\varepsilon_\tau = \frac{\pi}{n}(2l+1)$. Since $l \in \{-\frac{n}{2}, \ldots, \frac{n}{2} - 1\}$, $\omega\varepsilon_\tau \in [-\pi, \pi)$. Write

$$i\hat{\omega} - E = i\frac{\sin \omega\varepsilon_\tau}{\varepsilon_\tau} + \frac{1}{\varepsilon_\tau}(1 - \cos \omega\varepsilon_\tau) - E. \qquad (B.104)$$

If $|\omega\varepsilon_\tau| \geq \frac{\pi}{2}$, then $\cos \omega\varepsilon_\tau \leq 0$, so $1 - \cos \omega\varepsilon_\tau \geq 1$, and thus

$$|i\hat{\omega} - E| \geq \left|\frac{1}{\varepsilon_\tau}(1 - \cos \omega\varepsilon_\tau) - E\right| \geq \left|\frac{n}{\beta} - E\right| \geq \frac{n}{2\beta}. \qquad (B.105)$$

The function $x \to \frac{\sin x}{x}$ is decreasing on $[0, \pi/2]$, so $\inf_{[0,\pi/2]} \frac{\sin x}{x} = \frac{2}{\pi}$. Thus for $|\omega \epsilon_\tau| \leq \pi/2$,

$$|i\widehat{\omega} - E| \geq \left| \frac{\sin \omega \epsilon_\tau}{\omega \epsilon_\tau} \omega \right| \geq \frac{2}{\pi} |\omega|. \tag{B.106}$$

∎

Lemma B.21. *Let* $\xi = (\tau, \mathbf{x}, \sigma)$ *and* $\xi' = (\tau', \mathbf{x}', \sigma')$. *then* C *has a representation as a scalar product* $C(\xi, \xi') = \langle a_\xi \mid b_{\xi'} \rangle$ *with*

$$\|a_\xi\|^2 = \|b_{\xi'}\|^2 = \frac{1}{\beta} \sum_{\omega \in M_n} L^{-d} \sum_{\mathbf{p} \in \Lambda^*} \frac{1}{|i\widehat{\omega} - E(\mathbf{p})|}. \tag{B.107}$$

If

$$n \geq \max\{4, 2\beta \max_{\mathbf{p} \in \Lambda^*} |E(\mathbf{p})|\} \tag{B.108}$$

then

$$\|a_\xi\|^2 = \|b_{\xi'}\|^2 \leq \epsilon_\Gamma^{-1}(1 + \log n). \tag{B.109}$$

Proof. Let $\mathcal{H} = L^2(M_n \times \Lambda^* \times [0, 2\pi])$, with the scalar product

$$\langle a \mid b \rangle = \frac{1}{\beta} \sum_{\omega \in M_n} L^{-d} \sum_{\mathbf{p} \in \Lambda^*} \int_0^{2\pi} \frac{d\theta}{2\pi} \overline{a(\omega, \mathbf{p}, \theta)} b(\omega, \mathbf{p}, \theta). \tag{B.110}$$

Choose

$$
\begin{aligned}
a_\xi(\omega, \mathbf{p}, \theta) &= e^{-i\omega\tau - i\mathbf{p}\mathbf{x}} |i\widehat{\omega} - E(\mathbf{p})|^{-1/2} e^{-i\theta\sigma} \\
b_{\xi'}(\omega, \mathbf{p}, \theta) &= e^{-i\omega\tau' - i\mathbf{p}\mathbf{x}'} \frac{|i\widehat{\omega} - E(\mathbf{p})|^{-1/2}}{i\widehat{\omega} - E(\mathbf{p})} e^{-i\theta\sigma'},
\end{aligned} \tag{B.111}
$$

then $C(\xi, \xi') = \langle a_\xi \mid b_{\xi'} \rangle$ and (B.107) hold. On the right hand side of (B.107), sum over ω first, and split the sum into those ω for which $|\omega \epsilon_\tau| \leq \pi/2$ and the rest. By Lemma B.20 and by $L^{-d} \sum_{\mathbf{p}} 1 = \epsilon_\Gamma^{-1}$, we have

$$
\begin{aligned}
\|a_\xi\|^2 &\leq L^{-d} \sum_{\mathbf{p} \in \Lambda^*} \left(\frac{1}{\beta} \sum_{\substack{\omega \in M_n \\ |\omega \epsilon_\tau| \leq \pi/2}} \frac{\pi}{2|\omega|} + \frac{1}{\beta} \sum_{\substack{\omega \in M_n \\ |\omega \epsilon_\tau| \geq \pi/2}} \frac{2\beta}{n} \right) \\
&\leq \epsilon_\Gamma^{-1} \left(\frac{2}{n} \sum_{\substack{\omega \in M_n \\ |\omega \epsilon_\tau| \geq \pi/2}} 1 + \frac{1}{2} \sum_{l=-n/2}^{n/2-1} \frac{1}{|2l+1|} \right) \tag{B.112} \\
&\leq \epsilon_\Gamma^{-1}(1 + \log n).
\end{aligned}
$$

∎

The proof of the following lemma contains both the essential argument about convergence as $n \to \infty$ and a hint at the complications that arise in the full proof.

Lemma B.22. *For all* $\delta > 0$ *and all* $A \in \mathcal{L}(\mathcal{F}, \mathcal{F})$,

$$\int d\mu_C(\bar{\psi}, \psi)\left(e^{n^{-1-\delta}\sum_{k=0}^{n-1}A(\bar{\psi}(\tau_k), \psi(\tau_k))} - 1\right) \to 0 \tag{B.113}$$

as $n \to \infty$.

Proof. Since we want to prove something about the limit $n \to \infty$, we may restrict to those n that satisfy (B.108). By Theorem B.14 and Lemma B.21, it suffices to show that, for $q_n = \varepsilon_\Gamma^{-1}(1 + \log n)$,

$$X_n = \left\| \exp\left(n^{-1-\delta}\sum_{k=0}^{n-1}A(\bar{\psi}(\tau_k), \psi(\tau_k))\right) - 1 \right\|_{q_n} \to 0 \tag{B.114}$$

as $n \to \infty$. By the binomial theorem, applied in the form

$$\prod e^{\alpha_k} - 1 = \prod(1 + (e^{\alpha_k} - 1)) - 1 = \sum_{S \neq \emptyset} \prod_{k \in S}(e^{\alpha_k} - 1), \tag{B.115}$$

and the product inequality (B.45),

$$X_n \leq \sum_{\substack{S \subset \{0,\dots,n-1\} \\ S \neq \emptyset}} \prod_{k \in S} \left\| e^{n^{-1-\delta}A(\bar{\psi}(\tau_k), \psi(\tau_k))} - 1 \right\|_{q_n}. \tag{B.116}$$

$A(\bar{\psi}(\tau_k), \psi(\tau_k))$ is an element of the Grassmann algebra generated by the $|\Gamma|$ generators $(\psi_\xi, \bar{\psi}_\xi)_{\xi \in \Gamma}$. The maximal degree of such an element is $2|\Gamma|$. Thus, by definition of the norm, $\|A(\bar{\psi}(\tau_k), \psi(\tau_k))\| \leq \alpha q_n^{|\Gamma|}$ with a constant α that depends on Γ but not on n. Expansion of the exponential and the product inequality (B.45) imply

$$\prod_{k \in S} \left\| e^{n^{-1-\delta}A(\bar{\psi}(\tau_k), \psi(\tau_k))} - 1 \right\|_{q_n} \leq \exp\left(n^{-1-\delta}|S|\alpha q_n^{|\Gamma|}\right) - 1, \tag{B.117}$$

so

$$\begin{aligned} X_n &\leq \sum_{s=1}^{n-1}\binom{n-1}{s}\left(e^{n^{-1-\delta}\alpha|S|q_n^{|\Gamma|}} - 1\right)^s \tag{B.118} \\ &= \left(e^{n^{-1-\delta}\alpha|S|q_n^{|\Gamma|}}\right)^n - 1 = e^{\alpha n^{-\delta}q_n^{|\Gamma|}} - 1. \end{aligned}$$

This vanishes for $n \to \infty$ because $n^{-\delta}q_n^{|\Gamma|} \to 0$ in that limit. ∎

Note that for $\delta = 0$, this is not sufficient to prove convergence. This is the reason why one has to use some analyticity property to prove Theorem B.19. The deeper reason for the nonconvergence of the norm estimate for $\delta = 0$ is

that the bound by $\log n$ overestimates the actual behaviour of C. It is nontrivial to prove better estimates because the sum over ω is only conditionally convergent in the limit $n \to \infty$.

We now turn to the proof of Theorem B.19. Note that $E(\mathbf{p}) = 0$ is allowed in Lemma B.21, so the lemma applies to $C = \partial_\tau^{-1}$.

Theorem B.23. *Let*

$$\mathcal{R}_n = \int d\mu_C(\bar{\psi}, \psi) \left(\prod_{k=0}^{n-1} e^{-\frac{\beta}{n} A_k} - \prod_{k=0}^{n-1} (1 - \frac{\beta}{n} A_k) \right) \tag{B.119}$$

where A_k is an element of the Grassmann algebra generated by $(\bar{\psi}(\tau_k, \xi), \psi(\tau_k, \xi))_{\xi \in \Gamma}$. Thus $\|A_k\|_q \leq \alpha_k q^{|\Gamma|}$. Assume that there is an $\alpha > 0$ such that for all n, $\max_{k \leq n-1} |\alpha_k| \leq \alpha$. Let $\mathcal{R}_{n,p}$ be defined by $\mathcal{R}_n = \sum_{p=2}^{\infty} \beta^p \mathcal{R}_{n,p}$. Then for all $p \geq 2$, $\mathcal{R}_{n,p} \to 0$ as $n \to \infty$.

Proof. By Theorem B.14, it suffices to show that $\|\mathcal{R}_{n,p}\|_{q_n} \to 0$ as $n \to \infty$ for every fixed p, where $q_n = \varepsilon r^{-d}(1 + \log n)$. Let

$$B_k = \frac{1}{n^2} \sum_{p \geq 2} \frac{(-\beta A_k)^p}{p!} n^{2-p} = \sum_{p \geq 2} B_{k,p} \beta^p \tag{B.120}$$

then

$$\|B_{k,p}\|_q \leq \frac{1}{n^2} b_p q^{|\Gamma|} \tag{B.121}$$

with $b_p = \alpha^p / p!$. Let $\mathcal{N} = \{0, \dots, n-1\}$.

$$\mathcal{R}_n = \sum_{\substack{S \dot\cup T = \mathcal{N} \\ T \neq \emptyset}} \prod_{k \in S} (1 - \frac{\beta}{n} A_k) \prod_{k \in T} B_k$$

$$= \sum_{\substack{S \dot\cup T = \mathcal{N} \\ T \neq \emptyset}} \sum_{U \subseteq S} \prod_{k \in U} (-\frac{\beta}{n} A_k) \prod_{k \in T} \sum_{r_k \geq 2} B_{k,r_k} \beta^{r_k}. \tag{B.122}$$

Collecting powers of β, we have

$$\mathcal{R}_{n,p} = \sum{}' \prod_{k \in U} (-\frac{A_k}{n}) \prod_{k \in T} B_{k,r_k} \tag{B.123}$$

where \sum' denotes a sum over $S \dot\cup T = \mathcal{N}$, with $T \neq \emptyset$, over $U \subset S$, and over sequences $(r_k)_{k \in T}$, $r_k \geq 2$ for all $k \in T$, such that

$$\sum_{k \in T} r_k + |U| = p. \tag{B.124}$$

We now take the norm and use the product inequality, and the estimates for $\|A_k\|_q$ and $\|B_{k,r_k}\|_q$. With the notation $\gamma = |\Gamma|$, we then have

$$\|\mathcal{R}_{n,p}\|_{q_n} \leq \sum{}' \prod_{k \in U} \frac{\|A_k\|_{q_n}}{n} \prod_{k \in T} \|B_{k,r_k}\|_{q_n}$$

$$\leq \sum{}' \left(\tfrac{1}{n}\alpha q_n{}^\gamma\right)^{|U|} \prod_{k \in T} \frac{q_n{}^\gamma}{n^2} b_{r_k}. \tag{B.125}$$

By (B.124), $|U| \leq p$, so

$$\left(\tfrac{1}{n}\alpha q_n{}^\gamma\right)^{|U|} \leq q_n{}^{\gamma p} \left(\frac{\alpha}{n}\right)^{|U|}. \tag{B.126}$$

Inserting this, we can resum the r_k and U and get, with $b = \sum_{r \geq 2} b_r \leq e^\alpha - 1$,

$$\|\mathcal{R}_{n,p}\|_{q_n} \leq q_n{}^{\gamma p} \sum_{\substack{S \,\dot\cup\, T = \mathcal{N} \\ T \neq \emptyset}} \sum_{U \subset S} \left(\frac{\alpha}{n}\right)^{|U|} \left(\frac{q_n{}^\gamma}{n^2} b\right)^{|T|}$$

$$= q_n{}^{\gamma p} \sum_{\substack{S \,\dot\cup\, T = \mathcal{N} \\ T \neq \emptyset}} \left(1 + \frac{\alpha}{n}\right)^{|S|} \left(\frac{q_n{}^\gamma}{n^2} b\right)^{|T|}$$

$$= q_n{}^{\gamma p} \left[\left(1 + \frac{\alpha}{n} + \frac{q_n{}^\gamma}{n^2} b\right)^n - \left(1 + \frac{\alpha}{n}\right)^n\right]$$

$$= \left(1 + \frac{\alpha}{n}\right)^n q_n{}^{\gamma p} \left((1 + x_n)^n - 1\right) \tag{B.127}$$

where

$$x_n = \left(1 + \frac{\alpha}{n}\right)^{-1} \frac{q_n{}^\gamma}{n^2} b. \tag{B.128}$$

For $x \geq 0$, $\log(1 + x) \leq x$ and $e^x - 1 \leq xe^x$, so

$$(1 + x_n)^n - 1 \leq e^{n \log(1 + x_n)} - 1 \leq e^{n x_n} - 1 \leq n x_n e^{n x_n}. \tag{B.129}$$

Since $n x_n \leq \text{const } n^{-\delta}$ with $\delta > 0$,

$$0 \leq q_n{}^{\gamma p} \left((1 + x_n)^n - 1\right) \leq \text{const } q_n{}^{\gamma p} n^{-\delta} \to 0 \tag{B.130}$$

as $n \to \infty$, so $\mathcal{R}_{n,p}$ vanishes in that limit. ∎

We now only have to put all the pieces together to have a proof of Theorem B.19. Recall that $\zeta(\beta A)$ is an entire function of β. The $\tilde\zeta_n(\beta A)$ are also entire functions of β for all n. We know that $\tilde\zeta_n(\beta A) \to \zeta(\beta A)$ as $n \to \infty$. It follows by expanding in β, writing

$$\zeta(\beta A) = \sum_{p \geq 0} \zeta_p \beta^p, \quad \tilde\zeta_n(\beta A) = \sum_{p \geq 0} \tilde\zeta_{n,p} \beta^p, \tag{B.131}$$

that for all $p \geq 0$, $\tilde\zeta_{n,p} \to \zeta_p$ as $n \to \infty$. Now $\tilde\zeta_n(\beta A) = \zeta_n(\beta A) + \mathcal{R}_n$, and all summands are entire functions of β for any n. Let $\zeta_n(\beta A) = \sum_{p \geq 0} \zeta_{n,p} \beta^p$. By Theorem B.23,

$$\lim_{n \to \infty} \zeta_{n,p} = \lim_{n \to \infty} \tilde\zeta_{n,p} = \zeta_p \tag{B.132}$$

so the formal power series $\sum_{p \geq 0} z_p \beta^p$ with $z_p = \lim_{n \to \infty} \zeta_{n,p}$ converges and represents the function $\zeta(\beta A)$. Thus Theorem B.19 holds.

B.5.4 Nambu Formalism

It is useful for deriving the component form of the RGE to rename the Grassmann variables such that the distinction between ψ and $\bar{\psi}$ is in another index. This is a variant of the usual 'Nambu formalism'; see, e.g., [57]. Let

$$\Gamma = \Lambda \times \{-1,1\} \times \{1,2\}, \tag{B.133}$$

and denote $X = (x,\sigma,i) \in \Gamma$. For $x \in \Lambda$ and $\sigma \in \{-1,1\}$, the fields are defined as $\psi(x,\sigma,1) = \bar{\psi}_\sigma(x)$ and $\psi(x,\sigma,2) = \psi_\sigma(x)$. The antiperiodicity condition reads $\psi(x + \beta e_\tau, \sigma, i) = -\psi(x,\sigma,i)$ with e_τ the unit vector in τ-direction. The Grassmann algebra generated by the $(\psi(X))_{X\in\Gamma}$ is denoted by $\mathcal{A}_\Gamma[\psi]$. Given another set of Grassmann variables $(\eta(X))_{X\in\Gamma}$, the Grassmann algebra generated by the ψ and η is denoted by $\mathcal{A}_\Gamma[\psi,\eta]$. Furthermore, denote $\int_\Gamma dX\, F(X) = \sum_{i=1}^2 \sum_{\sigma\in\{-1,1\}} \int_\Lambda dx\, F(x,\sigma,i)$ and $\delta_\Gamma((x,\sigma,i),(x',\sigma',i')) = \delta_{ii'}\,\delta_{\sigma\sigma'}\,\delta_\Lambda(x,x')$, and define a bilinear form on $\mathcal{A}_\Gamma[\psi,\eta]$ by

$$(\psi,\eta)_\Gamma = \int_\Gamma dX\, \psi(X)\,\eta(X) = -(\eta,\psi)_\Gamma. \tag{B.134}$$

Then, for any operator a, $(\bar{\psi},a\psi) = \frac{1}{2}(\psi,\,A\,\psi)_\Gamma$, where, for $X = (x,\sigma,i)$ and $X' = (x',\sigma',i')$,

$$A(X,X') = \begin{cases} 0 & \text{if } i = i' \\ a(x,\sigma,x',\sigma') & \text{if } i = 1 \text{ and } i' = 2 \\ -a(x',\sigma',x,\sigma) & \text{if } i = 2 \text{ and } i' = 1. \end{cases} \tag{B.135}$$

In other words, when written as a matrix in the index i, A takes the form

$$\begin{pmatrix} 0 & a \\ -a^T & 0 \end{pmatrix} \tag{B.136}$$

with $(a^T)(x,\sigma,x',\sigma') = a(x',\sigma',x,\sigma)$ denoting the transpose of a. A is invertible if and only if a is invertible, and $A(X,X') = -A(X',X)$.

With this, $Z_{A,n_\tau} = \mathcal{N}_\Lambda \det a\, \tilde{Z}_\Lambda$ where $\tilde{Z}_\Lambda = \int d\mu_C(\psi) e^{-\lambda S_4(\psi)}$, where $C = A^{-1}$, and $d\mu_C$ is the linear functional ('Grassmann Gaussian measure') defined by $d\mu_C(\psi) = (\det a)^{-1}\, D_\Gamma\psi\, e^{\frac{1}{2}(\psi,A\psi)_\Gamma}$. The constant $\mathcal{N}_\Lambda \det a$ drops out of all correlation functions and can therefore be omitted. The 'measure' $d\mu_C$ is normalized, $\int d\mu_C(\psi) = 1$, and its characteristic function is

$$\int d\mu_C(\psi)\, e^{(\eta,\psi)_\Gamma} = e^{\frac{1}{2}(\eta,\,C\,\eta)_\Gamma}. \tag{B.137}$$

All moments of $d\mu_C$ can be obtained by differentiating (B.137) with respect to η and setting $\eta = 0$.

B.5.5 Matsubara Frequencies

To implement the antiperiodic boundary conditions in time direction, it is convenient to define the fields ψ on the torus $\mathbb{T}_2 = \varepsilon_\tau \mathbb{Z}/2\beta\mathbb{Z}$ and to require antiperiodicity with respect to translations by β, (4.57). We have $\mathbb{T}_2 = \mathbb{T} \cup (\mathbb{T} + \beta)$ with $\mathbb{T} = \{\tau \in \mathbb{T}_2 : \tau = \varepsilon_\tau k, \ k \in \{-\frac{n_\tau}{2}, \ldots, \frac{n_\tau}{2} - 1\}\}$. The dual to \mathbb{T}_2 is $\mathbb{T}_2^* = \frac{\pi}{\beta}\mathbb{Z}/2n_\tau\mathbb{Z} = \{\omega = \frac{\pi}{\beta}k : k \in \{-n_\tau, \ldots, n_\tau - 1\}\}$. The Fourier transform on \mathbb{T}_2 is

$$\tilde{f}(\omega) = \varepsilon_\tau \sum_{\tau \in \mathbb{T}_2} e^{-i\omega\tau} f(\tau), \quad f(\tau) = \frac{1}{2\beta} \sum_{\omega \in \mathbb{T}_2} e^{i\omega\tau} \tilde{f}(\omega). \tag{B.138}$$

If $f(\tau - \beta) = -f(\tau)$, then $\tilde{f}(\omega) = 0$ if $\frac{\omega\beta}{\pi}$ is even. In that case, with $\hat{f}(\omega) = \frac{1}{2}\tilde{f}(\omega)$,

$$f(\tau) = \frac{1}{\beta} \sum_{\omega \in M_{n_\tau}} e^{i\omega\tau} \hat{f}(\omega) \tag{B.139}$$

with M_{n_τ} given by (4.63). The orthogonality relations are

$$\int_{\mathbb{T}} d\tau \ e^{i(\omega_n \pm \omega_m)\tau} = \beta\delta_{mn}, \quad \frac{1}{\beta} \sum_{\omega \in M_{n_\tau}} e^{i\omega\tau} = \frac{1}{\varepsilon_\tau}(\delta_{\tau 0} - \delta_{\tau\beta}). \tag{B.140}$$

Lemma B.24. *Let* $\beta > 0$, $E_0 \geq 0$, *and* $n_\tau \geq 2\beta(E_0 + E_{max})$, *where* E_{max} *is defined in (4.74), and let* $\hat{\omega}$ *be defined as in (4.65). Then for all* $k \in B$ *and all* $\omega \in M_{n_\tau}$, $|i\hat{\omega} - E(k)| \leq E_0$ *implies* $|\omega| \leq \frac{\pi}{2}E_0$ *and* $|E(k)| \leq 2E_0$, *and*

$$\frac{1}{\beta} \sum_{\omega \in M_{n_\tau}} \mathbb{1}(|i\hat{\omega} - E(k)| \leq E_0) \leq E_0 \, \mathbb{1}(|E(k)| \leq 2E_0) . \tag{B.141}$$

Proof. By (4.65), $\text{Im} \ \hat{\omega} = \frac{1}{\varepsilon_\tau}(1 - \cos(\omega\varepsilon_\tau))$ and $\text{Re} \ \hat{\omega} = \frac{1}{\varepsilon_\tau}\sin(\omega\varepsilon_\tau)$. The condition $|i\hat{\omega} - E(k)| \leq E_0$ implies $|\text{Re} \ \hat{\omega}| \leq E_0$ and $|\text{Im} \ \hat{\omega} + E(k)| \leq E_0$. Thus $|\text{Im} \ \hat{\omega}| \leq E_0 + E_{max}$. Since $1 - \cos x \geq \frac{2}{\pi^2}x^2$,

$$(E_0 + E_{max})\varepsilon_\tau \geq \varepsilon_\tau|\text{Im} \ \hat{\omega}| = 1 - \cos(\omega\varepsilon_\tau) \geq \frac{2}{\pi^2}(\omega\varepsilon_\tau)^2, \tag{B.142}$$

so

$$|\omega\varepsilon_\tau| \leq \pi(\beta(E_0 + E_{max})(2n_\tau)^{-1})^{1/2} \leq \frac{\pi}{2}. \tag{B.143}$$

Since $\frac{\sin x}{x}$ is decreasing on $[0, \frac{\pi}{2}]$,

$$E_0 \geq |\text{Re} \ \hat{\omega}| \geq |\omega|\left|\frac{\sin(\omega\varepsilon_\tau)}{\omega\varepsilon_\tau}\right| \geq \frac{2}{\pi}|\omega|. \tag{B.144}$$

So $|\omega| \leq \frac{\pi}{2}E_0$. Since $1 - \cos x \leq \frac{1}{2}x^2$,

$$|\text{Im} \ \hat{\omega}| \leq \frac{1}{2\varepsilon_\tau}(\omega\varepsilon_\tau)^2 \leq \frac{\pi^2}{8}\beta E_0^2 \frac{1}{n_\tau} \leq E_0. \tag{B.145}$$

Thus $|\mathrm{Im}\,\hat{\omega} + E(\mathbf{k})| \leq E_0$ implies $|E(\mathbf{k})| \leq 2E_0$, hence $\mathbb{1}(|i\hat{\omega} - E(\mathbf{k})| \leq E_0)$ $\leq \mathbb{1}\left(|\omega| \leq \frac{\pi}{2}E_0\right)\mathbb{1}\left(|E(\mathbf{k})| \leq 2E_0\right)$. The summation over M_{n_τ} gives

$$\frac{1}{\beta}\sum_{\omega \in M_{n_\tau}} \mathbb{1}\left(|\omega| \leq \frac{\pi}{2}E_0\right) = \frac{1}{\beta}\sum_{n=-\frac{n_\tau}{2}}^{\frac{n_\tau}{2}} \mathbb{1}\left(|2n+1| \leq \frac{\beta E_0}{2}\right) \leq E_0. \qquad (B.146)$$

Note that the last inequality holds in particular if $\frac{\beta E_0}{2} < 1$, because then the sum is empty, hence the result zero, because $2n + 1$ is always odd. ∎

We end this appendix with a brief proof of (4.68), which concerns the limit $n_\tau \to \infty$. At temperature $T = \frac{1}{\beta} > 0$, the poles of the Fermi distribution $f_\beta(E) = (1+e^{\beta E})^{-1}$ in the complex E–plane are at $E = i\omega_n$, $\omega = \frac{\pi}{\beta}(2n+1) \in M$. For $n_\tau \to \infty$, the free propagator at positive temperature $T > 0$ is $C(\tau, E(\mathbf{p}))$, where

$$C(\tau, E) = \begin{cases} -e^{-\tau E}\left(1 - f_\beta(E)\right) & \text{if } \tau > 0 \\ e^{-\tau E}f_\beta(E) & \text{if } \tau \leq 0. \end{cases} \qquad (B.147)$$

Note that this obeys the antiperiodicity

$$C(\tau + \beta, E) = -C(\tau, E) \qquad (B.148)$$

so that C is periodic with period 2β. By (B.148) and because $e^{i\omega_n\beta} = -1$, the Fourier coefficients of C are

$$c_n(E) = \frac{2}{2\beta}\int_{-\beta}^{0} d\tau\, C(\tau, E)\, e^{i\omega_n\tau} = \frac{1}{\beta}\frac{1}{i\omega_n - E}. \qquad (B.149)$$

The Fourier series

$$C(\tau + \beta, E) = \frac{1}{\beta}\sum_{n \in \mathbb{Z}}\frac{e^{-i\omega_n\tau}}{i\omega_n - E} \qquad (B.150)$$

converges for $\tau \notin \beta\mathbb{Z}$. Thus (4.68) follows by taking the limit $\tau \uparrow 0$ of (B.150).

B.6 Feynman Graph Expansions

It is a natural question if there are any changes in the proofs of the connectedness theorems in Chap. 2 if we start from a fermionic theory instead of a bosonic one. In this section, we briefly discuss how the proofs given for the bosonic theory extend to fermions. We shall see that all essential formulas carry over unchanged, only that the variables are now Grassmann variables, and that the few formulas that do get changed get only trivial changes. In particular, we show that the proof that only connected graphs contribute to the effective action is unchanged even though the Grassmann algebra is non-commutative. In a nutshell, the reason for this is that the polymer activities

are elements of the even subalgebra, which is commutative, so that all results of Sect. 2.4 apply.

The main problem in generalizing the previous representations to the fermionic case is that the fermionic algebra is noncommutative whereas we have assumed, e.g. in the polymer expansions, that the activities are in a commutative algebra over \mathbb{C}. We now show that one can arrange everything such that only elements of the even subalgebra $\mathcal{A}_2[\psi, \eta]$ occur, so that the assumption that everything commutes still holds. The only place where the fermion signs appear is at the very end, when the derivatives are taken and evaluated at $\eta = 0$.

The definition of the fermionic effective action,

$$e^{G_{\text{eff}}(\psi, C, -\lambda V)} = \int d\mu_C(\phi) e^{-\lambda V(\phi + \psi)}, \tag{B.151}$$

is formally identical to the one for bosonic fields, only that now the Grassmann integral is implied in (B.151). Similarly, (2.117) carries over, only that the $v_k^{(m_k)}$ are now antisymmetric with respect to permutations, because no other part would contribute, by the permutation antisymmetry of the Wick ordered polynomials. Thus we have

$$G_{\text{eff}}(\psi, C, V) = \sum_{r=1}^{\infty} G_{\text{eff}}^{(r)}(\psi, C, V) \, \lambda^r \tag{B.152}$$

with

$$G_{\text{eff}}(\psi, C, V) = \mathcal{E}^T(V, \ldots, V, C, \psi) \tag{B.153}$$

and (see (2.121))

$$\mathcal{E}^T(V_1, \ldots, V_p; C_1, \psi) = \tag{B.154}$$

$$\sum_{m_1, \ldots m_p} \int \prod_{b \in B_0} dx_b \prod_{k=1}^{p} v_k^{(m_k)}(X_k) \left[\prod_{b \in B_0} \frac{\partial}{\partial \eta_b} E(\eta, C, C_1, \psi) \right]_{\eta=0}.$$

In the product over derivatives, the order is from small to large k, and for every k, i_k from 1 to m_k, and

$$E(\eta, C, C_1, \psi) = \left[\frac{\partial^p}{\partial \lambda_1 \ldots \partial \lambda_p} \log Z_F \right]_{\lambda=0} \tag{B.155}$$

with

$$Z_F = \int d\mu_{C_1}(\phi) \prod_{k=1}^{p} \left(1 + \lambda_k \Omega_C(e^{i(\eta^{(k)}, \phi + \psi)}) \right) \tag{B.156}$$

(the λ's are still ordinary complex variables, of course). Because of the simple algebraic nature of the Grassmann integral, there is not even any convergence question for this integral.

At this point we see that all factors are bilinear in the fermions, so all activities are in a commuting algebra. Expanding as in (2.126) and doing the Gaussian integrals, we get exactly the same formula as (2.127), only with $i\alpha$ replaced by η. Thus all activitis commute because they are in the even subalgebra, and Theorem 2.17 applies, hence (2.128)-(2.139) hold, with $i\alpha$ replaced by η. Thus the logarithm removes all disconnected graphs, exactly as in the bosonic case.

The only notable difference to the bosonic case arises when one now does the derivatives. While Lemma 2.13 would carry over to fermions unchanged if the coefficients in an expansion in η were \mathbb{C}-valued, the extra ψ-variables imply that the coefficients are actually in the Grassmann algebra (and in particular, they may have odd degree). The correct replacement of Lemma 2.13 is

Lemma B.25. *Let A and B be elements of the Grassmann algebra $\mathbb{C}\{\eta\}$, that is,*

$$A = \sum_I a_I \eta^I, \qquad B = \sum_I b_I \eta^I \qquad (B.157)$$

where $a_I, b_I \in \mathbb{C}$, and let $C(x)$ be the formal power series

$$C(x) = \sum_{r=0}^{\infty} c_r \lambda^r. \qquad (B.158)$$

Then

$$\left[A(\frac{\partial}{\partial \eta}) \, B(\eta) \, C((\eta, \psi)) \right]_{\eta=0} = \left[A(\eta) \, B(\frac{\partial_L}{\partial \eta}) \, C((\frac{\partial_L}{\partial \eta}, \psi)) \right]_{\eta=0} \qquad (B.159)$$

where $\frac{\partial_L}{\partial \eta}$ is the derivative with respect to η, acting to the left. Moreover

$$\left[A(\frac{\partial}{\partial \eta}) \, B(\eta) \right]_{\eta=0} = \left[B(\frac{\partial}{\partial \eta}) \, A(\eta) \right]_{\eta=0}. \qquad (B.160)$$

Proof. The proof is an easy exercise in Grassmann algebra and is left to the reader. ∎

The derivation of the graph rules is now easy. The place where the fermion signs arise is in the derivatives. Commuting them properly gives the usual fermionic sign factors. We leave these details to the reader.

We finally note that for the case of QED-type or Yukawa-type or phonon theories, where one has boson and fermion fields with interaction term $\phi\bar{\psi}\psi$, our results imply that all symmetry factors are equal to one.

B.7 The Thermodynamic Limit in Perturbation Theory

The propagator for the many-fermion system without cutoff is given in (4.64). Thus, for $k = (\omega, \mathbf{k}) \in \Lambda^*$ let

$$D_t(k) = \frac{1}{i\hat{\omega} - E(\mathbf{k})} \chi_1 \left(\epsilon_t^{-2} |i\hat{\omega} - E(\mathbf{k})|^2 \right) \qquad (B.161)$$

where $\hat{\omega}$ is defined in (4.65), χ_1 is given in (4.71), and define $C_t(k)$ similarly, with χ_1 replaced by $\chi_2 = 1 - \chi_1$. Then $C_t(k) + D_t(k) = (i\hat{\omega} - E(\mathbf{k}))^{-1}$ is independent of t. The functions C_t and D_t define operators $\tilde{D}_t(K, K')$ and $\tilde{C}_t(K, K')$ on the functions on Γ^* in the same way as \hat{C} defines C by (4.62). Denote $\dot{D}_t = \frac{\partial}{\partial t} D_t$, and let $\mathbb{1}(A) = 1$ if the event A is true and 0 otherwise.

Proposition B.26. D_t is a C^∞ function of t that vanishes identically if $t > \log \frac{\beta \epsilon_0}{2}$. If $t \leq \log \frac{\beta \epsilon_0}{2}$, then

$$\text{supp } D_t \quad \subset \quad \{k \in \Lambda^* : |i\hat{\omega} - E(\mathbf{k})| \leq \epsilon_t\}$$

$$\text{supp } \dot{D}_t \quad \subset \quad \{k \in \Lambda^* : \frac{1}{2}\epsilon_t \leq |i\hat{\omega} - E(\mathbf{k})| \leq \epsilon_t\}. \qquad (B.162)$$

Moreover

$$\left| \dot{D}_t(k) \right| \leq 4\epsilon_t^{-1} \mathbb{1}\left(|i\hat{\omega} - E(\mathbf{k})| \leq \epsilon_t\right) \leq 2\beta \, \mathbb{1}\left(|i\hat{\omega} - E(\mathbf{k})| \leq \epsilon_t\right), \qquad (B.163)$$

$$|D_t(k)| \leq \frac{\beta}{2} \mathbb{1}\left(|i\hat{\omega} - E(\mathbf{k})| \leq \epsilon_t\right),$$

$$\int_{\Lambda^*} dk \left| \dot{D}_t(k) \right| \leq 4\mathcal{V}_1 \quad \text{and} \quad \int_{\Lambda^*} dk \, |D_t(k)| \leq \mathcal{V}_1 \log \frac{\beta \epsilon_0}{2}, \qquad (B.164)$$

where \mathcal{V}_1 is the constant in (4.73).

Proof. $\chi_1(x) = 0$ if $x \geq 1$, so $D_t(k) \neq 0$ implies $|i\hat{\omega} - E(\mathbf{k})| \leq \epsilon_t \leq 1$. By Lemma B.24, this implies $|\omega| \leq \frac{\pi}{2}\epsilon_t$. Since $|\omega| \geq \frac{\pi}{\beta}$, $D_t \neq 0$ only for $t \leq \log \frac{\beta \epsilon_0}{2}$. Since $|i\hat{\omega} - E(\mathbf{k})| \geq |\text{Re } \hat{\omega}| \geq \frac{2}{\beta}$, the stated properties of D_t follow. The t-derivative gives $|\frac{\partial}{\partial t} D_t(k)| = 2\epsilon_t^{-2}|i\hat{\omega} - E(\mathbf{k})| \, |\chi_1'(\epsilon_t^{-2}|i\hat{\omega} - E(\mathbf{k})|^2)|$. Since $\chi_1'(x) = 0$ for $x \notin (\frac{1}{4}, 1)$, $\dot{D}_t(k) \neq 0$ implies $\frac{1}{2}\epsilon_t \leq |i\hat{\omega} - E(\mathbf{k})| \leq \epsilon_t$ which implies (B.163) and (B.162). (B.164) follows from these inequalities by Lemma B.24, by (4.73), and by $D_t = \int_t^{\log(\beta \epsilon_0/2)} ds \, \dot{D}_s$. ∎

The bounds (B.164) are crude because the restriction $|E(\mathbf{k})| \leq \epsilon_t$ was replaced by $|E(\mathbf{k})| \leq 2$ when (4.73) was applied. Better bounds would require a more detailed analysis of the behaviour as a function of L.

The proof of existence of the thermodynamic limit will proceed inductively in r. It will be an application of the dominated convergence theorem to the integrated RGE

$$I_{mr}(t \mid \underline{P}) = I_{mr}(0 \mid \underline{P}) + \frac{1}{2}A_m \int_0^t ds\, Q_{mr}(s \mid \underline{P}). \tag{B.165}$$

To this end, it is necessary to make the integration region independent of n_τ and L. Although $I_{mr}(t)$, given by (B.165), appears evaluated at $\underline{P} = (P_1, \dots, P_m) \in \Gamma^m$ on the RHS of (B.165) only, the integral defines the t-derivative of a function defined on $\Gamma_\infty^{*\,m}$, where $\Gamma_\infty^* = \mathbb{M} \times \mathcal{B} \times \{1, -1\} \times \{1, 2\}$ with $\mathcal{B} = \mathbb{G}^*$ the first Brillouin zone of the infinite lattice, and

$$\mathbb{M} = \{\omega_n = \frac{\pi}{\beta}(2n+1) : n \in \mathbb{Z}\} \tag{B.166}$$

the set of Matsubara frequencies in the limit $n_\tau \to \infty$. For a bounded function $F_m : \Gamma_\infty^{*\,m} \to \mathbb{C}$, let

$$|F_m|_0 = \sup_{\underline{P} \in \Gamma_\infty^{*\,m}} |F_m(\underline{P})| \tag{B.167}$$

Up to now, the dependence of $I_{mr}(t)$ on (n_τ, L) was not denoted explicitly. We now put it in a superscript and write $I_{mr}^{(n_\tau, L)}$. Let $\lim_{(n_\tau, L) \to \infty} = \lim_{L \to \infty} \lim_{n_\tau \to \infty}$.

Lemma B.27. *Let* $(I_{mr}^{(n_\tau, L)}(0))_{m, r, n_\tau, L}$ *be a family of bounded functions such that* $I_{mr}^{(n_\tau, L)}(0) = 0$ *if* $m > 2r + 2$, $\lim_{(n_\tau, L) \to \infty} I_{mr}^{(n_\tau, L)}(0) = I_{mr}(0)$ *exists and is a bounded function on* $\Gamma_\infty^{*\,m}$, *and* $\left| I_{mr}^{(n_\tau, L)}(0) \right| \leq K_{mr}^{(0)}$. *Let* $(I_{mr}^{(n_\tau, L)}(t))_{m, r, n_\tau, L}$ *be the solution to (B.165). Then, for all m and r, the function* $I_{mr}^{(n_\tau, L)}(t) = 0$ *if* $m > 2r + 2$, *and*

$$\frac{\partial}{\partial t} I_{mr}^{(n_\tau, L)}(t) = 0 \quad \text{for all } t > \log \frac{\beta}{2}, \tag{B.168}$$

there are bounded functions $I_{mr}(t) : \Gamma_\infty^{*\,m} \to \mathbb{C}$ *such that*

$$\lim_{(n_\tau, L) \to \infty} I_{mr}^{(n_\tau, L)}(t) = I_{mr}(t). \tag{B.169}$$

Let P_{mr} *be the polynomials defined recursively as*

$$P_{mr}(x) = 6^{1-r} |I_{mr}(0)| + \epsilon_0^{-1} \int d\kappa_{mr} i!\, x^{i-1} P_{m_1 r_1}(x) P_{m_2 r_2}(x) \tag{B.170}$$

(in particular, the coefficients of P_{mr} are independent of β), then for all n_τ, L, β, t

$$\left| I_{mr}^{(n_\tau, L)}(t) \right|_0 \leq (\epsilon_0 \beta)^{r-1} P_{mr}\left(4\mathcal{V}_1 \log \frac{\beta \epsilon_0}{2} \right). \tag{B.171}$$

Proof. Induction in r, with the statement of the Lemma as the inductive hypothesis. Let $r = 1$. Since $r_1 \geq 1$ and $r_2 \geq 1$, the right hand side of the equation is zero, so $I_{m1}^{(n_\tau,L)}(t,\underline{P}) = I_{m1}^{(n_\tau,L)}(0,\underline{P})$ for all t. Thus the statement follows from the hypotheses on $I_{mr}^{(n_\tau,L)}(0)$.

Let $r \geq 2$, and the statement hold for all $r' < r$. Let $m > 2r + 2$. Then $I_{mr}^{(n_\tau,L)}(0) = 0$. The five-tuple (m_1, r_1, m_2, r_2, i) contributes to the right hand side only if $r_1 + r_2 = r$, $m_1 + m_2 = m + 2i$, $i \geq 1$, and by the inductive hypothesis, only if $m_1 \leq 2r_1 + 2$ and $m_2 \leq 2r_2 + 2$. Thus $I_{mr}^{(n_\tau,L)}(t)$ can be nonzero only if $m = m_1 + m_2 - 2i \leq m_1 + m_2 - 2 \leq 2r_1 + 2 + 2r_2 + 2 - 2 = 2r + 2$. The integral appearing on the right hand side of (B.165) is a Riemann sum approximation to an integral over the (t, n_τ, L)-independent region $s \in [0, \infty)$, $k_j \in \mathbb{M} \times \mathcal{B}$. By the inductive hypothesis, the factors $I_{m_k r_k}^{(n_\tau,L)}$ have a limit (B.169) satisfying (B.171). Since for all t, \bar{D}_t is a bounded C^2 function, the same holds for the propagators (boundedness holds because $\beta < \infty$). Thus the integrand converges pointwise, and it suffices to show that it is bounded by an integrable function to get (B.169) and (B.171) by an application of the dominated convergence theorem. Let $\alpha = 4V_1 \log \frac{\beta\epsilon_0}{2}$, and let g be the function on $[0, \infty) \times (\mathbb{M} \times \mathcal{B})^{i-1}$ given by

$$g(s, \underline{k}) = \frac{2}{\epsilon_0 e^{-s}} \mathbb{1}\left(s \leq \log \frac{\beta\epsilon_0}{2}\right) \int d\kappa_{mr} \, i \, i!$$
$$P_{m_1 r_1}(\alpha)\beta^{r_1 - 1} P_{m_2 r_2}(\alpha)\beta^{r_2 - 1} \tag{B.172}$$
$$\prod_{j=2}^{i} \int_0^s ds_j \, \mathbb{1}\left(|\omega_j| \leq \tfrac{\pi}{2}\epsilon_0 e^{-s_j}\right) \mathbb{1}\left(|E(k_j| \leq 2\epsilon_0 e^{-s_j}\right).$$

The integrand on the right hand side of (B.165) is bounded by g by Proposition B.26, Lemma B.24, and the inductive hypothesis (B.171). Because g vanishes identically for $s > \log \frac{\beta\epsilon_0}{2}$, it is integrable. By (B.163) and (B.164),

$$\int_0^t ds \int g(\underline{k})dk_2 \dots dk_i \leq \beta^{r-1} \int d\kappa_{mr} i! \, \alpha^{i-1} P_{m_1 r_1}(\alpha) P_{m_2 r_2}(\alpha). \tag{B.173}$$

Thus (B.169) holds, and (B.171) holds with P_{mr} given by (B.170) (the factor 6^{1-r} comes from the assumption that $\beta\epsilon_0 \geq 6$). Since the right hand side of (B.165) vanishes for $t > \log \frac{\beta\epsilon_0}{2}$, (B.168) holds. ∎

As discussed, $I_{mr}^{(n_\tau,L)}(0)$ is not the initial interaction because at $t = 0$, the propagator $C_0 \neq 0$. The $I_{mr}^{(n_\tau,L)}(0)$ are obtained from the original interaction by Wick ordering with respect to C and integrating over all fields with covariance C_0. The existence of the limit $(n_\tau, L) \to \infty$ of the $I_{mr}^{(n_\tau,L)}(0)$ can be shown by a similar inductive proof (it is not trivial because in the limit $n_\tau \to \infty$, the absolute value of the propagator is not summable). An alternative proof is in Appendix D of [46].

B.8 Volume Improvement Bounds

In this Appendix, we prove the one-loop and two-loop volume bounds for the case $E(\mathbf{p}) = \mathbf{p}^2 - 1$, that is, for the spherical Fermi surface.

B.8.1 The One–Loop Volume Bound

Let

$$T(\varepsilon, \mathbf{q}) = \int\limits_{S^{d-1}} \frac{d\theta}{(2\pi)^d}\, \mathbb{1}\left(|E(\pi(0,\theta) + \mathbf{q})| \leq \varepsilon\right) . \tag{B.174}$$

Lemma B.28. *Let $E(\mathbf{p}) = \mathbf{p}^2 - 1$, $\mathbf{q} \in \mathbb{R}^d$ and $q = |\mathbf{q}|$. Then for all $d \geq 2$ and all $\varepsilon \leq 1/4$,*

$$T(\varepsilon, \mathbf{q}) \leq \begin{cases} 1 & \text{if } q \leq 2\varepsilon \\ \frac{\varepsilon}{q} & \text{if } 2\varepsilon \leq q \leq 1 \\ 2^d\, \frac{\varepsilon}{\sqrt{2-q}} & \text{if } 1 \leq q \leq 2 - 2\varepsilon \\ 2^d\, \varepsilon^{(d-1)/2} & \text{if } 2 - 2\varepsilon \leq q \leq 2 + \varepsilon \\ 0 & \text{otherwise.} \end{cases} \tag{B.175}$$

Proof. For $E(\mathbf{p}) = \mathbf{p}^2 - 1$, ρ and θ are simply polar coordinates, so $E(\pi(\rho, \theta) - \mathbf{q}) = \rho + q(q + 2\sqrt{1 + \rho}\,\cos\varphi)$, where φ is the angle between $\pi(\rho, \theta)$ and \mathbf{q}. For $d = 2$, we can choose coordinates such that $\varphi = \theta$; for $d \geq 3$, we write $d\theta = |\sin\varphi|^{d-2} d\varphi d\omega$, with ω varying over S^{d-2}/\mathbb{Z}_2, and $\varphi \in [0, 2\pi]$. The integrand is independent of ω and $\int d\omega/(2\pi)^{d-1} \leq 1$, so

$$T(\varepsilon, \mathbf{q}) \leq \int\limits_0^{2\pi} \frac{d\varphi}{2\pi} |\sin\varphi|^{d-2}\, \mathbb{1}\left(|q + 2\cos\varphi| \leq \tfrac{\varepsilon}{q}\right) . \tag{B.176}$$

If $q \leq \varepsilon$, use $T(\varepsilon, \mathbf{q}) \leq 1$. Let $2\varepsilon \leq q \leq 1$. The indicator function vanishes unless $\frac{1}{2}(-q - \tfrac{\varepsilon}{q}) \leq \cos\varphi \leq \frac{1}{2}(-q + \tfrac{\varepsilon}{q})$. Thus $|\cos\varphi| \leq 3/4$, and therefore $2\sin\varphi \geq 1$. Thus the Jacobian of a change of variables to $\gamma = q + 2\cos\varphi$ is bounded by one, and hence

$$T(\varepsilon, \mathbf{q}) \leq \int \frac{d\gamma}{2\pi}\, \mathbb{1}\left(|\gamma| \leq \tfrac{\varepsilon}{q}\right) \leq \tfrac{\varepsilon}{q}. \tag{B.177}$$

Let $q \geq 1$. We change variables from φ to $v = \cos\frac{\varphi}{2}$ and use $|\sin\theta| \leq 2v\sqrt{1 - v^2}$, to get

$$T(\varepsilon, \mathbf{q}) \leq \frac{2^{d-2}}{\pi} \int\limits_{-1}^1 dv\, |v|^{d-2}(1 - v^2)^{\frac{d-3}{2}}\, \mathbb{1}\left(|4v^2 + q - 2| \leq \varepsilon\right) . \tag{B.178}$$

If $q > 2 + \varepsilon$, $|4v^2 + q - 2| = 4v^2 + q - 2 > \varepsilon$ and the integrand is identically zero. If $|q - 2| \leq 2\varepsilon$, then $|4v^2 + q - 2| \leq \varepsilon$ implies $4v^2 \leq 3\varepsilon$ and therefore

$$T(\varepsilon, \mathbf{q}) \le \frac{2^{d-2}}{\pi} \int_{-\sqrt{\varepsilon}}^{\sqrt{\varepsilon}} dv\, |v|^{d-2}(1-v^2)^{\frac{d-3}{2}} \le \frac{2^{d-1}}{d-1}\varepsilon^{(d-1)/2}. \tag{B.179}$$

Finally, if $\delta = 2 - q \ge 2\varepsilon$, then $|4v^2 + q - 2| \le \varepsilon$ implies $\frac{1}{2}\sqrt{\delta - \varepsilon} \le 2|v| \le \frac{1}{2}\sqrt{\delta + \varepsilon}$. For $|x| \le \frac{1}{2}$, $|\sqrt{1+x} - 1| \le \sqrt{2}\,x$, so

$$|\,2|v| - \delta| \le \sqrt{\frac{2}{\delta}}\,\varepsilon. \tag{B.180}$$

and therefore

$$T(\varepsilon, \mathbf{q}) \le 2\frac{2^{d-2}}{\pi} \int_0^1 dv\, v^{d-2}(1-v^2)^{\frac{d-3}{2}} \mathbb{1}\left(|\,2|v| - \delta| \le \sqrt{\tfrac{2}{\delta}}\varepsilon\right)$$

$$\le \frac{2^{d-1}}{\pi} \int_0^1 \frac{dv}{\sqrt{1-v^2}} \mathbb{1}\left(|\,2|v| - \delta| \le \sqrt{\tfrac{2}{\delta}}\varepsilon\right). \tag{B.181}$$

Because $q \ge 1$, $\delta \le 1$, and thus the support of the integrand does not contain $v = 1$. Therefore $T(\varepsilon, \mathbf{q}) \le 2^d \frac{\varepsilon}{\sqrt{\delta}}$. ∎

Lemma B.28 implies Lemma 4.10 by Taylor expansion: using the coordinates ρ and θ, we have

$$V_1^{(\pm)}(\mathbf{q}, t, t') = \int_{-\epsilon_{t'}}^{\epsilon_{t'}} d\rho \int_{S^{d-1}} \frac{d\theta}{(2\pi)^d}\, J(\rho, \theta)\, \mathbb{1}\left(|E(\pi(\rho, \theta) + \mathbf{q})| \le \epsilon_t\right). \tag{B.182}$$

By (4.75) and because $\epsilon_{t'} \le \epsilon_t$,

$$|E(\pi(\rho, \theta) + \mathbf{q}) - E(\pi(0, \theta) + \mathbf{q})| \le \frac{2}{g_0}|\rho| \le \epsilon_t, \tag{B.183}$$

so

$$V_1^{(\pm)}(\mathbf{q}, t, t') \le 2|J|_0\, 2\epsilon_{t'}\, T\left((1 + \tfrac{2}{g_0}, \mathbf{q})\epsilon_t\right). \tag{B.184}$$

Lemma 4.10 now follows from Lemma B.28 by the observation that for $q \ge \sqrt{\epsilon_t}$, $\epsilon_t/q \le \sqrt{\epsilon_t}$.

Generalizing Lemma B.28 to the nonspherical case and taking into account that \mathcal{B} is a torus (so there may be more intersections of the two regions) is not difficult in principle, but it requires some care. The difference to the above proof is that the function $q + 2\cos\varphi$ is replaced by a 2π-periodic function f that, by strict convexity, is nonconstant. The second derivative at the critical points of f is proportional to the curvature of the Fermi surface at the corresponding point $\pi(0, \theta(\varphi))$ on \mathcal{S}.

B.8.2 The Two–Loop Volume Bound

Theorem B.29. *Let $k_0 \geq 2$, denote $\pi_0(\theta) = \pi(0, \theta)$, and let*

$$
\mathcal{W}(\varepsilon) = \sup_{\substack{q \in B \\ v_i \in \{\pm 1\}}} \int_{S^{d-1}} d\theta_1 \int_{S^{d-1}} d\theta_2 \; \mathbb{1}\left(|E(v_1 \pi_0(\theta_1) + v_1 \pi_0(\theta_2) + q)| \leq \varepsilon\right) .
$$

(B.185)

There is a constant $Q_V \geq 1$ such that for all $0 < \varepsilon \leq \epsilon_0$

$$
\mathcal{W}(\varepsilon) \leq Q_V \, \varepsilon \, \begin{cases} 1 + |\log \varepsilon| & \text{if } d = 2 \\ 1 & \text{if } d \geq 3. \end{cases}
$$

(B.186)

The constant Q_V depends on the curvature of the Fermi surface. Theorem B.29 is Theorem 1.2 of [46]. We prove it for the spherical Fermi surface, using Lemma B.28. The proof of Theorem B.29 for the nonspherical case is in Appendix B of [46].

Proof. We have

$$
\mathcal{W}(\varepsilon) = \sup_{p \in \mathbb{R}^d} \int_{S^{d-1}} \frac{d\theta}{(2\pi)^d} \, \mathcal{T}(\varepsilon, \mathbf{p} - \pi_0(\theta)).
$$

(B.187)

If p denotes the absolute value of \mathbf{p} and φ the angle between \mathbf{p} and π_0, we have $\xi(\varphi) = |\mathbf{p} - \pi_0(\theta)| = \sqrt{(1-p)^2 + 2p(1 - \cos\varphi)}$ and, by Lemma B.28, $\mathcal{W}(\varepsilon) = \sup_{p \geq 0} \mathcal{W}(\varepsilon, p)$ where

$$
\mathcal{W}(\varepsilon, p) = \int_0^{2\pi} d\varphi \, |\sin\varphi|^{d-2} \tilde{\mathcal{T}}(\varepsilon, \xi(\varphi))
$$

(B.188)

and

$$
\tilde{\mathcal{T}}(\varepsilon, \xi) = \begin{cases} 1 & \text{if } \xi \leq 2\varepsilon \\ \frac{\varepsilon}{q} & \text{if } 2\varepsilon \leq \xi \leq 1/M \\ K(M)\varepsilon & \text{if } 1/M \leq \xi \leq 2 - 1/M \\ 2^d \, \varepsilon / \sqrt{2 - \xi} & \text{if } -1/M \leq \xi - 2 \leq -2\varepsilon \\ 2^d \, \varepsilon^{(d-1)/2} & \text{if } -2\varepsilon \leq \xi - 2 \leq \varepsilon \\ 0 & \text{otherwise.} \end{cases}
$$

(B.189)

Here we take $M \geq 4$; Lemma B.28 implies that $K(M) = \max\{M, 2^d / \sqrt{M}\}$. Note that by definition of $\tilde{\mathcal{T}}$, the bound for \mathcal{W} follows immediately for the contribution from those φ for which $\xi(\varphi)$ is between $\frac{1}{M}$ and $2 - \frac{1}{M}$. The contributions that require more detailed bounds are the ones from the φ for which $\xi(\varphi) \approx 0$ and $\xi(\varphi) \approx 2$.

We split the proof in four cases:

1. $0 \leq p \leq 1 - \frac{1}{M}$

2. $1 - \frac{1}{M} \le p \le 1 + \frac{1}{M}$
3. $1 + \frac{1}{M} \le p \le 3 - \frac{2}{M}$
4. $3 - \frac{2}{M} \le p \le 3 + \frac{1}{M}$

(if $p \ge 3 + \frac{1}{M}$, $\mathcal{W}(\varepsilon, p) = 0$).

In Case 1, we have $\xi(\varphi) \ge 1 - p \ge \frac{1}{M}$, and $\xi(\varphi) \le 1 + p \le 2 - \frac{1}{M}$. Thus, by (B.189), $\mathcal{W}(\varepsilon, p) \le K(M)\varepsilon$.

Note that if $p \ge 1 + \frac{1}{M}$, then $\xi(\varphi) \ge p - 1 \ge \frac{1}{M}$, and thus $\xi(\varphi) \approx 0$ can occur only in Case 2. In Case 2, $\xi(\varphi) \approx 2$ also occurs, and the proof in Cases 3 and 4 is similar (in fact easier). Thus we shall do only Case 2 in detail.

We consider the contributions from the various regions given in the definition of $\tilde{\mathcal{T}}$ separately by inserting the corresponding indicator function in the integral. The integral with $\mathbb{1}\left(\xi(\varphi) \in (\frac{1}{M}, 2 - \frac{1}{M})\right)$ is bounded by $K(M)\varepsilon$ as in Case 1. Let

$$C_1 = \int_{-\pi}^{\pi} d\varphi |\sin\varphi|^{d-2} \frac{\varepsilon}{\xi(\varphi)} \, \mathbb{1}\left(2\varepsilon \le \xi(\varphi) \le \frac{1}{M}\right) . \tag{B.190}$$

The restriction $\xi(\varphi) \le \frac{1}{M}$ implies $2p(1 - \cos\varphi) \le \frac{1}{M}$. Since $p \ge 3/4$, this means that φ is near to zero. Changing variables to $v = \sin\frac{\varphi}{2}$, we have $\xi(\varphi) = \chi(v) = ((1-p)^2 + 4pv^2)^{1/2} \ge 2\sqrt{p}v$, and

$$C_1 \le 2^d \int_0^1 dv \, v^{d-2}(1-v^2)^{\frac{d-3}{2}} \frac{\varepsilon}{\chi(v)} \, \mathbb{1}\left(2\varepsilon \le \chi(v) \le \frac{1}{M}\right) . \tag{B.191}$$

For $d \ge 3$, we use $\frac{v^{d-2}}{\chi(v)} \le \frac{1}{2\sqrt{p}} v^{d-3} \le 1$, to get $C_1 \le 2^d \varepsilon$. For $d = 2$, we use that $\chi(v) \le \frac{1}{M}$ implies $v \le \frac{1}{M}$, so $(1-v^2)^{-1/2} \le 2$, to make a change of variables from v to $w = \chi(v)$, with Jacobian bounded by $2\sqrt{p}$, so

$$C_1 \le 16\sqrt{p} \int_{2\varepsilon}^{1/M} dw \, \frac{\varepsilon}{w} \le 32\varepsilon \, |\log(2M\varepsilon)|. \tag{B.192}$$

The contribution from those φ that have $\xi(\varphi) \le 2\varepsilon$,

$$C_2 = 2^d \int_0^1 dv \, v^{d-2}(1-v^2)^{\frac{d-3}{2}} \, \mathbb{1}\left((1-p)^2 + 4pv^2 \le 4\varepsilon^2\right) , \tag{B.193}$$

is nonzero only if $p - 1$ is of order ε and if $4pv^2 \le 4\varepsilon^2$, so

$$C_2 \le 2^{d+1} \int_0^{\varepsilon/\sqrt{p}} dv \, v^{d-2} = \frac{2^{d+1}}{d-1} \left(\frac{\varepsilon}{\sqrt{p}}\right)^{d-1} . \tag{B.194}$$

We now turn to the contribution from those φ with $\xi(\varphi) \approx 2$. Since $|p - 1| \leq \frac{1}{M}$, $(p-1)^2 + 2p(1 - \cos\varphi) \approx 4$ can hold only for $\varphi \approx \pi$. Thus we shift $\varphi \to \varphi' = \varphi + \pi$ and introduce $v = \sin\varphi'/2$. With $\eta(v) = \sqrt{(1-p)^2 + 4p(1-v^2)}$, we have to bound

$$C_3 = 2^d \int_0^1 dv\, v^{d-2}\, (1-v^2)^{\frac{d-3}{2}}\, \mathbb{1}\left(2\varepsilon \leq 2 - \eta \leq \tfrac{1}{M}\right) \frac{\varepsilon}{\sqrt{2-\eta}} \qquad (B.195)$$

and

$$C_4 = 2^d \int_0^1 dv\, v^{d-2}\, (1-v^2)^{\frac{d-3}{2}}\, \mathbb{1}\left(-\varepsilon \leq 2 - \eta \leq 2\varepsilon\right) 2^d \varepsilon^{\frac{d-1}{2}}. \qquad (B.196)$$

Now

$$2 - \eta = \frac{4 - \eta^2}{2 + \eta} \leq \frac{1}{2}(4 - \eta^2) = 2p(v^2 - a(p)) \qquad (B.197)$$

with $a(p) = (p-1)(p+3)/4p$. By a similar argument, we have $2 - \eta \geq \frac{2p}{1+p}(v^2 - a(p))$, and hence

$$\frac{1}{4}(v^2 - a) \leq 2 - \eta \leq 4(v^2 - a). \qquad (B.198)$$

Since $|a| \leq \frac{2}{M}$, the restriction on v imposed by the indicator function implies $(1 - v^2)^{-1/2} \leq 4$. Thus

$$C_3 \leq 2^{d+3}\varepsilon \int_0^1 dv\, v^{d-2} \frac{1}{\sqrt{v^2 - a}}\, \mathbb{1}\left(\tfrac{\varepsilon}{2} \leq v^2 - a \leq \tfrac{4}{M}\right) \qquad (B.199)$$

and

$$C_4 \leq 4^{d+1}\varepsilon^{\frac{d-2}{2}} \int_0^1 dv\, v^{d-2}\, \mathbb{1}\left(|v^2 - a| \leq 8\varepsilon\right). \qquad (B.200)$$

We first bound C_3. If $a > 0$, we change variables to $u = v^2 - a$. This gives $C_3 \leq 2^{d+2}\varepsilon I_3$ with

$$I_3 = \int_{\varepsilon/2}^{4/M} \frac{du}{\sqrt{u}}\, (a + u)^{\frac{d-3}{2}}. \qquad (B.201)$$

For $d \geq 3$, $(a + u)^{\frac{d-3}{2}} \leq 1$ and thus $I_3 \leq 2$. For $d = 2$,

$$I_3 = \int_{\varepsilon/2}^{4/M} \frac{du}{\sqrt{u}\sqrt{u + a}} \leq \int_{\varepsilon/2}^{4/M} \frac{du}{u} = \log\frac{8}{M\varepsilon}. \qquad (B.202)$$

If $a < 0$, let $\alpha = |a|$. With $u = v^2$, $C_3 \leq 2^{d+2} \varepsilon J_3$, where

$$J_3 = \int\limits_0^1 du\, u^{\frac{d-3}{2}} \frac{1}{\sqrt{u+\alpha}}\, \mathbb{1}\left(\tfrac{\varepsilon}{2} \leq u + \alpha \leq \tfrac{4}{M}\right). \tag{B.203}$$

For $d \geq 3$, we use that $u^{\frac{d-3}{2}} \leq 1$ and $\sqrt{u+\alpha} \geq \sqrt{u}$ to get $J_3 \leq 2$. For $d = 2$, we have to bound

$$\tilde{J}_3 = \int\limits_\nu^1 du \frac{1}{\sqrt{u}} \frac{1}{\sqrt{u+\alpha}}. \tag{B.204}$$

with $\nu = \max\{0, \tfrac{\varepsilon}{2} - \alpha\}$. Integration by parts gives

$$\tilde{J}_3 = \frac{1}{2} \frac{\sqrt{u}}{\sqrt{u+\alpha}} \Big|_\nu^1 + \frac{1}{4} \int\limits_\nu^1 du \frac{\sqrt{u}}{(u+\alpha)^{3/2}}. \tag{B.205}$$

Because $\alpha \geq 0$, $\sqrt{u}(u+\alpha)^{-1/2} \leq 1$, so the boundary term is at most one. The same argument implies that the remaining integral is bounded by

$$\int\limits_\nu^1 \frac{du}{u+\alpha} \leq \int\limits_{\varepsilon/2}^{1+\alpha} \frac{du}{u} \leq \log \frac{2 + 2|a|}{\varepsilon}. \tag{B.206}$$

Finally, we bound C_4. If $a \leq 0$, $v^2 - a \geq v^2$, so

$$C_4 \leq 4^{d+1} \varepsilon^{\frac{d-1}{2}} \int\limits_0^{\sqrt{8\varepsilon}} dv\, v^{d-2} \leq \text{const } \varepsilon^{d-1}. \tag{B.207}$$

If $0 < a < 16\varepsilon$, then $|v^2 - a| \leq 8\varepsilon$ implies $v^2 \leq 24\varepsilon$, and $C_4 \leq \text{const } \varepsilon$ follows in the same way as in (B.207). If $a \geq 16\varepsilon$, then we use that $v \geq 0$ to get

$$8\varepsilon \geq |(v - \sqrt{a})(v + \sqrt{a})| \geq \sqrt{a}\,|v - \sqrt{a}| \geq 4\sqrt{\varepsilon}\,|v - \sqrt{a}|, \tag{B.208}$$

that is, v must be in the interval $|v - \sqrt{a}| \leq 2\sqrt{\varepsilon}$. Then the integral over v is $\leq 2\sqrt{\varepsilon}$, and $C_4 \leq \text{const } \varepsilon$. ∎

Theorem 4.14 follows from Theorem B.29 by the same Taylor expansion in the ρ integration variables that was used in Sect. B.8.1 to derive Lemma 4.10 from Lemma B.28.

References

1. P.A.M. Dirac, *The Principles of Quantum Mechanics*, Second Edition, Oxford, Clarendon Press, 1947
2. R. P. Feynman, A. P. Hibbs, *Quantum Mechanics and Path Integrals*, McGraw Hill, 1965
3. B. Simon, *Functional Integration and Quantum Physics*, Academic Press, 1979
4. M. Reed, B. Simon, *Methods of Modern Mathematical Physics*, vol. II, Academic Press, 1975
5. G. Roepstorff, *Funktionalintegrale in der Quantenphysik*, Vieweg, Braunschweig, 1992
6. J. Glimm and A. Jaffe, *Quantum Physics: A Functional Integral Point of View*, Second Edition, Springer, 1987
7. V. Rivasseau, *From Perturbative to Constructive Renormalization*, Princeton University Press, 1991
8. S. Weinberg, *The Quantum Theory of Fields*, Cambridge University Press, 1995
9. Pierre J. van Baal, *A Course in Field Theory*, http://rulgm4.leidenuniv.nl/vanbaal/FTcourse.html
10. M. Aizenman, *Comm. Math. Phys.* **86** (1982) 1
11. J. Fröhlich, *Nucl. Phys.* **B 200** (1982), 281
12. R. Fernandez, J. Fröhlich, A. Sokal, *Random Walks, Critical Phenomena, and Triviality in Quantum Field Theory*, Springer, 1992
13. M. Lüscher, P. Weisz, *Nucl. Phys.* **B290** (1987) 25
14. K. Wilson, J. Kogut, *Phys. Reports* **12** (1974) 75
15. N.D. Mermin, H. Wagner, *Phys. Rev. Lett.* **17** (1966) 1133, R. Dobrushin, S.L. Shlosman, *Comm. Math. Phys.* **42** (1975) 31
16. D. Maison, H. Reeh, *Comm. Math. Phys.* **24** (1971) 67, S. Coleman, *Comm. Math. Phys.* **31** (1973) 259
17. R.B. Griffiths, *Phys. Rev.* **136** 437, *Jour. Math. Phys.* **8** (1967) 478,484, *Comm. Math. Phys.* **6** (1967) 121
18. D. Ruelle, *Statistical Mechanics*, W.A. Benjamin, 1969
19. R.B. Israel, *Convexity in the Theory of Lattice Gases*, Princeton University Press, 1978
20. E. Seiler, *Gauge Theories as a Problem of Constructive Quantum Field Theory and Statistical Mechanics*, Springer Lecture Notes in Physics 159, 1982
21. B. Simon, *The Statistical Mechanics of Lattice Gases*, Princeton University Press, 1993
22. A.C.D. van Enter, R. Fernàndez, A.D. Sokal, *J. Stat. Phys.* **72** (1993) 879
23. F. Wegner, in *Phase Transitions and Critical Phenomena*, vol. 6, edited by C. Domb and M. Green, Academic Press, 1976
24. L. R. Foulds, *Graph Theory Applications*, Springer, 1992
25. C. Borgs, J. Imbrie, *Commun. Math. Phys.* **145** (1992)235
26. G. Gallavotti, *Rev. Mod. Phys.* **57** (1985) 471

228 References

27. G. Gallavotti, F. Nicolò, *Comm. Math. Phys.* **100** (1985) 545; **101** (1985) 247
28. J. Feldman, T. Hurd, L. Rosen, J. Wright, *QED: A Proof of Renormalizability*, Springer Lecture Notes in Physics 312, (1988)
29. J. Polchinski, *Nucl. Phys.* **B 231** (1984), 269
30. C. Wieczerkowski, *Comm. Math. Phys.* **120** (1988), 149
31. G. Keller, C. Kopper, M. Salmhofer, *Helv. Phys. Acta*, **65** (1992) 32
32. G. Keller, C. Kopper, *Comm. Math. Phys.* **148** (1992) 445, *Phys. Lett.* **B273** (1991) 323
33. G. Keller, *Comm. Math. Phys.* **161** (1994) 311
34. C. Kopper, *Renormierungstheorie mit Flußgleichungen*, Aachen: Shaker, 1998
35. K. Gawedzki, A. Kupiainen, *Comm. Math. Phys.* **77** (1980) 33, *Comm. Math. Phys.* **99** (1985) 197
36. J. Feldman, J. Magnen, V. Rivasseau, R. Sénéor, *Comm. Math. Phys.* **98** (1985) 273 *Comm. Math. Phys.* **100** (1985) 23, *Comm. Math. Phys.* **109** (1987) 437
37. G. Mack, A. Pordt, *Comm. Math. Phys.* **97** (1985) 267
38. D. C. Brydges, H.-T. Yau, *Comm. Math. Phys.* **129**(1990) 351
39. D. C. Brydges, J. Dimock, T. Hurd, *Comm. Math. Phys.* **172** (1995) 143, mp_arc 96-538, mp_arc 96-681, and references therein
40. A. Abdesselam, V. Rivasseau, *Rev. Math. Phys.* **9** (1997) 123
41. T. Kennedy, E. Lieb, *Physica* **138A** (1986) 320
42. J. Hubbard, *Proc. R. Soc. London* **A 243** (1957) 336
43. N. W. Ashcroft, N. D. Mermin, *Solid State Physics*, Sunders College Publishing, 1976 Harcourt Brace College Publishers, New York
44. J. Feldman, E. Trubowitz, *Helv. Phys. Acta* **63** (1990) 157, ibid. **64** (1991) 213
45. J. Feldman, M. Salmhofer and E. Trubowitz, *J. Stat. Phys.* **84** (1996) 1209
46. J. Feldman, M. Salmhofer, and E. Trubowitz, *Comm. Pure Appl. Math.* **LI** (1998) 1133
47. J. Feldman, M. Salmhofer, and E. Trubowitz, *Regularity of the Moving Fermi Surface: The Full Selfenergy*, to appear in *Comm. Pure Appl. Math.*
48. M. Salmhofer, *Rev. Math. Phys.* **10** (1998) 553
49. P.B. Allen, *Physical Review* **B 13** (1976) 1416
50. W. Kohn and J. M. Luttinger, *Phys. Rev. Lett.* **15** (1965) 524
51. J. Feldman, H. Knörrer, R. Sinclair, E. Trubowitz, *Helv.Phys.Acta* **70**(1997)154
52. R. Sinclair, *Computational Evidence for Superconductivity in a Repulsive Model*, ETH Thesis No. 11660 (1996)
53. M. Salmhofer, *Comm. Math. Phys.* **194** (1998) 249
54. M. Salmhofer, *Phys. Lett.* **408 B** (1997) 245
55. J.Feldman, J.Magnen, V.Rivasseau, E.Trubowitz, *Helv.Phys.Acta* **65**(1992)679
56. J. Feldman, H. Knörrer, D. Lehmann, E. Trubowitz, in *Constructive Physics*, V. Rivasseau (ed.), Springer Lecture Notes in Physics, 1995
57. D. C. Brydges and J. Wright, *J. Stat. Phys.* **51** (1988) 435
58. G. Benfatto, G. Gallavotti, *J. Stat. Phys.* **59** (1990) 541
59. G. Benfatto, G. Gallavotti, A. Procacci, B. Scoppola, *Comm. Math. Phys.* **160** (1994) 93
60. F. Bonetto, V. Mastropietro, *Comm. Math. Phys.* **172** (1995) 57
61. J. Feldman, H. Knörrer, E. Trubowitz, *Comm. Math. Phys.* **195** (1998) 465
62. A. Abdesselam and V. Rivasseau, to appear
63. M. Disertori and V. Rivasseau, to appear
64. O. Bratteli, D.W. Robinson, *Operator Algebras and Quantum Statistical Mechanics II*, Springer 1981
65. Reed, M., Simon, B., *Methods of Modern Mathematical Physics*, vol. 1, Academic Press, New York 1980
66. U. Wolff, *Nucl. Phys.* **B225** (1983) 391–408

Index

Springer
and the
environment

At Springer we firmly believe that an
international science publisher has a
special obligation to the environment,
and our corporate policies consistently
reflect this conviction.
We also expect our business partners –
paper mills, printers, packaging
manufacturers, etc. – to commit
themselves to using materials and
production processes that do not harm
the environment. The paper in this
book is made from low- or no-chlorine
pulp and is acid free, in conformance
with international standards for paper
permanency.

 Springer